Production of Plant Secondary Metabolites Using Plant Tissue Culture and Bioreactor Culture Techniques

Production of Plant Secondary Metabolites Using Plant Tissue Culture and Bioreactor Culture Techniques

Editors

Kee-Yoeup Paek
Hosakatte Niranjana Murthy

 Basel • Beijing • Wuhan • Barcelona • Belgrade • Novi Sad • Cluj • Manchester

Editors

Kee-Yoeup Paek
Chungbuk National University
Cheongju
Republic of Korea

Hosakatte Niranjana Murthy
Karnatak University
Dharwad
India

Editorial Office
MDPI
St. Alban-Anlage 66
4052 Basel, Switzerland

This is a reprint of articles from the Special Issue published online in the open access journal *Plants* (ISSN 2223-7747) (available at: https://www.mdpi.com/journal/plants/special_issues/59NKDIZCSL).

For citation purposes, cite each article independently as indicated on the article page online and as indicated below:

Lastname, A.A.; Lastname, B.B. Article Title. *Journal Name* **Year**, *Volume Number*, Page Range.

ISBN 978-3-7258-0809-0 (Hbk)
ISBN 978-3-7258-0810-6 (PDF)
doi.org/10.3390/books978-3-7258-0810-6

© 2024 by the authors. Articles in this book are Open Access and distributed under the Creative Commons Attribution (CC BY) license. The book as a whole is distributed by MDPI under the terms and conditions of the Creative Commons Attribution-NonCommercial-NoDerivs (CC BY-NC-ND) license.

Contents

Baoyu Ji, Liangshuang Xuan, Yunxiang Zhang, Wenrong Mu, Kee-Yoeup Paek,
So-Young Park, et al.
Application of Data Modeling, Instrument Engineering and Nanomaterials in Selected Medid the Scientific Recinal Plant Tissue Culture
Reprinted from: *Plants* **2023**, *12*, 1505, doi:10.3390/plants12071505 1

Baoyu Ji, Liangshuang Xuan, Yunxiang Zhang, Guoqi Zhang, Jie Meng, Wenrong Mu, et al.
Advances in Biotechnological Production and Metabolic Regulation of *Astragalus membranaceus*
Reprinted from: *Plants* **2023**, *12*, 1858, doi:10.3390/plants12091858 23

Fengjiao Xu, Anjali Kariyarath Valappil, Ramya Mathiyalagan, Thi Ngoc Anh Tran,
Zelika Mega Ramadhania, Muhammad Awais and Deok Chun Yang
In Vitro Cultivation and Ginsenosides Accumulation in *Panax ginseng*: A Review
Reprinted from: *Plants* **2023**, *12*, 3165, doi:10.3390/plants12173165 37

Maria Titova, Elena Popova and Alexander Nosov
Bioreactor Systems for Plant Cell Cultivation at the Institute of Plant Physiology of the Russian Academy of Sciences: 50 Years of Technology Evolution from Laboratory to Industrial Implications
Reprinted from: *Plants* **2024**, *13*, 430, doi:10.3390/plants13030430 58

Ji-Hye Kim, Jong-Eun Han, Hosakatte Niranjana Murthy, Ja-Young Kim, Mi-Jin Kim,
Taek-Kyu Jeong and So-Young Park
Production of Secondary Metabolites from Cell Cultures of *Sageretia thea* (Osbeck) M.C. Johnst. Using Balloon-Type Bubble Bioreactors
Reprinted from: *Plants* **2023**, *12*, 1390, doi:10.3390/plants12061390 83

So Yeon Choi, Seong Sub Ku, Myung Suk Ahn, Eun Jin So, HyeRan Kim,
Sang Un Park, et al.
Metabolic Discrimination between Adventitious Roots and Standard Medicinal Part of *Atractylodes macrocephala* Koidz. Using FT-IR Spectroscopy
Reprinted from: *Plants* **2023**, *12*, 1821, doi:10.3390/plants12091821 99

Mei-Yu Jin, Miao Wang, Xiao-Han Wu, Ming-Zhi Fan, Han-Xi Li, Yu-Qing Guo, et al.
Improving Flavonoid Accumulation of Bioreactor-Cultured Adventitious Roots in *Oplopanax elatus* Using Yeast Extract
Reprinted from: *Plants* **2023**, *12*, 2174, doi:10.3390/plants12112174 113

Yashika Bansal, A. Mujib, Jyoti Mamgain, Yaser Hassan Dewir and Hail Z. Rihan
Phytochemical Composition and Detection of Novel Bioactives in Anther Callus of *Catharanthus roseus* L.
Reprinted from: *Plants* **2023**, *12*, 2186, doi:10.3390/plants12112186 129

Aicah Patuhai, Puteri Edaroyati Megat Wahab, Martini Mohammad Yusoff,
Yaser Hassan Dewir, Ali Alsughayyir and Mansor Hakiman
Plant Growth Regulator- and Elicitor-Mediated Enhancement of Biomass and Andrographolide Production of Shoot Tip-Culture-Derived Plantlets of *Andrographis paniculata* (Burm.f.) Wall. (Hempedu Bumi)
Reprinted from: *Plants* **2023**, *12*, 2953, doi:10.3390/plants12162953 146

Dipti Tonk, Abdul Mujib, Mehpara Maqsood, Mir Khusrau, Ali Alsughayyir and Yaser Hassan Dewir
Fungal Elicitation Enhances Vincristine and Vinblastine Yield in the Embryogenic Tissues of *Catharanthus roseus*
Reprinted from: *Plants* **2023**, *12*, 3373, doi:10.3390/plants12193373 . . . . **157**

Maria V. Titova, Dmitry V. Kochkin, Elena S. Sukhanova, Elena N. Gorshkova, Tatiana M. Tyurina, Igor M. Ivanov, et al.
Suspension Cell Culture of *Polyscias fruticosa* (L.) Harms in Bubble-Type Bioreactors—Growth Characteristics, Triterpene Glycosides Accumulation and Biological Activity
Reprinted from: *Plants* **2023**, *12*, 3641, doi:10.3390/plants12203641 . . . . **173**

Sumanta Das, Kaniz Wahida Sultana, Moupriya Mondal, Indrani Chandra and Ashwell R. Ndhlala
Unveiling the Dual Nature of Heavy Metals: Stressors and Promoters of Phenolic Compound Biosynthesis in *Basilicum polystachyon* (L.) Moench In Vitro
Reprinted from: *Plants* **2024**, *13*, 98, doi:10.3390/plants13010098 . . . . **192**

Review

Application of Data Modeling, Instrument Engineering and Nanomaterials in Selected Medid the Scientific Recinal Plant Tissue Culture

Baoyu Ji [1,2,†], Liangshuang Xuan [2,†], Yunxiang Zhang [1], Wenrong Mu [2], Kee-Yoeup Paek [3], So-Young Park [3], Juan Wang [1,*] and Wenyuan Gao [1,*]

1 School of Pharmaceutical Science and Technology, Tianjin University, Tianjin 300072, China
2 Shool of Pharmacy, Henan University of Chinese Medicine, Zhengzhou 450046, China
3 Department of Horticultural Science, Chungbuk National University, Cheongju 28644, Republic of Korea
* Correspondence: drwangjuan@tju.edu.cn (J.W.); pharmgao@tju.edu.cn (W.G.)
† These authors contributed equally to this work.

Abstract: At present, most precious compounds are still obtained by plant cultivation such as ginsenosides, glycyrrhizic acid, and paclitaxel, which cannot be easily obtained by artificial synthesis. Plant tissue culture technology is the most commonly used biotechnology tool, which can be used for a variety of studies such as the production of natural compounds, functional gene research, plant micropropagation, plant breeding, and crop improvement. Tissue culture material is a basic and important part of this issue. The formation of different plant tissues and natural products is affected by growth conditions and endogenous substances. The accumulation of secondary metabolites are affected by plant tissue type, culture method, and environmental stress. Multi-domain technologies are developing rapidly, and they have made outstanding contributions to the application of plant tissue culture. The modes of action have their own characteristics, covering the whole process of plant tissue from the induction, culture, and production of natural secondary metabolites. This paper reviews the induction mechanism of different plant tissues and the application of multi-domain technologies such as artificial intelligence, biosensors, bioreactors, multi-omics monitoring, and nanomaterials in plant tissue culture and the production of secondary metabolites. This will help to improve the tissue culture technology of medicinal plants and increase the availability and the yield of natural metabolites.

Keywords: bioreactors; intelligence model and optimization algorithm; multi-omics monitoring; nanomaterials; plant tissue culture; secondary metabolites

1. Introduction

The increment in the global demand for medicinal plant resources, a diversified use of plants, and the reduction in cultivated land have accelerated the lack of plant resources. Large-scale culture of plant cells and tissues is considered to be a suitable method to alleviate this situation. In the past, due to the differences in plant species and varieties, the immaturity of culture conditions, and the solidification of research thinking, the development of plant cell and tissue culture technology was relatively slow [1]. Although the growth of plant tissues can be easily observed at different stages of in vitro culture, the growth environment of different tissues is nonlinear and uncertain, and is also affected by many other factors [2]. The complex interaction of many factors makes the use of traditional statistical methods problematic for optimization, and requires a certain amount of processing. New technologies have benefited from the development of many fields such as information technology, engineering, instruments and the methods of analysis, and material science. In recent years, with breakthroughs in the biological mechanism of plant tissue and the continuous development of new technologies in various disciplines, the combination of theory and new technology has made rapid progress in the development of plant cell and tissue culture. Figure 1 shows the rapid progress in the development of plant cell and tissue culture with the help of the combination of theory and new technology. The development of various new technologies for plant tissue culture is aimed at achieving an efficient and high-volume production of plant biomass and effective secondary metabolites.

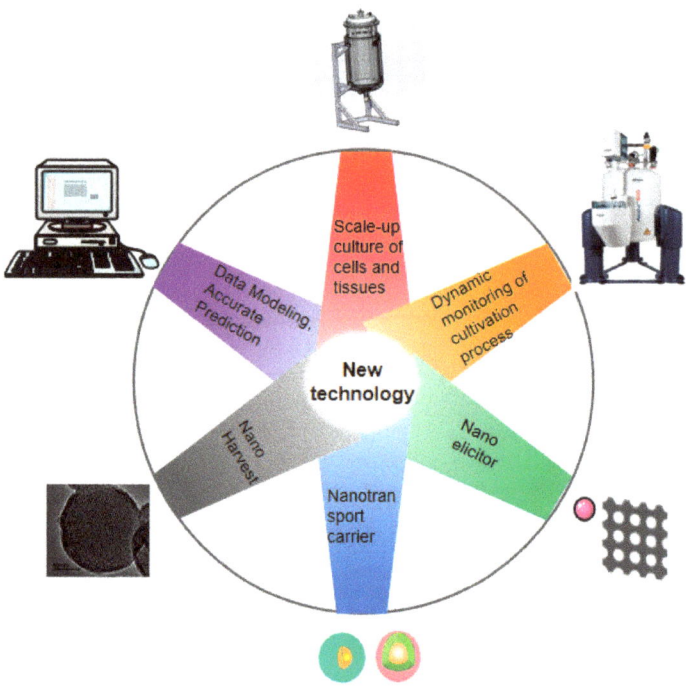

Figure 1. New technologies for plant tissue culture.

Whether for solid or liquid culture of plant tissue, the prediction and optimization of culture conditions have always been the focus of research. Data-driven modeling technology such as artificial intelligence (AI) [3], optimistic algorithm (OA) [4], and gene expression programming (GEP) can effectively be used for different purposes in plant tissue culture [5] such as the modeling, prediction, and optimization of plant genotypes, media, sterilization conditions, and plant growth regulators. Although we have these methods to optimize the culture condition, at present, the large-scale production of plant tissues can only be achieved by bioreactors. Bioreactors are engineering systems that support aerobic or anaerobic biochemical processes [6,7]. Various types of bioreactors have been designed and modified to improve the culture efficiency of different plant tissues. In addition, nuclear magnetic resonance-based metabolomics can be used as a tool to study the dynamics of plant cell metabolism and nutrition management [8]. Furthermore, nanomaterials also exhibit many special physical and chemical properties in plant tissue culture due to their unique structure and size [9]. Examples of the unique properties of nanoparticles include a very large specific surface area, high surface energy, and quantum confinement. These unusual properties may cause their environmental fates and behaviors to be very different to those of bulk congeners [10]. More and more nanomaterials are used in plant cell and tissue culture as elicitors [11,12]. Elicitors are often used in plant tissue culture to stimulate the defense system and the accumulation of secondary metabolites. In the past, biological elicitors and non-biological elicitors were more commonly used such as salicylic acid, methyl salicylate, bezoic acid, chitosan, bacterial, fungal, and algal [13,14].

In recent years, various types of nanomaterials have been used as new elicitors and have shown exciting effects in plant tissue culture [9]. Layered double hydroxide (LDH) [15] and chitosan [16] can be used as a transport carrier for various substances such as the transport of plant growth regulators, nutrients, etc., to promote the growth of plant tissue biomass and the contents of secondary metabolites. Engineered mesoporous silica nanoparticles (MSNPs) can be easily modified by specific functional groups, transported through the cell wall and cell membrane, and then enter the plant cells for the targeted binding of bioactive substances such as flavonoids [17]. After the combination, the MSNP-bioactive substance complex is carried out from the cell without causing significant harm to the hosts, achieving the sustainable harvesting of natural products. This paper summarizes the mechanism of medicinal plant tissue culture in recent years, and discusses the application of new technologies in plant cell and tissue culture.

2. Study on the Mechanism of Medicinal Plant Tissue Culture

2.1. The Generating Mechanism of Stem Cell, Callus, and Adventitious Roots

During the late period of plant embryonic development, stem cells are located in the apical niche of meristem, and stem cells will divide and differentiate to maintain the constitution of meristem, which can grow into different tissues and organs such as callus, adventitious roots (ARs), etc. [18,19]. Root apical meristem (RAM) determines the embryonic development of the underground part of plants. The brassinosteroid (BR) has been shown to regulate stem cell niche (SCN) in RAM. In the BR signal module in RAM, the loss of functional BR leads to the decrease in cell division rate and the frequency of distal stem cell renewal [20]. Rape steroids (BRAVO) produced by the expression of R2-R3MYB transcription factors at the quiescent center (QC) play an important role in counteracting the effects of BR on cell division [21]. QC-expressed gene Wuschel-related homeobox 5 (*WOX5*) is crucial in the identity and maintenance of QC and the mutant *WOX5* showed an aberrant differentiation pattern of stem cells [22].

Callus induction is a key step in plant suspension cell culture such as *Zea mays* L. [23], *Panax ginseng* [24], and *Abrus precatorius* Linn [25]. The high concentration of indole-3-acetic acid (IAA) promotes root formation while the high proportion of cytokinin is good for adventive shoot regeneration [26,27]. Although the cell mechanism of callus induction is complex, we know that the continuation of the cell division cycle is the key to callus induction [28]. Somatic embryogenesis (SE) can follow two different pathways called direct and indirect SE. Direct pathways occur when plant cells produce embryos without forming calluses. The indirect pathway requires an additional step in callus formation before embryonic development [29]. Abnormalities in SE are associated with the use of 2,4-D in most published protocols, an inflammatory auxin that disrupts the balance of endogenous auxin and the auxin polar transportation interfering with the embryo apical-basal polarity [30]. Mechanical trauma may be the main inducing factors for organ regeneration, which triggers callus formation by the dynamic hormone level and transcriptional changes [31]. Callus formation is derived from J0121-labeled cells (J0121-label is a marker of the pericycle cells of xylem in roots and near xylem in aerial tissues), and the pericycle cells are considered to be the main contributors to the formation of callus. Pericycle cells can form callus or lateral roots (LR), adventitious roots, and other organs under certain conditions. An important transcription factor involved in the differentiation of pericycle cells is Miniyo, which may lead to rapid cell differentiation or prevent cell differentiation after entering the nucleus of pericycle cells [32,33] (see part a in Figure 2 for a detailed pathway).

Adventitious root (AR) cultures show the characteristics of high root proliferation and biomass, and have the potential to synthesize specific bioactive compounds. Adventitious roots are a reliable source of natural chemicals due to their genetic and biosynthetic stability. ARs can be induced from various explants (such as leaves, roots, stems, petiole callus, etc.) in in vitro conditions. Factors affecting the morphogenesis of ARs include indole-3-acetic acid (IAA), nitrous oxide, and light [34]. Among them, IAA is the main growth-promoting hormone that triggers AR; the fine-tuned spatiotemporal interactions between hormones ultimately regulate IAA distribution and perception [35]. In many plant species, higher IAA concentration is required in the early stage of adventitious root development than in the later stage. Some downstream signaling pathways in IAA transduction are involved in the molecular mechanism of adventitious root formation [36]. SLR is an auxin signaling factor in the Aux/IAA protein family. The lack of a functional SLR in the slr-1 dominant mutant resulted in reduced auxin response, affecting LRs and callus development. An IAA signaling molecule, ALF4 may regulate the stability of IAA28 protein, and increased SLR levels in ALF4 mutants may lead to callus and lateral root formation [37]. LR and AR share key elements of genetic and hormonal regulatory networks, but are still influenced by different regulatory mechanisms. Light is an important environmental parameter affecting the development of AR and LR. Photoreceptors are involved in regulating the development of AR and LR [38]. Studies on Arabidopsis show that roots have photoreceptors for blue, red, and far red light [39] as well as a potential interaction between light and auxin in AR regulation [40]. The growth response of roots to light and the initiation of AR may be related to the change in the local endogenous auxin concentration [41] (see part b in Figure 2 for the detailed pathway).

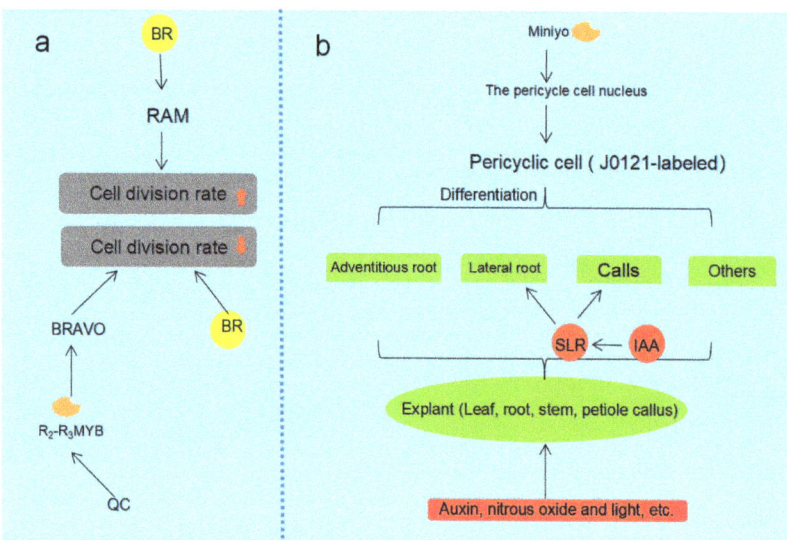

Figure 2. Effects of plant growth regulators. Note: (**a**) The effect of BR and transcription factor s on cell division rate. (**b**) Different plant tissues are differentiated from pericycle cells or induced by plant hormones. Yellow represents exogenous hormones, red represents exogenous hormones.

2.1.1. The Mechanism of Endogenous Plant Growth Regulators on Plant Tissue Growth and Plant Stress Resistance

Plant development can be manipulated by adding PGRs at specific stages of growth or maturation by adding plant hormones such as IAA, naphthaleneacetic acid (NAA), 6-furfurylaminopurine (KT), 6-benzylaminopurine (6-BA), ethylene, abscisic acid (ABA), brassinolide (BL), jasmonic acid (JA), salicylic acid (SA), etc. [42]. All of these can stimulate further developmental responses. In particular, IAA and cytokinin play a decisive role in cell growth and differentiation (especially meristem differentiation) as well as apical dominance [19]. IAA as well as ethylene and cytokinin plays an important role in the formation of lateral roots [43,44]. The same growth regulator causes different responses to different explants or species, indicating that plant hormone receptors may be specific. Each hormone can correspond to multiple receptors in plants, and the relationship between plant hormone receptors is very complex [45]. The interaction of different hormone synthesis is also complex, as IAA can promote the biosynthesis of gibberellin (GA) to promote fiber development, and GA may regulate fiber development downstream of IAA [46]. The function and crosstalk of common hormones are shown in Table 1.

Table 1. Function and crosstalk of common hormones.

Plant-Growth Regulator	Classification	Function	Hormone Crosstalk
Auxin	IBA	Root contact induction [47]	The hypocotyl and is enhanced by application of (IBA) combined with kinetin (Kin) [48]
	IAA	IAA affects plant growth and development including growth response, vascular development, leaf and flower initiation, root growth, and lateral root formation [49]	GA enhances auxin levels in stems by stimulating polar transport of IAA [50]
Cytokinin	6-BA	Involved in cell division [51]	2,4-dichlorophenoxyacetic acid (2,4-D), indoleacetic acid (IAA), and 6-BA promote the production of somatic cells [52]
Gibberellins	GA	Promote germination, growth and flowering, promote leaf expansion, but inhibit root growth [53]	GA and cytokinins antagonize many developmental processes including shoot and root elongation, cell differentiation, shoot regeneration in culture and meristem activity [49]
	SA	Depending on its concentration and plant growth conditions and developmental stages [54]	SA regulates IAA biosynthesis and transport. Low concentration of SA (50 μM) promotes adventitious roots and changes the structure of root apical meristem [55]
	ABA	Regulate seed dormancy and germination [56]	ABA, SA, and auxin can increase plant resistance to pathogens [57]
	Ethylene	Branch elongation and leaf abscission [58]	The signaling mechanisms of gibberellin, ethylene, and brassinolide may not evolve until mosses and vascular plants have evolved [59,60]
	Brassinolide	Can promote the growth of plant seedlings	Working with other PGRs alone affects plant growth and development and abiotic and biotic stress responses such as ABA, ethylene, SA, JA [61]

2.1.2. Effects of Plant Growth Regulators on the Synthesis of Secondary Metabolites in Medicinal Plants

Plant secondary metabolites are the main active ingredients of pharmaceuticals. Plant growth regulators have been used in recent years as elicitors to stimulate the production of secondary metabolites. Plant growth regulators stimulate the synthesis of different types of secondary metabolites such as terpenoids and phenols (coumarins, lignins, flavonoids, isoflavones), tannins, sulfur containing secondary metabolites (glucosinolates, phytoalexins), and nitrogen containing secondary metabolites (alkaloids, cyanogenic glucosides and non-protein amino acids) [62]. Some studies have shown that PGRs are not as effective as traditional elicitors in improving the plant secondary metabolism level [63]. In addition to increasing biomass accumulation, PGRs will also induce the quantitative modification of major volatile components [64].

At present, although the specific mechanism of PGRs has not been elucidated, some studies have confirmed some parts of the mechanism of plant hormone response. Studies have shown that secondary metabolic clusters such as terpenoids share a common gene evolutionary history with the major metabolic pathways of plant growth hormone synthesis [64], which may reflect that the types of secondary metabolic compounds could be related to plant hormones. Plants transduce the signal of the nitrogen state through a variety of signaling pathways. One of the pathways is to use cytokinins as messengers. Cytokinin-mediated signal transduction is related to the control of plant development, protein synthesis, and macronutrient acquisition, and can coordinate with nitrate components to change plant metabolism [65]. Calcium–hormone interaction regulates the expression of phenylalanine ammonia lyase (PAL), stilbene synthase (STS), dihydroflavonol reductase (DFR), and UDP-glucose, flavonoid 3-O-glucosyltransferase (UFGT), which will increase the content of classified components (anthocyanins) in grape suspension cells [66]. IAA level can be upregulated under the combined stress of drought and salt. The contents of flavonoids and phenolic compounds as well as the transcriptional level and activities of enzymes related to secondary metabolism are also increased [67]. In addition to regulating the growth and development of plants, melatonin can also enhance the adaptability of plants to biotic and abiotic stresses. Its strong antioxidant capacity can enhance the ability of antioxidant enzymes in plants, thereby increasing the content of secondary metabolites. Furthermore, melatonin can reduce the stress response of plants under low iron or high iron conditions and upregulate the photosynthetic rate and synthesis of phenols and flavonoids [68].

2.1.3. Biosensors Are Potential Tools for Plant Hormone Mechanism Research

A biosensor is an integrated receptor–transducer device that can convert a biological response into an electrical signal, biosensors for a wide range of applications such as health care and disease diagnosis, environmental monitoring, water and food quality monitoring, and drug delivery [69]. The interaction network of plant hormones is very complex. The visualization of the distribution and content of plant hormones in plant tissues can help people understand the basic laws of plant growth, and can also improve the yield and quality of plants by regulating the synthesis, transportation, and distribution of plant hormones in plants. The engineering of biosensors provides a new tool for the monitoring of phytohormones. The visualization and quantification of phytohormone distribution can be realized by using biosensors and other related technologies including fluorescent sensors, transcriptional reporters, degradation sensors, and luciferase [70].

Fluorescence biosensors are particularly effective for monitoring the spatial and temporal distribution of small signal molecules, which can be used to study the distribution and content of plant hormones in the subcellular [71]. The distribution of hormones in living tissues can be observed by the expression of the reporter gene. However, the relationship between the concentration of signal molecules and reporter molecules is complex, and it is necessary to use network dynamics to link the abundance of signals with the reporter's signal [71]. Genetically encoded biosensors are able to detect rapid changes in the concentration and distribution of plant hormones in living cells [70]. Transcriptional reporters, degron-based biosensors, and FRET-based direct biosensors have been developed to monitor auxin. Ole Herud-Sikimić et al. [72] reported a binding bag specifically designed for auxin. When auxin binds to the bag, the conformation of the auxin binding part will be changed to couple with the fluorescent protein, then the fluorescence resonance energy transfer signal will be generated so that the auxin level can be reflected. Other types of sensors have been described. This article is no longer detailed [71].

3. Application of Technology with Different Fields

3.1. Using Data-Driven Modeling Technology to Optimize Tissue Culture Scheme

In order to successfully and efficiently carry out plant tissue culture, we need to scientifically improve the culture conditions including light, temperature and humidity, appropriate medium composition, plant growth regulators, etc. However, the cultivation of plants of different species and different tissues of the same plant has great differences. It is very tedious and inefficient to screen the most suitable culture conditions through the experiments. For example, in the prediction of a medium formula, the mineral and hormone composition of the plant micropropagation medium is an important factor in the growth of explants, but the different plant species lead to different nutritional and hormone requirements. Blindly testing the types and proportions of various hormones and minerals is time-consuming, laborious, and costly work, and does not often succeed. Most of the optimization of plant tissue culture medium is improved on the basis of previous experiments. This method requires a lot of control experiments. Second, it is also necessary to effectively develop and optimize the new medium. Understanding the role of minerals and hormones and their interaction with other medium components and different plant tissues is critical to the successful in vitro culture of plant tissues [73,74]. However, the relationship between medium components is very complex, and there are even many unknown interactions. It is difficult to optimize the medium for different plant species and genotypes. Therefore, the development and application of a predictive modeling system for medium formulation and culture conditions can improve the efficiency of culture optimization [74]. Data-driven modeling is considered as an effective alternative to optimizing biological processes and nonlinear multivariate modeling [75]. Modeling systems are often combined with optimization algorithms to improve efficiency. Appropriate modeling techniques can not only predict the value of the software sensor, but also estimate the probability [75]. In recent years, modeling technology has developed rapidly. Common systems include the combination of an artificial intelligence (AI) model and optimization algorithm (OA) [76], and the combination of gene expression programming (GEP) and genetic algorithm (GA) [5].

3.1.1. Application of Artificial Intelligence (AI) Model and Optimization Algorithm (OA)

The modeling method of artificial intelligence (AI) is more effective than other modeling techniques and is a potential modeling tool for plant tissue culture [77]. AI and OA are widely used in different technical and scientific fields, and have been applied to improve different stages of plant tissue culture in recent years. Artificial intelligence tools build models from experimental and observational data. Using artificial intelligence models to model different steps of plant tissue culture is a very suitable and reliable method [75]. AI relies on the knowledge of mathematics and statistical equations and has engineering thinking and judgment capabilities. In recent years, various applications of AI and OA in the prediction of plant tissue culture medium and the optimization of growth and development have been reported. The steps of data modeling are: preprocessing of basic data (including principle component analysis and the data), network selection (parameter setting), training selection (error reduction) testing and interpretation of results, and the assessment of the developed model. Artificial intelligence tools include a variety of artificial neural networks, support vector machine (unsupervised learning artificial intelligence model for classification, clustering and regression analysis), and random forest (algorithms for classification and regression). All of which have the characteristics of simple design and high efficiency. The use of OA for optimal selection can significantly reduce the costs and time. This method relies on genetic algorithms and is especially suitable for plant tissue culture. The whole process takes into account the different culture stages of different plant tissues (such as embryo [75], callus [78], bud [79], and root [80]). The modeling, prediction, and optimization of plant genotypes, media, sterilization conditions, different types, and concentrations of plant growth regulators also need to be considered [75]. As shown in Figure 3, data-driven models are effectively used for different purposes in plant tissue culture. AI and OA have been effectively applied to predict and optimize the length and number of micro-buds [79,81] or roots [80], plant cell culture or hairy root biomass [82], and culture environmental conditions (such as temperature and sterilization [83,84]) to achieve the maximum productivity and efficiency as well as the classification of micro-buds and somatic embryos. Future AI–OA methods could also be used in the development of genetic engineering and genome editing.

3.1.2. Application of Gene Expression Programming (GEP) and Genetic Algorithm (GA)

Previous studies have focused on predicting the effects of various components of the medium and hormones on plant explants through traditional multi-layer perceptron neural network (MLPNN) and multiple linear regression (MLR) methods. Genetic programming (GP) is one of the most traditional and widely used evolutionary algorithms. GEP technology involves computer programs encoded by linear chromosomes of different sizes and shapes, which is an effective alternative to traditional GP. Jamshidi et al. [85] used GEP and M5′ model tree algorithms to predict the effects of medium components on the in vitro proliferation rate (PR), branch length (SL), branch necrosis (STN), and vitrification (Vitri). This method mainly includes modeling systems, multiple linear regression, radial basis function neural network, gene expression programming, optimization of GEP models, and genetic algorithm model optimization. In addition, they also compared the methods of combining GEP with the radial basis function neural network (RBFNN) and multiple linear regression (MLR), respectively, to predict the effects of minerals and certain hormones in pear rootstock medium on the proliferation index. The results showed that RBFNN and GEP showed a higher accuracy than MLR. At present, the application of GEP in plant tissue culture is not replacing GA, but GEP has shown more accurate results in its application, and its further development has great potential. For the application of artificial intelligence and gene expression programming in tissue culture, see Table 2.

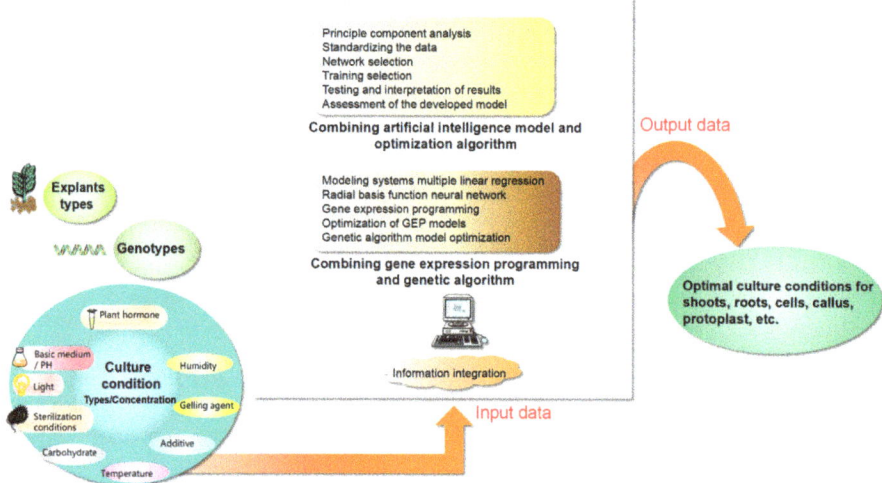

Figure 3. Overall roadmap for data-driven modeling technology.

Table 2. The application of artificial intelligence and gene expression programming in tissue culture.

Modeling	Plant	Optimal Results
Multilayer perceptron (MLP) as an artificial ANN and support vector regression (SVR)	*chrysanthemum*	The highest embryogenesis rate (99.09%) and the maximum number of somatic embryos per explant (56.24) can be obtained [3]
Artificial neural network–genetic algorithm	*Garnem*	The results showed that the optimized rooting medium was more effective than the other standard medium [74]
GEP and M5′ model tree	*Pear* rootstocks	Proliferation rate, shoot length, shoot tip necrosis, vitrification and quality index; GEP had a higher prediction accuracy than the M5′ model tree [74]
Adaptive neuro-fuzzy inference system–genetic algorithm	*Corylus avellana*	Cell culture-responsive taxol biosynthesis was modeled and predicted by cell extract, culture filtrate, and cell wall alone or in combination with methyl-β-cyclodextrin [86]
Image processing and ANN	*Lycopersicon esculentum* L.	Plant growth regulators, the concentration of gum Arabic (GA) additive, the cold pretreatment duration, and flower length on callus induction percentage and number of regenerated callus in an anther culture of tomato [78]
ANNs-GA	*Pistacia vera*	Gain insights, predict, and optimize the effect of several independent factors on four growth parameters [79]

At present, data-driven modeling technology has become a predictive tool for modeling complex biological research and plays an important role in plant tissue culture. Scientific prediction greatly reduces the workload of researchers and reduces the cost of in vitro culture. In the future, the development of modeling technology may make the in vitro culture of plant tissue become more automated and mechanized.

3.2. New Technologies to Help Bioreactor Engineering

3.2.1. Design of New Bioreactor

Many secondary metabolites of plants still cannot be obtained by artificial synthesis. The cell and tissue culture of medicinal plants is still the main source of secondary metabolite production [18]. Bioreactors can support the large-scale production of plant cells and tissues and has the advantages of stability, high efficiency, high yield, and low cost [6]. Bioreactors can be used to culture microorganisms or plant cells and tissues under monitored and controlled environmental conditions (such as pH, temperature, oxygen tension and nutrient supplement). Especially for medicinal plants with slow growth, a low bioactive component content, and those easily disturbed by environmental factors such as *Panax ginseng* and *Panax notoginseng*, the large-scale production of secondary metabolites by a bioreactor will be very competitive to solve existing problems such as the high cost of planting cultivation. By constructing the best environment for the effective growth and production of bioactive substances, the culture results can be significantly affected, which can be used to produce secondary metabolites, recombinant proteins, microbial fermentation, etc. From an economic point of view, the final purpose of the design of the bioreactor is to decrease the cost of planting production. To achieve this goal, a consideration of the cost of equipment, substrate cost, time, and the final total output is required [7].

Bioreactors should be able to control environmental conditions and allow the aseptic operation of different bioreactors to have a great influence on the state of secondary metabolites in plant tissue or cell culture. Agnieszka Szopa [87] compared the accumulation of phenolic acids and flavonoids in different types of bioreactors, and found that the conical bioreactor (CNB) was the best reactor for phenolic acid accumulation, and the nutrient sprinkle bioreactor (NSB) was the best for flavonoid accumulation. It can be seen that different reactors have different influences on the product selection. The appropriate design of the bioreactor will help to reduce costs, increase the productivity, and adaption to different kinds of cultures.

3.2.2. Types of Bioreactors

Typical plant cell and tissue culture bioreactors are made of glass or stainless steel [88]. The morphology, rheology, growth, and production behavior of the culture should be considered comprehensively when selecting the most suitable bioreactor type. According to the state of the growth environment, bioreactors can be divided into liquid phase bioreactors, gas phase bioreactors, and gas phase and liquid phase combined bioreactors. According to the structure of culture materials, they can be divided into suspension cell culture bioreactors and tissue culture bioreactors. Liquid phase reactors include mechanically driven reactors and gas-driven reactors. The stirred reactor will cause damage to the culture materials. For gas-driven reactors, the oxygen flux and shear force should be considered [89]. In recent years, reactors dedicated to plant tissue and cell culture (PTCC) have been continuously improved including airlift, tubular membrane, silicone-tubing aerated, slug bubble, disposable wave and orbitally shaken, spray, helical ribbon impeller, rotating wall vessel (RWV) bioreactors, and stirred tank reactors [6]. Stirred reactors, rotating drum reactors, airlift reactors, bubble columns, fluidized bed reactors, and packed bed reactors are the most commonly used industrial and commercial reactors [88]. The advantages and disadvantages of a variety of reactors have been summarized by our predecessors, so will not be elaborated here. Currently, one problem is the possibility of microbial and/or viral contamination in bioreactors. These can occur by accident at any step in the culture process, and the contamination can be bacteria, mycoplasma, viruses, parasites, and fungi.

In general, aseptic techniques, fungicides, and or antibiotics can be used to prevent contamination from occurring. The current work is focused on detecting bacterial contamination in bioreactors [90].

Stirred tank reactors (STRs) are currently the most functional bioreactors in PTCC. Airlift and bubble column systems were basically designed to improve the biological oxygen demand of the culture. Helical and double helical ribbon impellers appeared to be efficient for the suspension culture of high-density plants. An important advance in plant cell and tissue culture is the use of disposable bioreactors. A disposable reactor replaces the stainless-steel material in the device with plastic products, which greatly reduces the preparation time and cost of the reactor. Another advantage of disposable bioreactors is that they do not require cleaning, which will reduce the probability of contamination. At present, due to limited experience in the use of disposable reactors, a lack in the physical strength of plastic materials, and other reasons, this type of reactor is not widely used [91]. A large number of condition optimization experiments can be carried out simultaneously by using laboratory glassware or designing small reactors. The laboratory optimization of reactor culture conditions include temperature, pH value, oxygen supply, solid–liquid ratio, medium composition, and elicitor.

3.2.3. Dynamic Monitoring during Scale-Up Culture

A variety of detecting techniques for omics research such as high-performance liquid phase (HPLC), mass spectrum (MS), and nuclear magnetic resonance (NMR) can be used to determine the type and content of animal and plant metabolites. Some have been used to monitor robust and high-yield biological production processes. The conventional physical parameters that should be detected are pH, temperature, and dissolved oxygen. Sensors for bioreactor detection need to accurately measure the concentration of various nutrients and metabolites in the culture medium without affecting the normal production process. For nutrients and metabolites in reactors, spectroscopy is currently used for monitoring. Spectroscopy is particularly important for the biotechnology industry because it can be easily conducted and achieve online real-time measurement. The overview of spectroscopy is to study the interaction between matter and electromagnetic radiation. Analytical instruments include infrared spectroscopy, fluorescence spectroscopy, etc. Finally, statistical and mathematical techniques are used to analyze the chemical data of the monitored material. Each spectral method has its own advantages and disadvantages, which means that different spectral methods have different components and bioreactor matrices to analyze. Spectroscopy for bioreactor monitoring has been used in microbial fermentation, cell culture, etc. [92]. NMR is an important analytical technique in material science and medicine. NMR-based metabolomics is used to monitor the concentration of metabolites and can be used as a tool to study the dynamics of plant cell metabolism and nutrition management. The advantages of NMR include fast, reliable, and allow for online measurements of cell media, etc. [93]. Ninad Mehendale [94] proposed an NMR compatible platform for automatic real-time monitoring of biochemical reactions using a flow shuttle configuration for the real-time monitoring of biochemical reactions. Another advantage of the proposed low-cost platform is high spectral resolution. Nuclear magnetic resonance metabolomics can monitor metabolite reactions in different optimization processes.

3.3. Application of Nanomaterials in Plant Tissue Culture

Nanomaterials have a wide range of applications in biotechnology, and have played a great role in medical, agricultural, pharmaceutical, and other fields. They can be used in sensor development, agricultural chemical degradation, soil remediation, drug delivery, and so on. Various materials are used to manufacture NPs such as metal oxides, ceramics, silicates, magnetic materials, semiconductor quantum dots (QDs), lipids, and polymers. Dendritic macromolecules and emulsion encapsulated polymer nanomaterials have controlled and distressed release capabilities, and metal-based nanomaterials have a size dependence. Nanomaterial formulations increase system activity due to higher surface area, solubility, and smaller particle size [95]. In the field of plant tissue culture, nanomaterials are widely used as elicitor stimulation to increase the accumulation of secondary metabolites [96], or as a transport carrier for plant hormones [97] and other substances to specifically harvest effective substances in plant cells [17].

3.3.1. Nano-Elicitor

The Mechanism of Action of Nano-Elicitor

An elicitor stimulates the accumulation of secondary metabolites by stimulating the defense system in plant metabolism. Biological elicitors and abiotic elicitors are commonly used in plant tissue culture. Nanomaterials have been widely used in medicine, energy, biotechnology, and other fields due to their unique properties. In recent years, studies have found that using nanomaterials as elicitors can greatly increase the accumulation of secondary metabolites. The properties of nanomaterials can be precisely controlled by controlling the shape, size, and chemical composition of the materials. Nanoparticles can increase the production of reactive oxygen species and hydroxyl radicals that distort cell membranes, leading to changes in penetrability, facilitating the entry of nanoparticles into plant cells, and stimulating the production of secondary metabolites [98]. The process of nanomaterials acting on plant cells is shown in Figure 4. Controlled release of active ingredients is due to the slow release characteristics of nanomaterials, the combination of ingredients and materials, and the control of environmental conditions [95].

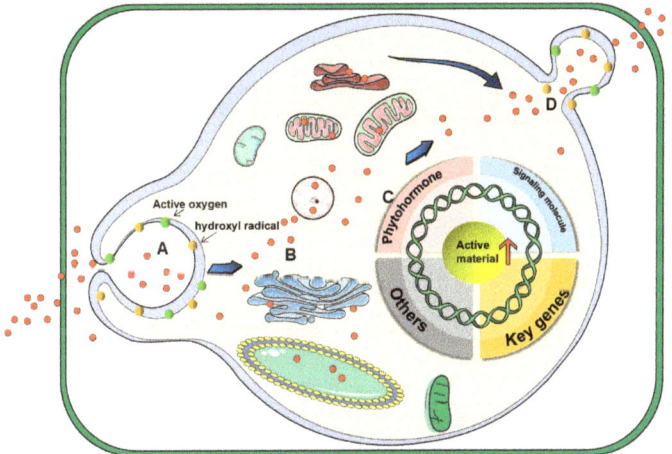

Figure 4. The process by which nano-elicitors act on plant cells. (**A**): The nano-elicitor increases reactive oxygen species and hydroxyl radicals, which can enhance the permeability of cell membranes. (**B**): The elicitor enters the cytoplasm and various organelles. (**C**): The elicitor stimulates intracellular hormones, related signaling molecules, key genes, and other components related to the synthesis of active substances, ultimately leading to an increase in the production of active substances. (**D**): Nano-elicitors are excluded from the cell by exocytosis.

Application of Nano-Elicitors

Common nanomaterials include multiwalled carbon nanotubes (MWCNTs), single walled carbon nanotubes (SWCNTs), graphene, and fullerene C70 [99]. However, not all nanomaterials can be used as plant tissue culture elicitors. In addition to the reported common metal nanomaterials (such as AgNPs, ZnNPs, CuNPs, TiNPs, etc.), there are other new nanomaterials used in plant tissue culture. Carbon nanotube nanomaterials and multi-walled carbon nanotubes can induce about twice the increase in the total phenol content (TPC) in thyme (seedlings); for flavonoids (TFC), this number would be 1.09-fold [100]. Studies have shown that the localization of graphene nanomaterial (GNS) inside the chloroplasts may be to activate photosynthetic pigments, thereby exhibiting a stimulating effect on fructose, sucrose, and starch. This will increase the pepper and eggplant yields, which will have no bad effect on the plants themselves. Nanomaterials can be combined with a variety of chemical functional groups to form composite nanomaterials, which indicates better influences on the content of secondary metabolites. The effects of silver nanoparticles, graphene, and their nanocomposite nanoparticles on stevia plants have been studied. Plants treated with nanocomposites were found to have more stevioside and rebaudioside [99]. At the same time, the effects of MgONPs, perlite, and composite nanomaterials made of MgONPs, perlite, and plant extracts on the volatile content of *Melissa officinalis* leaf were compared, and it was found that the composite elicitor could increase the content of the product by 0–17% [101].

Nano-elicitors play an important role in the production of many natural products. The discovery and preparation of nanomaterials are key technologies. Therefore, which nanomaterials can be used as potential elicitors for plant tissue culture? To answer this question, we conducted the following analyses. First, due to the application of new nanomaterials with the development of nanotechnology, more and more new nanomaterials have been created, so there is a need to combine biotechnology and engineering to achieve multidisciplinary cooperation and development. Second, although some composite nanomaterials have not been used in plant tissue culture, they have potential capabilities such as natural polysaccharide composite nanomaterials [102]. This material contains naturally extracted chemical components that are safer for plants, and this composite material generally has antibacterial and antioxidant functions that can reduce bacterial contamination in plant tissue culture and achieve a greener production model. Third, simple nanomaterials that bind to biological or abiotic elicitors [103] also have antibacterial ability, and it is also convenient to conduct parallel control experiments with common elicitors to obtain a better one. The reported effects of different types of nano-elicitors on plant tissues are shown in Table 3.

Table 3. The effects of different types of nano-elicitors on plant tissues.

Classification of Nanomaterials	Size/Concentration	Source of Plant Materials	Results
AgNPs	0, 25, 50, 100 nm, and 200 ppm	*Rosmarinus officinalis* L.	Increased carnosic acid (CA) levels by more than 11% [104].
FeNPs	75 mg/L	Hairy-root of *Dracocephalum kotschyi*	Rosmarinic acid (RA) increased by 9.7-fold, xanthomicrol, cirsimaritin, and isokaempferide increased by 11.87, 3.85, and 2.27-fold, respectively [105].
Chitosan encapsulated zinc oxide nanocomposite	10–50 nm,	*Capsicum annuum*	Photosynthetic pigments (about 50%), proline (about 2 times), proteins (about 2 times), antioxidant enzyme activity (about 2 times), PAL activity (about 2 times), soluble phenols (40%), and alkaloids (60%) [106].
Single-wall carbon nano tubes	25, 50, 100, 125, and 250 µg/mL	Callus of *Thymus daenensis*	Total phenolic (TPC) content increased by 1.290 ± 0.19 mgGAEg^{-1} DW, total flavonoid (TFC) content increased by 2.113 ± 0.05 mgGAEg^{-1} DW improved [107].
Graphene-based nanomaterials (GBNs)	50, 100, and 150 mg/L	*Ganoderma lucidum*	All GBNs increased the ganoderic acid (GA) content [108].
Magnetite nanoparticles (MNPs)	10.77, 20.5, 29.3 nm and 0.5, 1, 2 ppm	Callus of *Ginkgo biloba* L.	2 ppm + 10 nm MNPs increased the content of quercetin, kaempferol, p-coumarin, luding, caffeic acid, ginkgolide A, etc. [109].
SiO$_2$NPs	2 mM	*Crocus sativus* L.	Increased the content of crocin and activity of superoxide dismutase (SOD), catalase (CAT), and ascorbate peroxidase (APX) [110].
Nano-TiO$_2$	10, 60, and 120 mg/L	Callus of *Salvia tebesana*	The combination with methyl jasmonate increased the total phenols. O-diphenols, phenolic acid, flavonoid, flavane, flavonol, and proanthocyanidin were all increased [111].
TiO$_2$/perlite nanocomposites (NCs)	15.50–24.61 nm	Callus of *Hypericum perforatum*	Induced the production of hypericin, pseudohypericin, and volatile compounds [112].
MgO/perlite nanocomposites (NCs)	10–30 nm	*Melissa officinalis*	Elevated volatile compounds. The new compound rosmarinic acid was detected [101].

3.3.2. Nanomaterials as Transport Carriers

Nano-transport technology can transport some chemicals into cells in a targeted and efficient manner, which can effectively reduce the environmental pollution caused by plant growth. Currently, inorganic nanoparticles have been studied for drug delivery applications [113]. Nano-fertilizers and nano-insecticides have increased the economic benefits of agricultural products by 20–30% [114,115]. Nanomaterials for drug delivery such as calcium phosphate, gold, carbon materials, silicon oxide, iron oxide, and lactate dehydrogenase have multifunctional properties suitable for cell delivery such as wide availability, rich surface function, good biocompatibility, potential target delivery ability, and controllable drug release ability [116]. Plant development can be manipulated by adding plant growth regulators at a specific stage of growth or development, and further developmental responses can be stimulated by adding growth regulators such as auxin, cytokinin, gibberellin, ethylene, or abscisic acid. Auxin and cytokinin are the most important of these growth regulators. However, because these substances are greatly affected by the plant growth environment and the species itself, their content in plants is very unstable. On the other hand, PGRs are easily degraded by light, temperature, and other environmental factors, resulting in the loss of their activity, thereby affecting the growth of plant organs [97]. To promote the growth of plant tissue culture materials, we need to find a more efficient application model to ensure an even distribution of plant growth regulators in plant tissues. At the same time, we also need to supply plant growth regulators according to the needs of the plant tissue to achieve the rapid accumulation of plant tissue biomass [117]. Nano-carriers have great prospects for the application of plant growth regulators such as layered double hydroxide (LDH) [15] and chitosan nanoparticles [16].

Layered Double Hydroxide Nanomaterials as Transport Carriers

LDH has certain advantages as transport carriers. The hydroxide component of LDH can effectively prevent external enzymes and oxygen from destroying the loaded drugs in the interlayer. The intercalated structure makes the drugs carried by LDH have a controlled release effect. LDH-drug mixed nanoparticles can achieve cell localization by adjusting the zeta potential. LDH material has low cytotoxicity [116], and its size should be kept below 150–200 nm for endocytosis [118]. LDH has been used as a gene and in drug delivery in recent years [119] such as in an antioxidant carrier, fertilizer carrier [119], plant nutriment [15], pesticide slow release carrier [120], herbicide carrier [121], and plant hormone transport carriers.

Ilya Shlar et al. [116] found that LDH could be used as the carrier of IAA and acted on the plants of *Vigna radiata* (L.) Wilczek. The molecules of IAA are effectively inserted into LDH through the co-precipitation effect. The insertion can protect IAA molecules from enzymatic degradation, and the intercalation can achieve a sustained release performance. The results showed that the IAA–LDH complex treatment increased the rooting efficiency by 4.6 times and improved the biological activity of IAA to promote adventitious root development. Hussein et al. [122] found that the use of zinc-aluminum layered double hydroxide nanocomposites could achieve the controlled release of plant growth regulator NAA in the film. Yanfang Liu et al. [123] used the ion exchange method to insert sodium naphthalene acetate (NAA, a plant growth regulator) into layered double hydroxide Mg/Al-LDH, so that NAA showed a significant release effect. Vander A. de Castro et al. [117] used an alginate polymer (a substance with good biodegradability) to encapsulate zinc-aluminum LDH and NAA to act on plant seeds. The results showed that the alginate film containing ZnAl–NAA–LDH enhanced the root area, fresh root material, and shoot length of the plants. Inas H. Hafez et al. [124] constructed a gibberellic acid (GA) nanohybrid system using inorganic magnesium aluminum layered double hydroxide (LDH) as raw material. The biodegradation process of intercalated GA is characterized by a long soil storage period and slow degradation rate. Shifeng Li et al. [125] prepared β-naphthoxyacetic acid (BNOA) layered double hydroxides (LDHs) by the co-precipitation method and discussed its release mechanism. They believed that with the cleavage of the LDH nanolayer structure, the nanohybrid of BNOA-LDHs has good controlled release characteristics, and pH is the key factor.

Chitosan Nanoparticles as a Transport Carrier of Substances in Plant Tissue Culture

Compared with synthetic polymers, natural polymers have good biocompatibility and biodegradability. The structure of some natural polymers is similar to that of biological macromolecules, and they are more easily recognized, utilized, and metabolized by organisms [126]. Chitosan is obtained by the acetylation of chitin after alkali treatment. The structure of the amino group and carboxyl group can make chitosan functionalized such as carboxymethylation, etherification, esterification, crosslinking copolymerization, and other modifications [127]. Research has found that chitosan has good biodegradability, bioadhesion, ecological safety, and high biocompatibility due to its high charge density, the interaction of amine and carboxyl groups, and the existence of hydrogen bonds [128]. With these properties, chitosan can be used in biotechnology, medicine, food, agriculture, environment, and other fields [129]. Chitosan nanocarrier systems have also been developed for use in plant growth regulators as chitosan molecules can slowly release plant growth regulators. There are many benefits of controlled release such as protecting plant hormones from the environment and protecting plant cells from the risk of explosive release [16].

Chitosan nanomaterials have the characteristics of promoting plant growth and increasing the content of secondary metabolites. Farhad Mirheidari et al. [130] applied IAA, GA3, and chitosan nanofibers (alone or in combination) to *Roselle* plants. The results showed that IAA + GA3 + CNF (800 + 800 + 100 mg/L) stimulated the growth parameters of rose plants and promoted the content of healthy phytochemicals (ascorbic acid, β-carotene, anthocyanin) and the improvement in plant antioxidant capacity. This indicates that the combined application of plant growth regulators and chitosan nanofibers (CNFs) has an important impact on various growth parameters and metabolite status, and reflects the positive effect of chitosan nanofibers on plants. In plant tissue culture, chitosan is mostly used as a transport carrier for plant hormones. Salicylic acid-chitosan nanoparticles ensure their continuous availability in plants by slowly releasing SA. The seedling index was 1.6 times higher than that of the control group, and the chlorophyll (a,b) content (1.46 times), ear length, grain number per ear, and grain weight per pot were also increased [131]. Plant hormones such as SA can fight toxic symptoms caused by heavy metal stress such as chlorophyll synthesis and biomass loss [132]. Using SA-loaded chitosan nanoparticles to act on plants can promote the absorption of SA by plants. Studies have shown that salicylic acid nanoparticles (SANP) improve plant growth and phytoremediation efficiency under arsenic stress, and enhance the ability of plants to resist arsenic stress [133]. As an essential element for plant growth, Cu may also cause heavy metal toxicity to plants. Any organic compound that can affect plant development can be seen as a plant growth regulator. It has been proven that the complex of 1-hydroxy-1-methoxycarbonyl-copper (Cu(II)) can be used as a new plant growth regulator. This regulator can realize the controlled release by delivering the Cu(II) complex to plants using chitosan-coated calcium alginate microcapsules [134]. Some chitosan nanoparticles can also be used as carriers of bacteria. J. J. Perez et al. [135] developed a novel, green, low-cost chitosan-starch hydrogel as a delivery vehicle for plant growth-promoting bacteria—*Azospirillum*. The results showed that the release of bacteria in saline was gradual and could be used as a bio-nano fertilizer. Different chitosan nanomaterials as plant tissue culture material transport carriers are shown in Table 4.

Table 4. Different chitosan nanomaterials as transport carriers in plant tissue culture.

Chitosan Nanomaterials and Carrier Materials	Dimensions (MD)	Results
Novel chitosan/alginate microcapsules simultaneously loaded with copper(II)cations and trichoderma viride	-	Chitosan/alginate microcapsules may incorporate both viral spores and chemical bioactivators without inhibiting their activity [134].
Deoxycholic acid carboxymethyl chitosan (DACMC) loaded with rotenone	91.3–140.0 nm	The in vitro release data of rotenone-loaded DACMC followed the Ritger and Peppas Case II transport mechanism. Highlights the potential of DACMC to reduce the use of organic solvents in the production of water-insoluble pesticides [136].
Alginate/chitosan nanoparticles encapsulated GA3	472–503 nm	Nanoparticles can improve the biological activity of gibberellic acid and have good application prospects in agriculture. Good performance and time stability [137].
Nanocarriers of plant growth regulator gibberellic acid (GA3) composed of alginate/chitosan (ALG/CS) and chitosan/tripolyphosphate (CS/TPP)	450 ± 10 nm	ALG/CS-GA3 nanoparticles have higher stability and efficiency in increasing the leaf area and chlorophyll and carotenoid content [97].
SA-CS NPs	368.7 ± 0.05 nm	SA-CS NPs can significantly affect the source activity by slowly releasing SA to manipulate various physiological and biochemical reactions of wheat plants [131].
SA-CS NPs	-	The results showed that the activity of antioxidant defense enzymes in maize increased, and the balance of reactive oxygen species (ROS) and the deposition of cell wall lignin increased, which had a positive effect on disease control and *maize* plant growth. It is a potential biological promoter [138].
Silica or chitosan encapsulated salicylic acid (SA) capsules	9.6–11.0 mm, 7.2–8.5 mm	In the in vitro system, the plants treated with a low proportion capsule had the best antifungal effect. At the same time, the capsule treated plants had higher levels of root and rosette development than the free SA treated plants [139].
Chitosan nanoparticles loaded with indole-3-acetic acid (IAA)	149–183 nm	CNPs-IAA can be applied to the hydroponic crop of crocantela variety *Latuca sativa* L. and has beneficial effects on plant growth, increasing the number of lettuce leaves by 30.9% [140].

3.3.3. Nano-Assisted Harvesting Technology

Plant secondary metabolites are synthesized and accumulated in plant cells, but traditional recovery methods generally dry plant tissues, which is destructive to the tissue. In addition to destroying expensive transgenic plant cell cultures, the activity of unstable biomolecules may be lost during the solvent extraction process, resulting in reduced yields [141]. Nano-harvesting refers to the use of nanoparticles to combine and carry active molecules away from plant cells. This method can not only obtain secondary metabolites, but also protect the integrity of the original cells and is a sustainable method for harvesting metabolites from plant tissues. Engineered mesoporous silica nanoparticles (MSNPs) have the characteristics of high surface area, unique size, shape, pore structure, and surface functionalization. Such properties allow MSNPs to be easily modified by specific functional groups to target binding bioactive materials through cell barriers. Mesoporous silica nanoparticles designed with an amine function can bind to active substances of flavonoids and carry them out of the cell [17]. Silica nanoparticles were found to obtain flavonoids from plant cultures without significant harm to host plants [142]. After the harvest of nanoparticles, the roots were found to be re-synthesized by regulation [142].

The cellular uptake and excretion mechanisms of MSNPs are important in designing novel biomolecule separation and delivery applications. Positively charged MSNP particles have been shown to facilitate penetration into the membranous and cell walls. The amine-functionalized MSNPs spontaneously entered and exited the plant cells through dynamic exchange for 20 ± 5 min. Ti-functionalized weakly charged MSNPs were absorbed and excreted through a thermal activation mechanism, whereas amine-modified positively charged particles were absorbed and excreted mainly through direct cell penetration. Particle size and surface properties (charge) are the two most critical factors for cellular uptake, but the extent of uptake depends on the type of plant. Excretion mechanisms depend on the cell type, nanoparticle size, shape, and surface modification, and the surface charge may be a decisive factor in controlling transport and excretion [17].

The nano-capture of secondary metabolites is a new biotechnology for the separation and transfer of active biological molecules in living tissues. At present, nano-harvesting technology is only applied to the capture of flavonoids, but through the modification of nanomaterials, other active ingredients are also expected to be extracted.

4. Conclusions

The cultivation technology of plant cells, tissues, and organs is a modern biological means to solve the shortage of medicinal resources and the shortage of wild plant resources. The ultimate goal of plant tissue culture is to achieve large-scale biomass production and the accumulation of precious secondary metabolites. The mechanism of plant tissue growth and development is constantly being updated, and the technology in various fields is gradually being integrated into tissue culture technology. For example, data-driven modeling technology takes artificial intelligence modeling and optimization algorithm as the means of prediction, accurately and efficiently analyzes the cultivation conditions, develops new cultivation methods, and systematically and scientifically optimizes the prediction. Data-driven modeling is one of the most applicable methods and plays an important role as a predictive tool for modeling complex biological research. From another aspect, the amplification of the bioreactor culture and improvement in the dynamic monitoring system are conducive to the realization of the industrial automatic production of culture. Designing different bioreactor types will not only help to improve and maintain high productivity, but also reduce the process costs. Suitable types of reactors should be selected for different cultures. Nanotechnology has been widely used in plant tissue culture. However, nanotechnology is not limited to the content introduced in this article. Its mode of action is more extensive and can be applied in various processes of plant tissue culture. The role of nanomaterials largely depends on their chemical and mineral composition, size, and sometimes the shape and concentration of application. Technology in various fields with continuous update, and more innovative ways will be applied to the plant tissue culture process. In summary, the application of new technologies in intelligent algorithms, instrument engineering, nanomaterials, and other fields to adapt to the standardization of commercial applications has had a positive impact on plant tissue culture and the industrial production of secondary metabolites.

Author Contributions: B.J., L.X., J.W. and W.G. designed and supervised the research; B.J. and L.X. analyzed the documents; Y.Z. and W.M. designed the pictures and tables assistance; B.J., L.X. and Y.Z. supervised and completed the writing; K.-Y.P. and S.-Y.P. reviewed the manuscript and gave valuable opinions. W.G. agrees to serve as the author responsible for contact and communication. All authors have read and agreed to the published version of the manuscript.

Funding: This research was funded by the National Key Research and Development Program (2020YFA0907903), the ability establishment of the sustainable use for valuable Chinese medicine resources (2060302) and the Scientific Research Transformation Foundation of Wenzhou Safety (Emergency) Institute of Tianjin University.

Data Availability Statement: No new data were created or analyzed in this study. Data sharing is not applicable to this article.

Conflicts of Interest: The authors declare no conflict of interest.

References

1. Chandran, H.; Meena, M.; Barupal, T.; Sharma, K. Plant Tissue Culture as a Perpetual Source for Production of Industrially Important Bioactive Compounds. *Biotechnol. Rep.* **2020**, *26*, e00450. [CrossRef] [PubMed]
2. Espinosa-Leal, C.A.; Puente-Garza, C.A.; García-Lara, S. In Vitro Plant Tissue Culture: Means for Production of Biological Active Compounds. *Planta* **2018**, *248*, 1–18. [CrossRef] [PubMed]
3. Hesami, M.; Naderi, R.; Tohidfar, M.; Yoosefzadeh-Najafabadi, M. Development of Support Vector Machine-Based Model and Comparative Analysis with Artificial Neural Network for Modeling the Plant Tissue Culture Procedures: Effect of Plant Growth Regulators on Somatic Embryogenesis of Chrysanthemum, as a Case Study. *Plant Methods* **2020**, *16*, 112. [CrossRef] [PubMed]
4. Geipel, K.; Song, X.; Socher, M.L.; Kümmritz, S.; Püschel, J.; Bley, T.; Ludwig-Müller, J.; Steingroewer, J. Induction of a Photomixotrophic Plant Cell Culture of Helianthus Annuus and Optimization of Culture Conditions for Improved A-Tocopherol Production. *Appl. Microbiol. Biotechnol.* **2014**, *98*, 2029–2040. [CrossRef]
5. Jamshidi, S.; Yadollahi, A.; Arab, M.M.; Soltani, M.; Eftekhari, M.; Sabzalipoor, H.; Sheikhi, A.; Shiri, J. Combining Gene Expression Programming and Genetic Algorithm as a Powerful Hybrid Modeling Approach for Pear Rootstocks Tissue Culture Media Formulation. *Plant Methods* **2019**, *15*, 136. [CrossRef]
6. Valdiani, A.; Hansen, O.K.; Nielsen, U.B.; Johannsen, V.K.; Shariat, M.; Georgiev, M.I.; Omidvar, V.; Ebrahimi, M.; Dinanai, E.T.; Abiri, R. Bioreactor-Based Advances in Plant Tissue and Cell Culture: Challenges and Prospects. *Crit. Rev. Biotechnol.* **2019**, *39*, 20–34. [CrossRef]
7. Khanahmadi, M.; Paek, K.Y. Bioreactor Technology for Sustainable Production of Valuable Plant Metabolites: Challenges and Advances. In *Crop Improvement: Sustainability Through Leading-Edge Technology*; Springer: Cham, Switzerland, 2017; pp. 169–189.
8. Kim, H.K.; Choi, Y.H.; Verpoorte, R. Nmr-Based Plant Metabolomics: Where Do We Stand, Where Do We Go? *Trends Biotechnol.* **2011**, *29*, 267–275. [CrossRef]
9. Kim, D.H.; Gopal, J.; Sivanesan, I. Nanomaterials in Plant Tissue Culture: The Disclosed and Undisclosed. *RSC Adv.* **2017**, *7*, 36492–36505. [CrossRef]
10. Ma, X.; Geiser-Lee, J.; Deng, Y.; Kolmakov, A. Interactions between Engineered Nanoparticles (Enps) and Plants: Phytotoxicity, Uptake and Accumulation. *Sci. Total. Environ.* **2010**, *408*, 3053–3061. [CrossRef] [PubMed]
11. Zhang, H.; Demirer, G.S.; Zhang, H.; Ye, T.; Goh, N.S.; Aditham, A.J.; Cunningham, F.J.; Fan, C.; Landry, M.P. DNA Nanostructures Coordinate Gene Silencing in Mature Plants. *Proc. Natl. Acad. Sci. USA* **2019**, *116*, 7543–7548. [CrossRef] [PubMed]
12. Khan, M.A.; Khan, T.; Riaz, M.S.; Ullah, N.; Ali, H.; Nadhman, A. Plant Cell Nanomaterials Interaction: Growth, Physiology and Secondary Metabolism. In *Comprehensive Analytical Chemistry*; Elsevier: Amsterdam, The Netherlands, 2019; pp. 23–54.
13. Patel, H.; Krishnamurthy, R. Elicitors in Plant Tissue Culture. *J. Pharmacogn. Phytochem.* **2013**, *2*, 60–65.
14. Alcalde, M.A.; Perez-Matas, E.; Escrich, A.; Cusido, R.M.; Palazon, J.; Bonfill, M. Biotic Elicitors in Adventitious and Hairy Root Cultures: A Review from 2010 to 2022. *Molecules* **2022**, *27*, 5253. [CrossRef] [PubMed]
15. Wu, H.; Zhang, H.; Li, X.; Zhang, Y.; Wang, J.; Wang, Q.; Wan, Y. Optimized Synthesis of Layered Double Hydroxide Lactate Nanosheets and Their Biological Effects on *Arabidopsis* Seedlings. *Plant Methods* **2022**, *18*, 17. [CrossRef] [PubMed]
16. Mujtaba, M.; Khawar, K.M.; Camara, M.C.; Carvalho, L.B.; Fraceto, L.F.; Morsi, R.E.; Elsabee, M.Z.; Kaya, M.; Labidi, J.; Ullah, H. Chitosan-Based Delivery Systems for Plants: A Brief Overview of Recent Advances and Future Directions. *Int. J. Biol. Macromol.* **2020**, *154*, 683–697. [CrossRef]
17. Khan, M.A.; Fugate, M.; Rogers, D.T.; Sambi, J.; Littleton, J.M.; Rankin, S.E.; Knutson, B.L. Mechanism of Mesoporous Silica Nanoparticle Interaction with Hairy Root Cultures During Nanoharvesting of Biomolecules. *Adv. Biol.* **2021**, *5*, 2000173. [CrossRef]
18. Ramawat, K.G. An Introduction to the Process of Cell, Tissue, and Organ Differentiation, and Production of Secondary Metabolites. In *Plant Cell and Tissue Differentiation and Secondary Metabolites: Fundamentals and Applications*; Springer: Cham, Switzerland, 2021; pp. 1–22.
19. Su, Y.-H.; Liu, Y.-B.; Zhang, X.-S. Auxin–Cytokinin Interaction Regulates Meristem Development. *Mol. Plant* **2011**, *4*, 616–625. [CrossRef]
20. González-García, M.-P.; Vilarrasa-Blasi, J.; Zhiponova, M.; Divol, F.; Mora-García, S.; Russinova, E.; Caño-Delgado, A.I. Brassinosteroids Control Meristem Size by Promoting Cell Cycle Progression in *Arabidopsis* Roots. *Development* **2011**, *138*, 849–859. [CrossRef]
21. Vilarrasa-Blasi, J.; González-García, M.-P.; Frigola, D.; Fàbregas, N.; Alexiou, K.G.; Lopez-Bigas, N.; Rivas, S.; Jauneau, A.; Lohmann, J.U.; Benfey, P.N. Regulation of Plant Stem Cell Quiescence by a Brassinosteroid Signaling Module. *Dev. Cell* **2014**, *30*, 36–47. [CrossRef] [PubMed]
22. Sarkar, A.K.; Luijten, M.; Miyashima, S.; Lenhard, M.; Hashimoto, T.; Nakajima, K.; Scheres, B.; Heidstra, R.; Laux, T. Conserved Factors Regulate Signalling in *Arabidopsis* Thaliana Shoot and Root Stem Cell Organizers. *Nature* **2007**, *446*, 811–814. [CrossRef]
23. Vasil, V.; Vasil, I.K. Plant Regeneration from Friable Embryogenic Callus and Cell Suspension Cultures of *Zea mays* L. *J. Plant Physiol.* **1986**, *124*, 399–408. [CrossRef]
24. Le, K.-C.; Jeong, C.-S.; Lee, H.; Paek, K.-Y.; Park, S.-Y. Ginsenoside Accumulation Profiles in Long-and Short-Term Cell Suspension and Adventitious Root Cultures in Panax Ginseng. *Hortic. Environ. Biotechnol.* **2019**, *60*, 125–134. [CrossRef]
25. Karwasara, V.S.; Tomar, P.; Dixit, V.K. Influence of Fungal Elicitation on Glycyrrhizin Production in Transformed Cell Cultures of *Abrus precatorius* Linn. *Pharmacogn. Mag.* **2011**, *7*, 307.

26. Osterc, G.; Petkovšek, M.M.; Stampar, F. Quantification of Iaa Metabolites in the Early Stages of Adventitious Rooting Might Be Predictive for Subsequent Differences in Rooting Response. *J. Plant Growth Regul.* **2016**, *35*, 534–542. [CrossRef]
27. Guo, D.-P.; Zhu, Z.-J.; Hu, X.-X.; Zheng, S.-J. Effect of Cytokinins on Shoot Regeneration from Cotyledon and Leaf Segment of Stem Mustard (*Brassica juncea* Var. *Tsatsai*). *Plant Cell Tissue Organ Cult.* **2005**, *83*, 123–127. [CrossRef]
28. Ikeuchi, M.; Sugimoto, K.; Iwase, A. Plant Callus: Mechanisms of Induction and Repression. *Plant Cell* **2013**, *25*, 3159–3173. [CrossRef] [PubMed]
29. Horstman, A.; Bemer, M.; Boutilier, K. A Transcriptional View on Somatic Embryogenesis. *Regeneration* **2017**, *4*, 201–216. [CrossRef] [PubMed]
30. Garcia, C.; de Almeida, A.-A.F.; Costa, M.; Britto, D.; Valle, R.; Royaert, S.; Marelli, J.-P. Abnormalities in Somatic Embryogenesis Caused by 2, 4-D: An Overview. *Plant Cell Tissue Organ Cult.* **2019**, *137*, 193–212. [CrossRef]
31. Goldman, J.A.; Poss, K. Gene Regulatory Programmes of Tissue Regeneration. *Nat. Rev. Genet.* **2020**, *21*, 511–525. [CrossRef]
32. Beeckman, T.; De Smet, I. Pericycle. *Curr. Biol.* **2014**, *24*, R79–R378. [CrossRef]
33. Bustillo-Avendaño, E.; Ibáñez, S.; Sanz, O.; Barros, J.A.S.; Gude, I.; Perianez-Rodriguez, J.; Micol, J.L.; Del Pozo, J.C.; Moreno-Risueno, M.A.; Pérez-Pérez, J.M. Regulation of Hormonal Control, Cell Reprogramming, and Patterning During De Novo Root Organogenesis. *Plant Physiol.* **2018**, *176*, 1709–1727. [CrossRef]
34. Ho, T.-T.; Ha, T.M.N.; Nguyen, T.K.C.; Le, T.D. Pilot-Scale Culture of Adventitious Root for the Production of Pharmacology Active from Medicinal Plants: A Mini Review. *BIO Web Conf.* **2021**, *40*, 03003. [CrossRef]
35. Pacifici, E.; Polverari, L.; Sabatini, S. Plant Hormone Cross-Talk: The Pivot of Root Growth. *J. Exp. Bot.* **2015**, *66*, 1113–1121. [CrossRef] [PubMed]
36. Gonin, M.; Bergougnoux, V.; Nguyen, T.D.; Gantet, P.; Champion, A. What Makes Adventitious Roots? *Plants* **2019**, *8*, 240. [CrossRef] [PubMed]
37. Perez-Garcia, P.; Moreno-Risueno, M. Stem Cells and Plant Regeneration. *Dev. Biol.* **2018**, *442*, 3–12. [CrossRef]
38. Bellini, C.; Pacurar, D.I.; Perrone, I. Adventitious Roots and Lateral Roots: Similarities and Differences. *Annu. Rev. Plant Biol.* **2014**, *65*, 639–666. [CrossRef]
39. Jung, J.K.H.; McCouch, S. Getting to the Roots of It: Genetic and Hormonal Control of Root Architecture. *Front. Plant Sci.* **2013**, *4*, 186. [CrossRef]
40. Sorin, C.; Bussell, J.D.; Camus, I.; Ljung, K.; Kowalczyk, M.; Geiss, G.; McKhann, H.; Garcion, C.; Vaucheret, H.; Sandberg, G. Auxin and Light Control of Adventitious Rooting in *Arabidopsis* Require Argonaute1. *Plant Cell* **2005**, *17*, 1343–1359. [CrossRef]
41. Correa-Aragunde, N.; Graziano, M.; Chevalier, C.; Lamattina, L. Nitric Oxide Modulates the Expression of Cell Cycle Regulatory Genes During Lateral Root Formation in Tomato. *J. Exp. Bot.* **2006**, *57*, 581–588. [CrossRef]
42. Kupke, B.M.; Tucker, M.R.; Able, J.A.; Porker, K.D. Manipulation of Barley Development and Flowering Time by Exogenous Application of Plant Growth Regulators. *Front. Plant Sci.* **2022**, *12*, 3171. [CrossRef]
43. Du, Y.; Scheres, B. Lateral Root Formation and the Multiple Roles of Auxin. *J. Exp. Bot.* **2018**, *69*, 155–167. [CrossRef]
44. Aloni, R.; Aloni, E.; Langhans, M.; Ullrich, C.I. Role of Cytokinin and Auxin in Shaping Root Architecture: Regulating Vascular Differentiation, Lateral Root Initiation, Root Apical Dominance and Root Gravitropism. *Ann. Bot.* **2006**, *97*, 883–893. [CrossRef] [PubMed]
45. Phillips, G.C.; Garda, M. Plant Tissue Culture Media and Practices: An Overview. *Vitr. Cell. Dev. Biol.-Plant* **2019**, *55*, 242–257. [CrossRef]
46. Zhu, L.; Jiang, B.; Zhu, J.; Xiao, G. Auxin Promotes Fiber Elongation by Enhancing Gibberellic Acid Biosynthesis in Cotton. *Plant Biotechnol. J.* **2022**, *20*, 423. [CrossRef] [PubMed]
47. Ludwig-Müller, J.R. Indole-3-Butyric Acid in Plant Growth and Development. *Plant Growth Regul.* **2000**, *32*, 219–230. [CrossRef]
48. Fattorini, L.; Hause, B.; Gutierrez, L.; Veloccia, A.; Della Rovere, F.; Piacentini, D.; Falasca, G.; Altamura, M.M. Jasmonate Promotes Auxin-Induced Adventitious Rooting in Dark-Grown *Arabidopsis* Thaliana Seedlings and Stem Thin Cell Layers by a Cross-Talk with Ethylene Signalling and a Modulation of Xylogenesis. *BMC Plant Biol.* **2018**, *18*, 182. [CrossRef]
49. Greenboim-Wainberg, Y.; Maymon, I.; Borochov, R.; Alvarez, J.; Olszewski, N.; Ori, N.; Eshed, Y.; Weiss, D. Cross Talk between Gibberellin and Cytokinin: The *Arabidopsis* Ga Response Inhibitor Spindly Plays a Positive Role in Cytokinin Signaling. *Plant Cell* **2005**, *17*, 92–102. [CrossRef]
50. Björklund, S.; Antti, H.; Uddestrand, I.; Moritz, T.; Sundberg, B. Cross-Talk between Gibberellin and Auxin in Development of *Populus* Wood: Gibberellin Stimulates Polar Auxin Transport and Has a Common Transcriptome with Auxin. *Plant J.* **2007**, *52*, 499–511. [CrossRef]
51. Liu, Y.; Zhang, M.; Meng, Z.; Wang, B.; Chen, M. Research Progress on the Roles of Cytokinin in Plant Response to Stress. *Int. J. Mol. Sci.* **2020**, *21*, 6574. [CrossRef]
52. Wu, Y.; Dor, E.; Hershenhorn, J. Strigolactones Affect Tomato Hormone Profile and Somatic Embryogenesis. *Planta* **2017**, *245*, 583–594. [CrossRef]
53. Hedden, P.; Sponsel, V. A Century of Gibberellin Research. *J. Plant Growth Regul.* **2015**, *34*, 740–760. [CrossRef]
54. Rivas-San Vicente, M.; Plasencia, J. Salicylic Acid Beyond Defence: Its Role in Plant Growth and Development. *J. Exp. Bot.* **2011**, *62*, 3321–3338. [CrossRef] [PubMed]

55. Pasternak, T.; Groot, E.P.; Kazantsev, F.V.; Teale, W.; Omelyanchuk, N.; Kovrizhnykh, V.; Palme, K.; Mironova, V.V. Salicylic Acid Affects Root Meristem Patterning Via Auxin Distribution in a Concentration-Dependent Manner. *Plant Physiol.* **2019**, *180*, 1725–1739. [CrossRef] [PubMed]
56. Arc, E.; Sechet, J.; Corbineau, F.; Rajjou, L.; Marion-Poll, A. Aba Crosstalk with Ethylene and Nitric Oxide in Seed Dormancy and Germination. *Front. Plant Sci.* **2013**, *4*, 63. [CrossRef] [PubMed]
57. Denancé, N.; Sánchez-Vallet, A.; Goffner, D.; Molina, A. Disease Resistance or Growth: The Role of Plant Hormones in Balancing Immune Responses and Fitness Costs. *Front. Plant Sci.* **2013**, *4*, 155. [CrossRef] [PubMed]
58. Broekaert, W.F.; Delauré, S.L.; De Bolle, M.F.C.; Cammue, B.P.A. The Role of Ethylene in Host-Pathogen Interactions. *Annu. Rev. Phytopathol.* **2006**, *44*, 393–416. [CrossRef]
59. Rensing, S.A.; Lang, D.; Zimmer, A.D.; Terry, A.; Salamov, A.; Shapiro, H.; Nishiyama, T.; Perroud, P.-F.; Lindquist, E.A.; Kamisugi, Y. The Physcomitrella Genome Reveals Evolutionary Insights into the Conquest of Land by Plants. *Science* **2008**, *319*, 64–69. [CrossRef]
60. Vandenbussche, F.; Fierro, A.C.; Wiedemann, G.; Reski, R.; Van Der Straeten, D. Evolutionary Conservation of Plant Gibberellin Signalling Pathway Components. *BMC Plant Biol.* **2007**, *7*, 65. [CrossRef]
61. Choudhary, S.P.; Yu, J.-Q.; Yamaguchi-Shinozaki, K.; Shinozaki, K.; Tran, L.-S.P. Benefits of Brassinosteroid Crosstalk. *Trends Plant Sci.* **2012**, *17*, 594–605. [CrossRef]
62. EL Sabagh, A.; Islam, M.S.; Hossain, A.; Iqbal, M.A.; Mubeen, M.; Waleed, M.; Reginato, M.; Battaglia, M.; Ahmed, S.; Rehman, A. Phytohormones as Growth Regulators During Abiotic Stress Tolerance in Plants. *Front. Agron.* **2022**, *4*, 4. [CrossRef]
63. Khan, T.; Abbasi, B.H.; Khan, M.A. The Interplay between Light, Plant Growth Regulators and Elicitors on Growth and Secondary Metabolism in Cell Cultures of *Fagonia indica*. *J. Photochem. Photobiol. B Biol.* **2018**, *185*, 153–160. [CrossRef]
64. Monfort, L.E.F.; Bertolucci, S.K.V.; Lima, A.F.; de Carvalho, A.A.; Mohammed, A.; Blank, A.F.; Pinto, J.E.B.P. Effects of Plant Growth Regulators, Different Culture Media and Strength Ms on Production of Volatile Fraction Composition in Shoot Cultures of *Ocimum basilicum*. *Ind. Crop. Prod.* **2018**, *116*, 231–239. [CrossRef]
65. Sakakibara, H.; Takei, K.; Hirose, N. Interactions between Nitrogen and Cytokinin in the Regulation of Metabolism and Development. *Trends Plant Sci.* **2006**, *11*, 440–448. [CrossRef] [PubMed]
66. Martins, V.; Garcia, A.; Costa, C.; Sottomayor, M.; Gerós, H. Calcium-and Hormone-Driven Regulation of Secondary Metabolism and Cell Wall Enzymes in Grape Berry Cells. *J. Plant Physiol.* **2018**, *231*, 57–67. [CrossRef] [PubMed]
67. Ibrahim, W.; Zhu, Y.-M.; Chen, Y.; Qiu, C.-W.; Zhu, S.; Wu, F. Genotypic Differences in Leaf Secondary Metabolism, Plant Hormones and Yield under Alone and Combined Stress of Drought and Salinity in Cotton Genotypes. *Physiol. Plant.* **2019**, *165*, 343–355. [CrossRef] [PubMed]
68. Jahan, M.; Guo, S.; Baloch, A.R.; Sun, J.; Shu, S.; Wang, Y.; Ahammed, G.J.; Kabir, K.; Roy, R. Melatonin Alleviates Nickel Phytotoxicity by Improving Photosynthesis, Secondary Metabolism and Oxidative Stress Tolerance in Tomato Seedlings. *Ecotoxicol. Environ. Saf.* **2020**, *197*, 110593. [CrossRef]
69. Naresh, V.; Lee, N. A Review on Biosensors and Recent Development of Nanostructured Materials-Enabled Biosensors. *Sensors* **2021**, *21*, 1109. [CrossRef]
70. Isoda, R.; Yoshinari, A.; Ishikawa, Y.; Sadoine, M.; Simon, R.; Frommer, W.B.; Nakamura, M. Sensors for the Quantification, Localization and Analysis of the Dynamics of Plant Hormones. *Plant J.* **2021**, *105*, 542–557. [CrossRef]
71. Wells, D.M.; Laplaze, L.; Bennett, M.J.; Vernoux, T. Biosensors for Phytohormone Quantification: Challenges, Solutions, and Opportunities. *Trends Plant Sci.* **2013**, *18*, 244–249. [CrossRef]
72. Herud-Sikimić, O.; Stiel, A.C.; Kolb, M.; Shanmugaratnam, S.; Berendzen, K.W.; Feldhaus, C.; Höcker, B.; Jürgens, G. A Biosensor for the Direct Visualization of Auxin. *Nature* **2021**, *592*, 768–772. [CrossRef]
73. Lotfi, M.; Mars, M.; Werbrouck, S. Optimizing Pear Micropropagation and Rooting with Light Emitting Diodes and Trans-Cinnamic Acid. *Plant Growth Regul.* **2019**, *88*, 173–180. [CrossRef]
74. Arab, M.M.; Yadollahi, A.; Eftekhari, M.; Ahmadi, H.; Akbari, M.; Khorami, S.S. Modeling and Optimizing a New Culture Medium for In Vitro Rooting of G× N15 Prunus Rootstock Using Artificial Neural Network-Genetic Algorithm. *Sci. Rep.* **2018**, *8*, 9977. [CrossRef]
75. Hesami, M.; Naderi, R.; Yoosefzadeh-Najafabadi, M.; Rahmati, M. Data-Driven Modeling in Plant Tissue Culture. *J. Appl. Environ. Biol. Sci.* **2017**, *7*, 37–44.
76. Hesami, M.; Jones, A.M.P. Application of Artificial Intelligence Models and Optimization Algorithms in Plant Cell and Tissue Culture. *Appl. Microbiol. Biotechnol.* **2020**, *104*, 9449–9485. [CrossRef]
77. Arab, M.M.; Yadollahi, A.; Shojaeiyan, A.; Ahmadi, H. Artificial Neural Network Genetic Algorithm as Powerful Tool to Predict and Optimize In Vitro Proliferation Mineral Medium for G× N15 Rootstock. *Front. Plant Sci.* **2016**, *7*, 1526. [CrossRef] [PubMed]
78. Niazian, M.; Shariatpanahi, M.E.; Abdipour, M.; Oroojloo, M. Modeling Callus Induction and Regeneration in an Anther Culture of Tomato (*Lycopersicon esculentum* L.) Using Image Processing and Artificial Neural Network Method. *Protoplasma* **2019**, *256*, 1317–1332. [CrossRef] [PubMed]
79. Nezami-Alanagh, E.; Garoosi, G.-A.; Maleki, S.; Landín, M.; Gallego, P.P. Predicting Optimal In Vitro Culture Medium for Pistacia Vera Micropropagation Using Neural Networks Models. *Plant Cell Tissue Organ Cult.* **2017**, *129*, 19–33. [CrossRef]
80. Mridula, M.R.; Nair, A.S.; Kumar, K.S. Genetic Programming Based Models in Plant Tissue Culture: An Addendum to Traditional Statistical Approach. *PLoS Comput. Biol.* **2018**, *14*, e1005976. [CrossRef]

81. Prasad, A.; Prakash, O.; Mehrotra, S.; Khan, F.; Mathur, A.K.; Mathur, A. Artificial Neural Network-Based Model for the Prediction of Optimal Growth and Culture Conditions for Maximum Biomass Accumulation in Multiple Shoot Cultures of Centella Asiatica. *Protoplasma* **2017**, *254*, 335–341. [CrossRef]
82. Duan, Y.; Jiang, Y.; Ye, S.; Karim, A.; Ling, Z.; He, Y.; Yang, S.; Luo, K. Ptrwrky73, a Salicylic Acid-Inducible Poplar Wrky Transcription Factor, Is Involved in Disease Resistance in *Arabidopsis* Thaliana. *Plant Cell Rep.* **2015**, *34*, 831–841. [CrossRef]
83. Mansouri, A.; Fadavi, A.; Mortazavian, S.M.M. An Artificial Intelligence Approach for Modeling Volume and Fresh Weight of Callus—A Case Study of Cumin (*Cuminum cyminum* L.). *J. Theor. Biol.* **2016**, *397*, 199–205. [CrossRef]
84. Ivashchuk, O.A.; Fedorova, V.I.; Shcherbinina, N.V.; Maslova, E.V.; Shamraeva, E.O. Microclonal Propagation of Plant Process Modeling and Optimization of Its Parameters Based on Neural Network. *Drug Invent. Today* **2018**, *10*, 3170–3175.
85. Jamshidi, S.; Yadollahi, A.; Arab, M.M.; Soltani, M.; Eftekhari, M.; Shiri, J. High Throughput Mathematical Modeling and Multi-Objective Evolutionary Algorithms for Plant Tissue Culture Media Formulation: Case Study of Pear Rootstocks. *PLoS ONE* **2020**, *15*, e0243940. [CrossRef] [PubMed]
86. Farhadi, S.; Salehi, M.; Moieni, A.; Safaie, N.; Sabet, M.S. Modeling of Paclitaxel Biosynthesis Elicitation in Corylus Avellana Cell Culture Using Adaptive Neuro-Fuzzy Inference System-Genetic Algorithm (Anfis-Ga) and Multiple Regression Methods. *PLoS ONE* **2020**, *15*, e0237478. [CrossRef]
87. Szopa, A.; Kokotkiewicz, A.; Bednarz, M.; Jafernik, K.; Luczkiewicz, M.; Ekiert, H. Bioreactor Type Affects the Accumulation of Phenolic Acids and Flavonoids in Microshoot Cultures of *Schisandra chinensis* (Turcz.) Baill. *Plant Cell Tissue Organ Cult.* **2019**, *139*, 199–206. [CrossRef]
88. Su, R.; Sujarani, M.; Shalini, P.; Prabhu, N. A Review on Bioreactor Technology Assisted Plant Suspension Culture. *Asian J. Biotechnol. Bioresour. Technol.* **2019**, *5*, 1–13. [CrossRef]
89. Eibl, R.; Eibl, D. Design of Bioreactors Suitable for Plant Cell and Tissue Cultures. *Phytochem. Rev.* **2008**, *7*, 593–598. [CrossRef]
90. Phelan, K.; May, K.M. Basic Techniques in Mammalian Cell Tissue Culture. *Curr. Protoc. Cell Biol.* **2015**, *66*, 1.1.1–1.1.22. [CrossRef]
91. Lihua, F. Preparation and Properties of Nanoscaled Silver/Natural Polymer Antibacterial Sols. Master's Dissertation, Beijing Forestry University, Beijing, China, 2012.
92. Abu-Absi, N.R.; Martel, R.P.; Lanza, A.M.; Clements, S.J.; Borys, M.C.; Li, Z.J. Application of Spectroscopic Methods for Monitoring of Bioprocesses and the Implications for the Manufacture of Biologics. *Pharm. Bioprocess.* **2014**, *2*, 267–284. [CrossRef]
93. Wang, R.C.C.; Campbell, D.A.; Green, J.R.; Čuperlović-Culf, M. Automatic 1d 1h Nmr Metabolite Quantification for Bioreactor Monitoring. *Metabolites* **2021**, *11*, 157. [CrossRef] [PubMed]
94. Mehendale, N.; Jenne, F.; Joshi, C.; Sharma, S.; Masakapalli, S.K.; MacKinnon, N. A Nuclear Magnetic Resonance (Nmr) Platform for Real-Time Metabolic Monitoring of Bioprocesses. *Molecules* **2020**, *25*, 4675. [CrossRef]
95. Ghormade, V.; Deshpande, M.V.; Paknikar, K.M. Perspectives for Nano-Biotechnology Enabled Protection and Nutrition of Plants. *Biotechnol. Adv.* **2011**, *29*, 792–803. [CrossRef] [PubMed]
96. Lala, S. Nanoparticles as Elicitors and Harvesters of Economically Important Secondary Metabolites in Higher Plants: A Review. *IET Nanobiotechnol.* **2021**, *15*, 28–57. [CrossRef] [PubMed]
97. Santo Pereira, A.E.; Silva, P.M.; Oliveira, J.L.; Oliveira, H.C.; Fraceto, L.F. Chitosan Nanoparticles as Carrier Systems for the Plant Growth Hormone Gibberellic Acid. *Colloids Surf. B Biointerfaces* **2017**, *150*, 141–152. [CrossRef] [PubMed]
98. Nokandeh, S.; Ramezani, M.; Gerami, M. The Physiological and Biochemical Responses to Engineered Green Graphene/Metal Nanocomposites in *Stevia rebaudiana*. *J. Plant Biochem. Biotechnol.* **2021**, *30*, 579–585. [CrossRef]
99. Ghorbanpour, M.; Hadian, J. Multi-Walled Carbon Nanotubes Stimulate Callus Induction, Secondary Metabolites Biosynthesis and Antioxidant Capacity in Medicinal Plant *Satureja khuzestanica* Grown In Vitro. *Carbon* **2015**, *94*, 749–759. [CrossRef]
100. Samadi, S.; Saharkhiz, M.J.; Azizi, M.; Samiei, L.; Ghorbanpour, M. Multi-Walled Carbon Nanotubes Stimulate Growth, Redox Reactions and Biosynthesis of Antioxidant Metabolites in *Thymus daenensis* Celak. In Vitro. *Chemosphere* **2020**, *249*, 126069. [CrossRef]
101. Rezaei, Z.; Jafarirad, S.; Kosari-Nasab, M. Modulation of Secondary Metabolite Profiles by Biologically Synthesized Mgo/Perlite Nanocomposites in *Melissa officinalis* Plant Organ Cultures. *J. Hazard. Mater.* **2019**, *380*, 120878. [CrossRef]
102. Torres, F.G.; Troncoso, O.P.; Pisani, A.; Gatto, F.; Bardi, G. Natural Polysaccharide Nanomaterials: An Overview of Their Immunological Properties. *Int. J. Mol. Sci.* **2019**, *20*, 5092. [CrossRef]
103. Soraki, R.K.; Gerami, M.; Ramezani, M. Effect of Graphene/Metal Nanocomposites on the Key Genes Involved in Rosmarinic Acid Biosynthesis Pathway and Its Accumulation in Melissa Officinalis. *BMC Plant Biol.* **2021**, *21*, 260. [CrossRef]
104. Hadi Soltanabad, M.; Bagherieh-Najjar, M.B.; Mianabadi, M. Carnosic Acid Content Increased by Silver Nanoparticle Treatment in Rosemary (*Rosmarinus officinalis* L.). *Appl. Biochem. Biotechnol.* **2020**, *191*, 482–495. [CrossRef]
105. Nourozi, E.; Hosseini, B.; Maleki, R.; Mandoulakani, B.A. Iron Oxide Nanoparticles: A Novel Elicitor to Enhance Anticancer Flavonoid Production and Gene Expression in Dracocephalum Kotschyi Hairy-Root Cultures. *J. Sci. Food Agric.* **2019**, *99*, 6418–6430. [CrossRef] [PubMed]
106. Asgari-Targhi, G.; Iranbakhsh, A.; Ardebili, Z.O.; Tooski, A.H. Synthesis and Characterization of Chitosan Encapsulated Zinc Oxide (Zno) Nanocomposite and Its Biological Assessment in Pepper (*Capsicum annuum*) as an Elicitor for In Vitro Tissue Culture Applications. *Int. J. Biol. Macromol.* **2021**, *189*, 170–182. [CrossRef] [PubMed]

107. Samadi, S.; Saharkhiz, M.J.; Azizi, M.; Samiei, L.; Karami, A.; Ghorbanpour, M. Single-Wall Carbon Nano Tubes (Swcnts) Penetrate *Thymus daenensis* Celak. Plant Cells and Increase Secondary Metabolite Accumulation In Vitro. *Ind. Crop. Prod.* **2021**, *165*, 113424. [CrossRef]
108. Darzian Rostami, A.; Yazdian, F.; Mirjani, R.; Soleimani, M. Effects of Different Graphene-Based Nanomaterials as Elicitors on Growth and Ganoderic Acid Production by *Ganoderma lucidum*. *Biotechnol. Prog.* **2020**, *36*, e3027. [CrossRef]
109. El-Saber, M.M.; Diab, M.; Hendawey, M.; Farroh, K. Magnetite Nanoparticles Different Sizes Effectiveness on Growth and Secondary Metabolites in *Ginkgo biloba* L. Callus. *Egypt. J. Chem.* **2021**, *64*, 4523–5432. [CrossRef]
110. Tavakoli, F.; Rafieiolhossaini, M.; Ravash, R. Effects of Peg and Nano-Silica Elicitors on Secondary Metabolites Production in *Crocus sativus* L. *Russ. J. Plant Physiol.* **2021**, *68*, 931–940. [CrossRef]
111. Shoja, A.A.; Çirak, C.; Ganjeali, A.; Cheniany, M. Stimulation of Phenolic Compounds Accumulation and Antioxidant Activity in In Vitro Culture of *Salvia tebesana* Bunge in Response to Nano-TiO_2 and Methyl Jasmonate Elicitors. *Plant Cell Tissue Organ Cult.* **2022**, *149*, 423–440. [CrossRef]
112. Ebadollahi, R.; Jafarirad, S.; Kosari-Nasab, M.; Mahjouri, S. Effect of Explant Source, Perlite Nanoparticles and TiO_2/Perlite Nanocomposites on Phytochemical Composition of Metabolites in Callus Cultures of Hypericum Perforatum. *Sci. Rep.* **2019**, *9*, 12998. [CrossRef]
113. Xu, Z.P.; Zeng, Q.H.; Lu, G.Q.; Yu, A.B. Inorganic Nanoparticles as Carriers for Efficient Cellular Delivery. *Chem. Eng. Sci.* **2006**, *61*, 1027–1040. [CrossRef]
114. DeRosa, M.C.; Monreal, C.; Schnitzer, M.; Walsh, R.; Sultan, Y. Nanotechnology in Fertilizers. *Nat. Nanotechnol.* **2010**, *5*, 91. [CrossRef]
115. Kah, M.; Kookana, R.S.; Gogos, A.; Bucheli, T.D. A Critical Evaluation of Nanopesticides and Nanofertilizers against Their Conventional Analogues. *Nat. Nanotechnol.* **2018**, *13*, 677–684. [CrossRef] [PubMed]
116. Xu, Z.P.; Lu, G.Q.M. Layered Double Hydroxide Nanomaterials as Potential Cellular Drug Delivery Agents. *Pure Appl. Chem.* **2006**, *78*, 1771–1779. [CrossRef]
117. de Castro, V.A.; Duarte, V.G.O.; Nobre, D.A.C.; Silva, G.H.; Constantino, V.R.L.; Pinto, F.G.; Macedo, W.R.; Tronto, J. Plant Growth Regulation by Seed Coating with Films of Alginate and Auxin-Intercalated Layered Double Hydroxides. *Beilstein J. Nanotechnol.* **2020**, *11*, 1082–1091. [CrossRef] [PubMed]
118. Leong, K.W. Polymeric Controlled Nucleic Acid Delivery. *MRS Bull.* **2005**, *30*, 640–646. [CrossRef]
119. Ladewig, K.; Xu, Z.P.; Lu, G.Q. Layered Double Hydroxide Nanoparticles in Gene and Drug Delivery. *Expert Opin. Drug Deliv.* **2009**, *6*, 907–922. [CrossRef]
120. Wang, C.; Zhu, H.; Li, N.; Wu, Q.; Wang, S.; Xu, B.; Wang, Y.; Cui, H. Dinotefuran Nano-Pesticide with Enhanced Valid Duration and Controlled Release Properties Based on a Layered Double Hydroxide Nano-Carrier. *Environ. Sci. Nano* **2021**, *8*, 3202–3210. [CrossRef]
121. Rebitski, E.P.; Darder, M.; Aranda, P. Layered Double Hydroxide/Sepiolite Hybrid Nanoarchitectures for the Controlled Release of Herbicides. *Beilstein J. Nanotechnol.* **2019**, *10*, 1679–1690. [CrossRef]
122. bin Hussein, M.Z.; Zainal, Z.; Yahaya, A.H.; Foo, D.W.V. Controlled Release of a Plant Growth Regulator, A-Naphthaleneacetate from the Lamella of Zn–Al-Layered Double Hydroxide Nanocomposite. *J. Control. Release* **2002**, *82*, 417–427. [CrossRef]
123. Liu, Y.; Song, J.; Jiao, F.; Huang, J. Synthesis, Characterization and Release of a-Naphthaleneacetate from Thin Films Containing Mg/Al-Layered Double Hydroxide. *J. Mol. Struct.* **2014**, *1064*, 100–106. [CrossRef]
124. Hafez, I.H.; Berber, M.R.; Minagawa, K.; Mori, T.; Tanaka, M. Design of a Multifunctional Nanohybrid System of the Phytohormone Gibberellic Acid Using an Inorganic Layered Double-Hydroxide Material. *J. Agric. Food Chem.* **2010**, *58*, 10118–10123. [CrossRef]
125. Li, Y.; Shen, Y.; Xiao, M.; Liu, D.; Fan, L. Synthesis and Controlled Release Properties of B-Naphthoxyacetic Acid Intercalated Mg–Al Layered Double Hydroxides Nanohybrids. *Arab. J. Chem.* **2019**, *12*, 2563–2571. [CrossRef]
126. Thomas, S.; Ninan, N.; Mohan, S.; Francis, E. *Natural Polymers, Biopolymers, Biomaterials, and Their Composites, Blends, and Ipns*; CRC Press: Boca Raton, FL, USA, 2012.
127. Rafique, A.; Zia, K.M.; Zuber, M.; Tabasum, S.; Rehman, S. Chitosan Functionalized Poly (Vinyl Alcohol) for Prospects Biomedical and Industrial Applications: A Review. *Int. J. Biol. Macromol.* **2016**, *87*, 141–154. [CrossRef] [PubMed]
128. Jiménez-Gómez, C.P.; Cecilia, J.A. Chitosan: A Natural Biopolymer with a Wide and Varied Range of Applications. *Molecules* **2020**, *25*, 3981. [CrossRef] [PubMed]
129. Bakshi, P.S.; Selvakumar, D.; Kadirvelu, K.; Kumar, N.S. Chitosan as an Environment Friendly Biomaterial—A Review on Recent Modifications and Applications. *Int. J. Biol. Macromol.* **2020**, *150*, 1072–1083. [CrossRef]
130. Mirheidari, F.; Hatami, M.; Ghorbanpour, M. Effect of Different Concentrations of Iaa, Ga_3 and Chitosan Nano-Fiber on Physio-Morphological Characteristics and Metabolite Contents in Roselle (*Hibiscus sabdariffa* L.). *S. Afr. J. Bot.* **2022**, *145*, 323–333. [CrossRef]
131. Kadam, P.M.; Prajapati, D.; Kumaraswamy, R.V.; Kumari, S.; Devi, K.A.; Pal, A.; Sharma, S.K.; Saharan, V. Physio-Biochemical Responses of Wheat Plant Towards Salicylic Acid-Chitosan Nanoparticles. *Plant Physiol. Biochem.* **2021**, *162*, 699–705. [CrossRef]
132. Karimi, N.; Shayesteh, L.S.; Ghasmpour, H.; Alavi, M. Effects of Arsenic on Growth, Photosynthetic Activity, and Accumulation in Two New Hyperaccumulating Populations of Isatis Cappadocica Desv. *J. Plant Growth Regul.* **2013**, *32*, 823–830. [CrossRef]
133. Souri, Z.; Karimi, N.; Sarmadi, M.; Rostami, E. Salicylic Acid Nanoparticles (Sanps) Improve Growth and Phytoremediation Efficiency of *Isatis cappadocica* Desv., under as Stress. *IET Nanobiotechnol.* **2017**, *11*, 650–655. [CrossRef]

134. Vincekovic, M.; Jalsenjak, N.; Topolovec-Pintaric, S.; Đermic, E.; Bujan, M.; Juric, S. Encapsulation of Biological and Chemical Agents for Plant Nutrition and Protection: Chitosan/Alginate Microcapsules Loaded with Copper Cations and Trichoderma Viride. *J. Agric. Food Chem.* **2016**, *64*, 8073–8083. [CrossRef]
135. Perez, J.J.; Francois, N.J.; Maroniche, G.A.; Borrajo, M.P.; Pereyra, M.A.; Creus, C.M. A Novel, Green, Low-Cost Chitosan-Starch Hydrogel as Potential Delivery System for Plant Growth-Promoting Bacteria. *Carbohydr. Polym.* **2018**, *202*, 409–417. [CrossRef] [PubMed]
136. Aljafree, N.F.A.; Kamari, A. Synthesis, Characterisation and Potential Application of Deoxycholic Acid Carboxymethyl Chitosan as a Carrier Agent for Rotenone. *J. Polym. Res.* **2018**, *25*, 133. [CrossRef]
137. Pereira, A.E.S.; da Silva, P.M.; de Melo, N.F.; Rosa, A.H.; Fraceto, L.F. Alginate/Chitosan Nanoparticles as Sustained Release System for Plant Hormone Gibberellic Acid. *Colloids Surf B Biointerfaces* **2017**, *150*, 141–152. [CrossRef]
138. Kumaraswamy, R.V.; Kumari, S.; Choudhary, R.C.; Sharma, S.S.; Pal, A.; Raliya, R.; Biswas, P.; Saharan, V. Salicylic Acid Functionalized Chitosan Nanoparticle: A Sustainable Biostimulant for Plant. *Int. J. Biol. Macromol.* **2019**, *123*, 59–69. [CrossRef] [PubMed]
139. Sampedro-Guerrero, J.; Vives-Peris, V.; Gomez-Cadenas, A.; Clausell-Terol, C. Improvement of Salicylic Acid Biological Effect through Its Encapsulation with Silica or Chitosan. *Int. J. Biol. Macromol.* **2022**, *199*, 108–120. [CrossRef] [PubMed]
140. Valderrama, A.; Lay, J.; Flores, Y.; Zavaleta, D.; Delfín, A.R. Factorial Design for Preparing Chitosan Nanoparticles and Its Use for Loading and Controlled Release of Indole-3-Acetic Acid with Effect on Hydroponic Lettuce Crops. *Biocatal. Agric. Biotechnol.* **2020**, *26*, 101640. [CrossRef]
141. Do, B.H.; Ryu, H.-B.; Hoang, P.; Koo, B.-K.; Choe, H. Soluble Prokaryotic Overexpression and Purification of Bioactive Human Granulocyte Colony-Stimulating Factor by Maltose Binding Protein and Protein Disulfide Isomerase. *PLoS ONE* **2014**, *9*, e89906. [CrossRef]
142. Slowing, I.; Vivero-Escoto, J.; Zhao, Y.; Kandel, K.; Peeraphatdit, C.; Trewyn, B.; Lin, V. Exocytosis of Mesoporous Silica Nanoparticles from Mammalian Cells: From Asymmetric Cell-to-Cell Transfer to Protein Harvesting. *Small* **2011**, *7*, 1526–1532. [CrossRef]

Disclaimer/Publisher's Note: The statements, opinions and data contained in all publications are solely those of the individual author(s) and contributor(s) and not of MDPI and/or the editor(s). MDPI and/or the editor(s) disclaim responsibility for any injury to people or property resulting from any ideas, methods, instructions or products referred to in the content.

Review

Advances in Biotechnological Production and Metabolic Regulation of *Astragalus membranaceus*

Baoyu Ji [1,2,†], Liangshuang Xuan [2,†], Yunxiang Zhang [1,†], Guoqi Zhang [1], Jie Meng [1], Wenrong Mu [2], Jingjing Liu [3], Kee-Yoeup Paek [4], So-Young Park [4], Juan Wang [1,*] and Wenyuan Gao [1,*]

1. School of Pharmaceutical Science and Technology, Tianjin University, Tianjin 300072, China; xlpxlp@aliyun.com (B.J.); yunxiang_zhang@tju.edu.cn (Y.Z.); 18768153215@163.com (J.M.)
2. School of Pharmacy, Henan University of Chinese Medicine, Zhengzhou 450046, China; m18203675687@163.com (L.X.)
3. School of Graduate, Tianjin University of Traditional Chinese Medicine, Tianjin 301617, China
4. Department of Horticultural Science, Chungbuk National University, Cheongju 28644, Republic of Korea
* Correspondence: drwangjuan@tju.edu.cn (J.W.); pharmgao@tju.edu.cn (W.G.)
† These authors contributed equally to this work.

Abstract: Legume medicinal plants *Astragalus membranaceus* are widely used in the world and have very important economic value, ecological value, medicinal value, and ornamental value. The bioengineering technology of medicinal plants is used in the protection of endangered species, the rapid propagation of important resources, detoxification, and the improvement of degraded germplasm. Using bioengineering technology can effectively increase the content of secondary metabolites in *A. membranaceus* and improve the probability of solving the problem of medicinal plant resource shortage. In this review, we focused on biotechnological research into *A. membranaceus*, such as the latest advances in tissue culture, including callus, adventitious roots, hairy roots, suspension cells, etc., the metabolic regulation of chemical compounds in *A. membranaceus*, and the research progress on the synthetic biology of astragalosides, including the biosynthesis pathway of astragalosides, microbial transformation of astragalosides, and metabolic engineering of astragalosides. The review also looks forward to the new development trend of medicinal plant biotechnology, hoping to provide a broader development prospect for the in-depth study of medicinal plants.

Keywords: *Astragalus membranaceus*; biotechnology; tissue culture; metabolic regulation; synthetic biology

1. Introduction

Astragalus membranaceus is the dry root of *Astragalus membranaceus* (Fisch.) Bge. var. *mongholicus* (Bge.) Hsiao or *Astragalus membranaceus* (Fisch.) Bge. There are more than 200 kinds of drugs which use *A. membranaceus* as a raw material. *A. membranaceus* has high utilization value, and has been used in food, dietary tonics, medicine, cosmetics (anti-aging), health care products and so on [1]. There are various chemical compounds in *A. membranaceus*, including flavonoids, polysaccharides, saponins, amino acids, sterols, and alkaloids [2]. The main active ingredients are *Astragalus* saponin (AST) and calycasin-7-glycoside (CG). According to scientific research, *A. membranaceus* has potential clinical applications for various conditions, such as heart failure [3], anemia, pneumonia, kidney disease, cancer, diabetes [4], skin problems, and reproductive system problems [5], improving immune responses [6], osteoporosis prevention and treatment, and radiation protection [7]. *Astragalus* saponins have significant medicinal value and can be used to produce cancer vaccines [8], health products, and cosmetics [9]. More than 50 saponins have been isolated from *Astragalus membranaceus* [5]. Astragalosides mainly include astragalosides I–IV, isoastragalosides I, II, IV, acetyl astragalosides, and soybean saponins [7]. The *Chinese Pharmacopoeia* stipulates that the content of astragaloside IV in *A. membranaceus* should not be less than 0.080%, and the content of calycosin-7-glucoside should be more than 0.020%. It can be seen that the content of astragaloside IV in the *A. membranaceus* root is very low.

With the improvement in people's living standards, the demand for *Astragalus membranaceus* and *Astragalus* saponins is increasing at home and abroad [10]. *Astragalus* saponins are mainly extracted from *A. membranaceus*. As one of the most popular herbal medicines in the world, *A. membranaceus* is mainly derived from wild and cultivated resources in China. Wild plants of *A. membranaceus* are mainly produced in Inner Mongolia, Gansu, Shanxi, and Ningxia [3]. Wild resources have been over-exploited, seriously damaging the ecological environment. Cultivated plants are mainly derived from Shanxi Hunyuan, Gansu Longxi, Inner Mongolia Niuyingzi, and Shaanxi Zizhou [11]. The contents of active substances in cultivated plants are unstable and vary with changes in environmental conditions. Furthermore, infection, disease and pesticide application have also reduced the quality of cultivated *A. membranaceus*. It takes 3~4 years to collect cultivated *A. membranaceus* [12], which has increased the scarcity of medicinal plant resources and the probability of extinction of endangered medicinal plant resources [13,14]. It is of great significance to apply and develop new biotechnology to supply and replace traditional cultivation methods.

Bioengineering technology has played an important role in solving the problem of shortages of traditional Chinese medicine resources. The use of cell engineering, genetic engineering, bioreactor engineering, and biochemical engineering can quickly produce high-quality seedlings, modify the metabolic pathways of medicinal plants, and obtain active ingredients more easily, while saving the use of wild medicinal materials, which also greatly reduces production costs [15].

1.1. Astragalus Tissue Culture

Astragalus originates from the roots of *Astragalus membranaceus* (Fisch.) Ege. var. mongholicus (Ege.) Hsiao or *Astragalus membranaceus* (Fisch.) Ege. Its main chemical components are saponins and flavonoids. In recent years, plant tissue culture technology has provided conditions for the resource development of *Astragalus membranaceus*; see Table 1.

Table 1. *Astragalus* plant tissue culture research.

Culture Material	Culture Medium	Cultivation Conditions	Result
Callus	MS + 0.5 mg/L 6-BA + 2 mg/L 2.4-D + 0.1 mg/LVC+3% sucrose + 0.7% agar	The culture temperature was 25 °C, 24 h dark culture, pH value was 5.8	The induction rate can reach 100% [16]
Suspension cell	MS + 0.1 mg/L (NAA) + 1.0 mg/L (6- BA) + 1.5% (w/v) sucrose and 0.8% (w/v) agar After three weeks, MS + 0.5 mg/L (IAA) + 1.5% (w/v) sucrose and 0.8% (w/v) agar	Seed germination, callus induction and subculture were carried out in a growth chamber illuminated with fluorescent light (ca. 1400 mol m^{-2} s^{-1}) over a 16/8 day and night at 25 ± 2 °C	The seedlings developed fragile callus within 2 weeks [17]
Adventitious root	MS +7 mg/L(IBA)+ 30 g/L sucrose + 7 g/L agar	In dark, at 25 ± 2 °C, for 6 weeks of culture	Adventitious roots were successfully induced [18]

1.1.1. Establishment of Regeneration Plant System of *Astragalus*

The leaves of *Astragalus* were explants for callus induction. The induction rate of the leaves on the MS + 2.0 mg/L 6-BA + 2.0 mg/L NAA medium reached 83.3%, and the hypocotyl induction rate reached 100% on the MS + 2.0 mg/L 6-BA + 2.0mg/L NAA medium [19].

1.1.2. Research Progress on the Culture of Adventitious Roots of *Astragalus*

The induction of adventitious roots from stem of *Astragalus* needs callus induction first. The optimal concentrations of KT (Kinetin) and NAA are 2 mg/L and 1 mg/L, respectively. The high concentration of NAA can promote an increase in the root number, root length and number of fibrous roots, but inhibits the root thickness. Therefore, the optimal medium for inducing adventitious roots from the callus is MS + 2 mg/L KT + 2 mg/L NAA [20]. The adventitious root culture has the highest growth rate by inoculation root length 1.5 cm, inoculation volume 30 g (fresh weight)/bioreactor, and aeration volume 0.1 vvm (air volume /culture volume/min). The total polysaccharide content of adventitious root is higher than 1 yr old roots and 3 yr old plants; the total saponins and flavonoids contents are basically the same as 3-year-old roots and higher than 1-year-old roots [12].

1.1.3. Research Progress of the Hairy Roots Culture of *Astragalus*

Some researchers have screened the hairy root line AMRRL VI and AMHRL II with high saponins and isoflavone content, respectively, and optimized the culture temperature of 27.8 °C, inoculation volume of 1.54%, sucrose concentration of 3.24% and culture time of 36 days [21]. The yield of a 30 L culture of *Astragalus* hairy roots for 20 days is similar to a 250 mL and 1 L shake flask culture, but higher than the 10 L bioreactor [22]. Hairy roots can grow rapidly and have genetic and biochemical stability, high productivity, and adaptability for large-scale systems [23].

1.1.4. Research Progress on the Culture of *Astragalus* Suspension Cells

The best explant for the growth of *Astragalus* suspension cells is the hypocotyl, and the optimal culture conditions are MS + 1.5 mg/L 2,4-D + 2.0 mg / L 6-BA + 3% sucrose, pH:5.5–6.0. The initial inoculation volume is 50 g/L, and the subculture inoculation volume is 37 g/L. The growth characteristics of *Astragalus* suspension cells have been studied; the growth period is 25 d. In detail, the first ten days are the lag period, and 11–19 d is the logarithmic phase. The cell biomass reached its maximum at 19 days, 20–22 days for the stationary phase, and 23–25 days for the decay period. The cell growth curve was drawn. This was the first time that the *Astragalus* cell suspension system has been established. The growth curve of the suspension cell culture was nearly an "s" shape, and its growth cycle was about 14 days. The best harvest time for the cell culture to produce *Astragalus* polysaccharides, saponins and flavonoids was the 15th day [24].

1.1.5. Culture of *Astragalus* Protoplasts

The preparation of protoplasts of *Astragalus* leaves is far easier for *Astragalus* callus, and it can obtain a large number of highly viable protoplasts. In this method, the explant can be hydrolyzed by a mixed enzyme of 2% cellulase + 0.5% hemicellulose + 0.5% pectinase for 12 h. Then, the high-quality *Astragalus* protoplasts will be separated more efficiently [25].

1.2. Astragalus Metabolism Control Research

Plants and plant cells, tissues and organs in vitro, with physiological and morphological responses to microbial, physical, and chemical factors, are called elicitors. Arousal is the process by which plants induce or enhance the synthesis of secondary metabolites to ensure their persistence and competitiveness [26]. The application of elicitors is the focus of current research, and has been considered as one of the most effective methods to improve the synthesis of secondary metabolites in medicinal plants [27]. At present, some hypotheses about the mechanisms of the elicitor-promoting plant secondary metabolism have been raised. As an exogenous signal, an elicitor can cause the recognition of receptor sites on the plant cell membrane. The combination of the two causes physiological and biochemical chain reactions in the cell membrane and in the cell, which changes the permeability of the cell membrane to specific molecules, and changes the chemical signal molecules, enzyme activity, and gene expression, which are related to the synthesis of secondary metabolites. All of these will eventually lead to the synthesis and accumulation of active ingredients [28].

The elicitor can be a compound that stimulates any type of physiological abnormality in plants [29].

Elicitors can be divided into abiotic types, which are compounds that stimulate any type of physiological abnormality in plants [26], physical induction, such as trauma, waterlogging, high temperature, light, etc., and biological type (bacteria, fungi and their extracts, such as *Saccharomyces cerevisiae*, *Pichia pastoris*, chitosan, cell wall polysaccharides, and plant endophytes) [27].

For a study on the metabolism control of *Astragalus*, see Table 2. The effects of inducing factors on the secondary metabolites of *Astragalus* are shown in Table 3.

Table 2. Study on metabolism control of *Astragalus*.

Culture Material	Activity Component	Influence Factor	Metabolic Regulation	Result
Suspension cells	PAL Activity and Total Phenol	Yeast extract (10 g/L) was added for 36 h on the 13th day of culture	PAL activity induced by yeast extract was positively correlated with total phenol accumulation	Increased PAL activity and total phenol content [30]
Astragalus	*Astragalus* saponin	—	HMGR, FPS, SE, and CAS are the main regulatory genes	Regulated the synthesis of *Astragalus* saponin [31]
Hairy roots	Saponins and Isoflavone	Regulation of Methyl Jasmonate (MJ), Acetylsalicylic Acid (ASA) and Salicylic Acid (SA)	MVD, IDI, FPS, SS, CHI, IFS	It is revealed that MVD, IDI, FPS and SS are key enzyme genes that MJ induces and which regulate the saponin biosynthesis pathway CHI and IFS are the key factors of the isoflavone biosynthesis pathway [21]
Astragalus	Cyclo*Astragalus* phenol and *Astragalus* phenol	Endophytic fungi	—	Endophytic fungi were found to transform sapogenins (Cyclonosterol and *Astragalus* cresol) [23]
Hairy roots of *Astragalus*	ASTS, MAO rhzomorphand CG	100 μM methyl jasmonate (MeJA) treatment	2127 genes were up-regulated by MeJA and 1247 genes were down-regulated by MeJA	The accumulation of ASTS, MAO rhizomorph and CG in hairy roots treated with MeJA increased significantly [32]
Astragalus	Genistein -7-O-β-D-glucoside (CGs)	Low temperature stress, light dependence	CHS, CHR, CHI, IFS, and I3'H PAL1, C4H	Temperature fluctuations up-regulated the transcription of CHS, CHR, CHI, IFS, and I3'H, but had different effects on the transcription of PAL1 and C4H of phenylpropanoid pathway in leaves [33]
Hairy roots	Hairy cephalosporins (CA) and formononetin (FO)	AMHRCs were co-cultured with immobilized aspergillus niger (IAN) for 54 h	—	The CA and FO biosynthetic pathway gene expression was significantly up-regulated, thereby increasing the production of CA and FO [34]
Root	Isoflavone	After 10 days of UV-B treatment (λ = 313 nm, 804 j/m)	—	UV-B radiation significantly induced isoflavone synthesis [35]
Hairy roots	Isoflavone	Ultraviolet light (UV-A, UV-B and UV-C) irradiation	PAL, C4H	86.4 kJ/m (2) UV-B upregulated the transcription and expression of all genes involved in the isoflavone biosynthesis pathway of AMHRCs [36]

Table 3. Effect of Inducer on Secondary Metabolites of *Astragalus*.

Culture Material	Active Ingredient	Influence Factor	Increase Multiples
Astragalus adventitious roots	Calycosin isoflavone glycoside	Hydrogen peroxide, the L-phenylalanine	The culture treated with hydrogen peroxide and L-phenylalanine was 8.6 times higher than that treated with hydrogen peroxide alone [37]
Astragalus adventitious roots	Calycosin isoflavone glycoside	Drought stress, methyl jasmonate, and L-phenylalanine	The three combinations induced the highest CG content, 3.12 times higher than that of the field plants [18]
Astragalus hair root	*Astragalus* saponin I, *Astragalus* saponin and *Astragalus* methyside	Methyl jasmonate	It reached 2.98, 2.85, 2.30, and 1.57 times in the control group, respectively [32]
Astragalus hair root	*Astragalus* methylside	Chitosan	It was 2.1 times higher than that in the control group [38]

1.2.1. *Astragalus* Saponins

In order to increase the production of astragaloside IV (AGIV) in hairy root cultures (AMHRCs) of *Astragalus*, a new method combining deacetylated biocatalysis and the induction of immobilized penicillium canescens (IPC) was proposed. The AMHRCs and IPC were co-cultured for 60 h, and the content of AGIV was 14.59 times higher than the control group. In addition, under the induction of IPC, the expression of AGIV biosynthetic pathway-related genes was significantly up-regulated. This method provides an effective and sustainable approach for the large-scale production of AGIV [39]. The accumulation of *Astragalus* metabolic components is affected by many factors. Yeast extract (10g/L) was added on the 13th day of the *Astragalus* suspension cell culture. The PAL activity was upregulated after 36h, and the total phenol content reached the maximum after 48 h. The maximum PAL activity and total phenol content were 2.88 times and 2.12 times higher than the control, respectively. Therefore, the PAL activity induced by yeast extract was positively correlated with the accumulation of total phenols. It is believed that after adding yeast extract, the early defense response of *Astragalus* cells to biological and abiotic environmental stressors is up-regulated. Therefore, the phenolic branch of secondary metabolism is up-regulated [30]. Hexanal can significantly affect the growth of the *Astragalus* adventitious root at solid culture condition, 10 μmol/L n-hexanal has an inhibitory effect, and 50 μmol/L has a promoting effect; 50 μmol/L n-hexanal in a liquid culture system can promote the synthesis of saponins in *Astragalus* adventitious roots [40].

Methyl jasmonate (MJ), acetylsalicylic acid (ASA) and salicylic acid (SA) can be used to induce and regulate the biosynthesis of saponins and isoflavones in the hairy root of *Astragalus*. Under the condition of an MJ concentration of 157.4 μM and an induction time of 18.4 h, the total saponin yield was 2.07 times that of the control. It was also revealed that *MVD*, *IDI*, *FPS*, and *SS* are the key enzyme genes in the pathway of MJ-induced regulation of astragaloside biosynthesis [21]. To study the response of the hairy root of *Astragalus* to 100 μM methyl jasmonate (MeJA), DESeq analysis showed that 2127 genes were up-regulated and 1247 genes were down-regulated, and among the 2,127 up-regulated genes 17 were new astragaloside biosynthesis genes and 7 were isolated new genes for the biosynthesis of isoflavones and isoflavone glucoside-7-O-β-D-glucoside. The accumulation of ASTS, Mucor, and CG in the hairy roots treated with MeJA increased significantly [32].

Exogenous transcription factors also have a regulatory effect on the synthesis of astragalosides. Li et al. (2022) [41] found that Arabidopsis MYB12, anthocyanin pigment 1 (PAP1), and maize leaf color (LC) transcription factors had regulatory effects on the synthesis of astragaloside metabolites. The overexpression of LC led to the accumulation of astragaloside I-IV. The accumulation of astragaloside I and IV was higher. The overexpression of MYB12 increased the accumulation of astragaloside I in transgenic hairy roots, followed by astragaloside IV. The overexpression of PAP1 increased the synthesis of astragaloside I and IV. Several key genes in the biosynthesis pathway of valeric acid, especially 3-hydroxy-3-methylglutaryl coenzyme A reductase (HMGR1, HMGR2 and HMGR3), were differentially up-regulated in response to these transcription factors, which could lead to the synthesis of astragaloside IV in the hairy roots of membranous *Astragalus membranaceus*.

Real-time fluorescence quantitative PCR analysis of eight key enzymes in the synthetic pathway of astragaloside acetyl-CoA acetyltransferase(AATC), 3-hydroxy-3-methylglutaryl-CoA synthase (HMGS), 3-hydroxy-3-Methylglutaryl coenzyme A reductase (HMGR), isoprene pyrophosphate isomerase (IDI), farnesyl pyrophosphate synthase (FPS), squalene synthase (SS), squalene epoxy Enzyme (SE), and Cycloaltinane synthase (CAS) gene expression levels, found that in HMGR, FPS, SE, and CAS genes under the water regulation of *Astragalus* saponins content, the regulation effect was obviously the main regulatory gene [31]. Real-time quantitative polymerase chain reaction (QRT-PCR) technology was used to study the expression levels of genes related to the AST biosynthetic pathway in *Astragalus* plant seedling roots (SRS), adventitious roots (ARs), and hairy roots (HRs). It was found that genes involved in the AST biosynthetic pathway had the lowest transcription levels in ARs, and showed similar patterns in HRs and SRs. ARs and CG had higher phenylalanine concentrations, and phenylalanine was the predecessor of the phenylpropane biosynthesis pathway [42]. Therefore, both biological and abiotic factors can affect the metabolic regulation of the active ingredient of *Astragalus*.

1.2.2. *Astragalus* Flavone

Isoflavonoid and Calycosin-7-glucoside (CGs) are accumulated in the whole plant of *Astragalus*, and mainly concentrated in the leaves. It was found that after 10 days of the UV-B treatment of *Astragalus* root, the change in isoflavone content in the *Astragalus* root was positively correlated with the expression of isoflavone biosynthesis related genes, and UV-B radiation significantly induced the isoflavone synthesis, which also provided a feasible heuristic strategy for understanding the accumulation of isoflavones in *Astragalus* [35]. Using ultraviolet light (UV-A, UV-B and UV-C) to promote the accumulation of isoflavones in AMHRCs, it was found that 86.4 kJ/m (2) UV-B had the best effect of promoting the production of isoflavones in AMHRCs at 34d. UV-B up-regulated the transcriptional expression of all genes involved in the isoflavone biosynthesis pathway. According to the results, PAL and C4H are two potential key genes that control isoflavone biosynthesis [36]. Using MJ, ASA, and SA signaling molecules to induce and regulate the biosynthesis of isoflavone secondary metabolic active components in *Astragalus* hairy roots, it was found that the yield of total isoflavones with an MJ concentration of 283 µM and induction time of 33.75 h was 9.71 times that of the blank control group. It was also revealed that CHI and IFS are the key enzyme genes in the pathway of MJ-induced regulation of *Astragalus* isoflavone biosynthesis [21]. *Astragalus* hairy root cultures (AMHRCs) were co-cultured with immobilized food-grade fungi to increase the production of trichomes (CA) and formononetin (FO), and 34-day-old AMHRCs were immobilized with *immobilized Aspergillus niger* (IAN) for 54 h of co-cultivation. In the highest accumulation of CA and FO, IAN induction can promote the generation of endogenous signaling molecules involved in plant defense responses, resulting in the significant upregulation of CA and FO biosynthetic pathway gene expressions, thereby increasing the production of CA and FO [34].

By studying the changes in CGs content and the expression of related genes under different conditions, including phenylalanine ammonia lyase (PAL1), cinnamic acid 4-hydroxylase (C4H), chalcone synthase (CHS), chalcone reductase (CHR), chalcone isomerase (CHI), isoflavone synthase (IFS), and isoflavone 3'-hydroxylase (I3'H) expression, the effects of different conditions on CGs biosynthesis were found. These seven genes' expression levels showed a light-dependent manner under low temperature stress, but they showed different expression patterns when *Astragalus* plants were transferred from 16 °C to 2 °C or 25 °C or 2 °C (maintained for 24 h) to 25 °C. There were different effects on the transcription of PAL1 and C4H in the phenylpropanoid pathway in leaves. The expression of PAL1 changed markedly, which was consistent with the change in CGs content. PAL enzyme activity seemed to be the limiting factor in determining CGs levels. The PAL enzyme inhibitor L-alpha-aminooxy-beta-phenylpropionic acid almost completely blocked the accumulation of CGs at low temperatures. This confirmed that PAL1, as an intelligent gene switch, directly controls the accumulation of CGs in a light-dependent manner during low temperature treatment [33]. PAL may be a key point for flux into flavonoid biosynthesis in the genetic control of secondary metabolisms in *Astragalus Mongholicus* [43]. A new AmCHR gene was cloned from *Astragalus*, which is a new member of the CHR gene family. It is speculated that the expression of AmCHR is closely related to the accumulation of isoflavone glucoside in *Astragalus* [44]. Chalcone synthase (CHS) is a key enzyme and rate-limiting enzyme for the biosynthesis of flavonoids. The AnCHS gene was cloned from *Astragalus*, and proved that the expression of AnCHS and the accumulation of *Astragalus* isoflavones are closely related [45].

1.2.3. *Astragalus* Polysaccharide

Astragalus polysaccharide is the main component of *Astragalus*, which is composed of hexuronic acid, glucose, fructose, rhamnose, arabinose, galacturonic acid, and glucuronic acid, etc. It can be used as an immune promoter or regulator, and has anti-virus, anti-tumor, anti-aging, anti-radiation, anti-stress, and anti-oxidation effects, among others. The pathway of secondary metabolite biosynthesis can be explained by the identification of candidate genes and important regulatory factors. The best inducer for screening the *Astragalus* polysaccharide metabolic pathway is silver nitrate solution, the best treatment site is the underground part, the best treatment time is 6–9 days, and the best polysaccharide content change detection site is the *Astragalus* root; 36 unigenes were found in the metabolic pathway of the polysaccharides metabolized [46]. Studies have found that methyl jasmonate, salicylic acid, IAA, and NAA promote the accumulation of *Astragalus* hairy root polysaccharides and total saponins, and IBA and 2,4-D have a negative effect on the accumulation of *Astragalus* hairy root polysaccharides and total saponins [47]. This provides theoretical guidance and technical support for regulating secondary metabolic pathways and increasing the content of *Astragalus* secondary metabolites by genetic engineering.

The study on the regulation of secondary metabolites of *Astragalus* is shown in Figure 1.

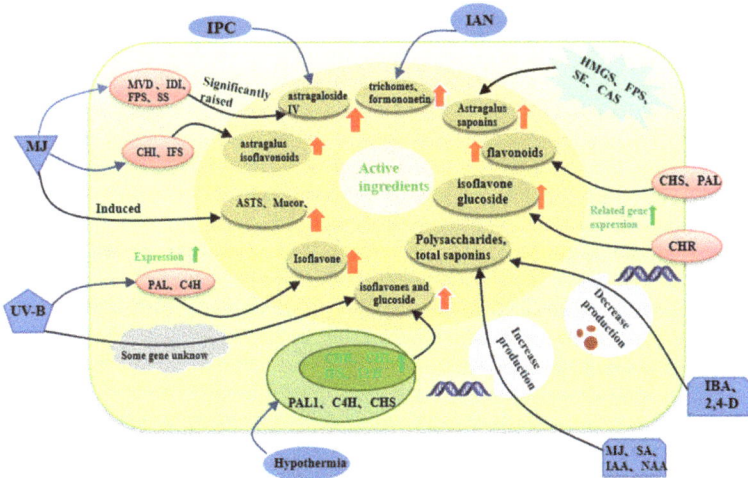

Figure 1. Study on the regulation of secondary metabolites of *Astragalus*. Note: The blue boxes represent biological and abiotic elicitors, the pink boxes represent genes, and the brown boxes represent secondary metabolite types.

3-hydroxy-3-methylglutaryl coenzyme A reductase (HMGR), farnesyl pyrophosphate synthase (FPS), squalene epoxy enzyme (SE), cycloaltinane synthase (CAS), methyl jasmonate(MJ), salicylic acid(SA), chalcone synthase (CHS), chalcone Reductase (CHR), chalcone isomerase (CHI), isoflavone synthase (IFS) and isoflavone 3′-hydroxylase (I3′H), phenylalanine ammonia (PAL), cinnamic acid 4-hydroxylase (C4H), farnesyl pyrophosphate synthase (FPS), squalene synthase (SS), isoprene pyrophosphate isomerase (IDI), mevalonate-5-diphosphate decarboxylase (MVD).

1.3. Research Progress on Synthetic Biological Pathways of Astragalosides

1.3.1. Research Progress on the Biosynthesis Pathway of Astragalosides

Astragalosides are the main active ingredient of *Astragalus*; they belong to triterpenoid saponins and are extremely important secondary metabolites in *Astragalus*. They are similar to the biosynthesis of other triterpenoid saponins. The biosynthetic pathway of astragalosides in plants includes the mevalonate (MVA) pathway and the 2-C-methl-D-erythritol-4-phospate (MEP) pathway. These two pathways ultimately produce the precursor isopentenyl pyrophosphate (IPP). IPP is catalyzed by farnesyl pyrophosphate synthase (FPS) to produce farnesyl diphosphate (FPP), FPP is catalyzed by squalene synthase (SS) to produce squalene, squalene is catalyzed by squalene epoxidase (SE) to produce 2,3-oxidosqualene, OS), and 2,3-oxidosqualene is catalyzed by cycloartenol synthase (CAS) to produce cycloartenol, which is the precursor of triterpene saponins. The cycloastragenol biosynthesis pathway can be seen in Figure 2. The functions of cytochrome P450 (CYP450), glycosyltransferases, and other genes required for the downstream synthesis pathway of astragalosides are being analyzed [48].

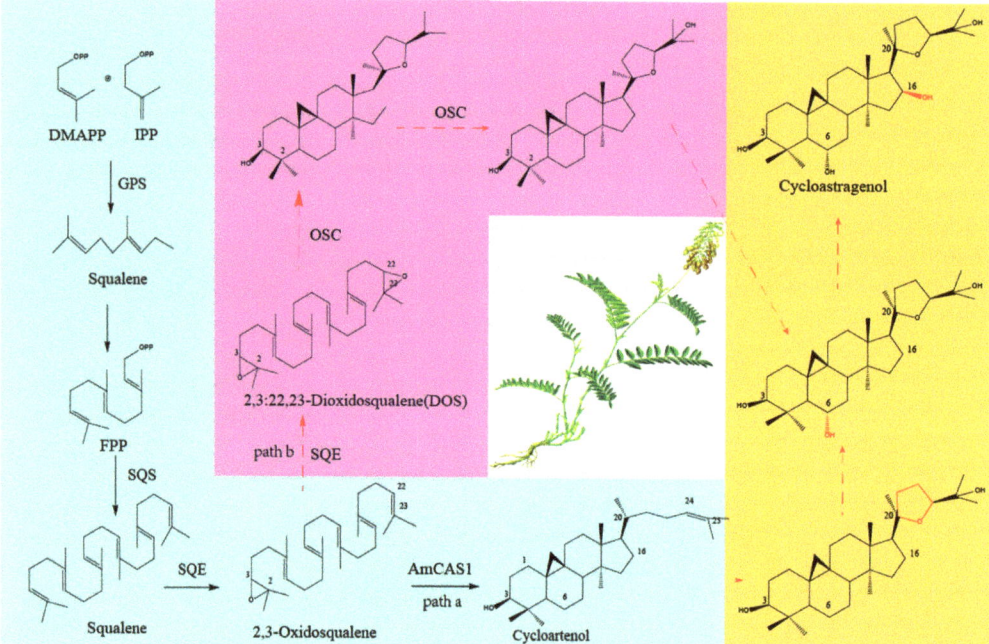

Figure 2. Cycloastragenol biosynthesis pathway. Note: The blue part represents the resolved pathway, and the orange and purple parts represent the speculative pathway. The red dotted line represents the speculative path.

Cycloartenol synthase (CAS) has a typical 9, 19—cyclopropane moiety. Astragalosides are gradually generated by cycloastragenol-type saponins in the downstream pathway. Cycloastragenol, which was further synthesized from cycloartenol, has a 20, 24-epoxy ring, and C6, C16, and C25 hydroxyl groups. Then, structurally diverse glycosides synthesize different astragalosides through a variety of glycosylation modes. This includes xylose, glucose, and glycosyl parts of single-chain, double-chain, triple-chain, or branched chains. The sites of action are 3-OH, 6-OH, 25-OH, and 20-OH. CAS converts the 2,3-oxidosqualene skeleton into a chair–boat–chair conformation, and then several specific CYP450s may catalyze the conversion of cycloartenol into cycloastragenol. CYP450 constitutes an important, highly differentiated sequence superfamily, which can be divided into 10 families. In these families, the CYP72 family is involved in the catabolism of isoprenoid hormones. The CYP71 family modified shikimic acid products and intermediates. The CYP85 family is involved in the modification of cyclic terpenes and sterols in the brassinolide pathway. Further studies have identified CYP93E1 from Leguminosae (CYP71 family) and CYP88D6 from *Glycyrrhiza* (CYP85 family), both of which are involved in the biosynthesis of saponins. Therefore, CYP71, CYP72, and CYP85 families may be involved in the biosynthesis of astragalosides [49].

Chen et al. (2023) [50] found that AmOSC3 is acycloartenol synthase expressed in both aerial and underground parts. It is related to the synthesis of astragalosides (cycloartane-type) in the roots. Jing Chen et al. [49] found a high-quality CAS transcript by transcriptome analysis of *A.mongholicus*, and detected 22 CYP450s related to the synthesis pathway of astragalosides. Seven transcripts belong to the CYP71 family (P57867.0, P5215.0, P7366.0, P50274.0, P26377.0, P71378.0, and P60800.0). One transcript belongs to the CYP72 family (P71249.0). Nine transcripts belong to the CYP85 family (P52746.0, P50482.0, P60763.0, P51105.0, P65633.0, P32417.0, P69412.0, P69412.2, and P43338.0). In addition, 25 UGT genes were detected to be related to the downstream synthesis pathway of astragaloside. Liu Lu et al. [51], based on transcriptome data, screened fourteen CYP450 genes with up-regulated expression, of which nine genes, four genes, and one gene were relatively highly expressed in roots, leaves, and stems, respectively. Duan et al. (2023) [52] found, for the first time, that AmCAS1 (a cycloartenol synthase) can catalyze the conversion of 2,3-oxidosqualene to cycloartenol, by transcriptome and phylogenetic analysis. Four glycosyltransferases were screened out, which were AmUGT15, AmUGT14, AmUGT13, and AmUGT7. They can catalyze the biosynthesis of cycloastragenol saponins, and the catalytic functions are 3-O-xylosylation, 3-O-glucosylation, 25-O-glucosylation/25-O-xylosylation, and 20-O-glucosylation, respectively. Among them, AmUGT15 can catalyze the conversion of cycloastragenol-6-O-D-glucoside to astragaloside IV or catalyze cycloastragenol into cycloastragenol-3-O-D-xyloseside; AmUGT13 can catalyze the conversion of astragaloside IV to Astragaloside VII or cycloastragenol-3-O-D-xylside to isostragaloside IV; AmUGT7 catalyzes the conversion of cycloastragenol-3-O-D-glucoside to cycloaraloside; and AmUGT14 can catalyze the conversion of cycloastragenol to cycloastragenol-3-O-D-glucoside. The Astragaloside biosynthesis pathway can be seen in Figure 3.

1.3.2. *Astragalus* Saponin Metabolism Engineering Research

A pair of special primers were designed with the cloned starch grains and starch synthase (GBSS) gene sequence from the hairy roots of *Astragalus* to construct the expression vector PbI-GBSS. It was confirmed that the enzyme activity of GBSS gene-transformed *Escherichia coli* was 20% higher than that of the untransformed strain [53]. The total RNA of *Astragalus* leaves was extracted by reverse transcription to synthesize cDNA, which was cloned into pPIC9K by in vitro homologous recombination technology to construct the pPIC9K-PAL expression vector. Then, it was transformed into *Pichia pastoris* GS115 to obtain a relatively pure phenylpropane amino acid ammonia lyase. The content of phenylalanine ammonia lyase after purification accounted for 11.54% of the total protein, and the highest specific activity reached 4270 U/mg [54]. At present, there has not been much research on the synthetic biology of *Astragalus*, which still needs to be explored continuously.

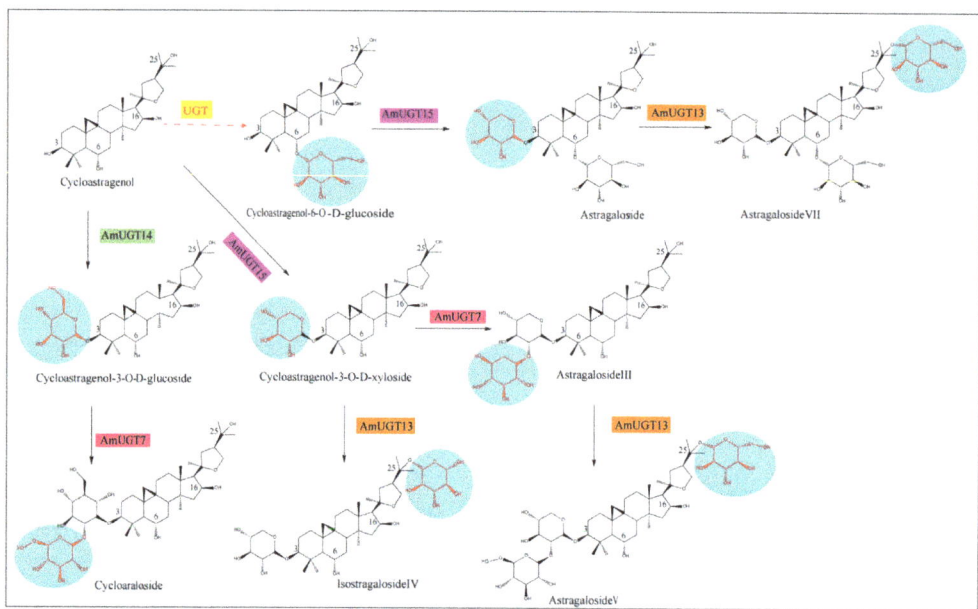

Figure 3. Astragaloside biosynthesis pathway. Note: The blue part represents different glycosyl groups, the pink, green and orange parts represent different glycosyltransferases. The red dotted line represents the speculative path.

1.3.3. Biotransformation of *Astragalosides*

The biotransformation methods of astragalosides include microbial transformation and enzymatic hydrolysis. In recent years, the biotransformation of cycloastragenol saponins has achieved many research results.

Microbial Transformation

The microbial transformation of astragalosides mainly uses enzymes in microorganisms to convert raw materials into required target products through complex and special metabolic pathways. Fungi are commonly used microorganisms for the biotransformation of astragalosides. Eight different yeast strains, such as *Aspergillus niger*, *Aspergillus oryzae* and white rot fungi, were selected for the biotransformation of astragaloside IV. It was found that *Aspergillus niger* had the strongest biotransformation activity. After transformation, the content of astragaloside IV increased by 10.7 times and reached 2.326 mg/g [55]. Li Ye [56] selected the strain *Absidia corymbifera* AS2 to transform astragalosides to ASI. This strain enhanced ASI production approximately fourfold when cultures were supplemented with 5 g/L of crude. Bedir et al., (2015) [57] studied the microbial transformation of *Astragalus* derived sapogenins, namely, *Cycloastragenol*, *astragenol*, and *Cyclocanthogenol*, by *Cunninghamella blakesleeana* NRRL 1369 and Glomerella fusarioides ATCC 9552. The unique enzyme system of both fungi resulted in hydroxylation, cyclization, dehydrogenation, and oxidation reactions. Under the deacetylation of fungal endophyte Penicillium canescens, which were isolated from *pigeon pea*, a novel and highly efficient biotransformation method of astragalosides to astragaloside IV in Radix Astragali was investigated [51]. Meng et al.(2018) [53] studied the biotransformation characteristics of astragaloside components in human intestinal flora and obtained four products and determined their structures: astragaloside I (AS-I), astragaloside II (AS-II), astragaloside III (AS-III), and astragaloside A (AS-IV). A novel biotechnology approach of combining deacetylation biocatalysis with the elicitation of IPC in *Astragalus membranaceus* hairy root cultures (AMHRCs) was proposed for the elevated production of astragaloside IV (AG IV). The highest AG IV accumulation was achieved in 36-day-old AMHRCs co-cultured with IPC for 60 h, which resulted in the enhanced production of AG IV by 14.59-fold in comparison with that in the control (0.193 ± 0.007 mg/g DW).

Enzymolysis

The most commonly used enzymes are β-glucosidase and β-xylosidase. Commercial β-glucosidase can hydrolyze astragaloside IV into cycloastragenol-6-O-β-D-glucoside; β-Glucosidases and β-xylosidases are two categories of enzymes that could cleave out nonreducing, terminal β-D-glucosyl and β-D-xylosyl residues with the release of D-glucose and Dxylose [58]. Qi Li et al.(2019) [58] used the purified thermostable and sugar-tolerant enzymes from *Dictyoglomus thermophilum* to hydrolyze ASI synergistically, which provided a specific, environmentally friendly and cost-effective way to produce CAG. An acetyl esterase from *A. corymbifera* AS2 was purified and its catalytic pathways were investigated, which showed unique enzymatic characteristics and enabled the clarification of the biotransformation pathways of astragalosides [59]. A novel β-glucosidase from *Phycicoccus* sp. Soil748 (Bgps) was discovered, possessing the efficient conversion rate for cycloastragenol-6-O-β-D-glucoside (CMG) into Cycloastragenol (CA). The optimum temperature and pH value of Bgps were determined as 45 °C and 7.0 [60]. It can be seen that the strains with high conversion efficiency were excavated from a variety of microorganisms, and the highly specific and efficient enzymes were isolated and purified. After large-scale expression in vitro, the development of commercial invertase is helpful for large-scale production [48].

2. Conclusions

As the leading industry in the world today, bioengineering technology has a high utilization rate in many industries such as medicine, agriculture, the chemical industry, and food. It has a significant role in promoting economic development, and also provides valuable technology for the development of medicine and information resources. Using bioengineering technology to produce medicinal plant active ingredients is an effective way to solve the problem of Chinese medicine resource shortages. With the development of synthetic biology, metabolic engineering, and protein engineering, the application of new bioengineering technology for medicinal plant resources' reasonable utilization and in-depth development has great development potential. Based on the traditional literature and modern biotechnology research, this article summarizes the tissue culture and metabolic regulation of active ingredients and the synthetic biology of *Astragalus*.

However, the synthesis mechanisms of some biologically active ingredients are still unclear, and there is a lack of information on the biosynthetic pathways and the complex regulatory mechanisms of the biosynthesis of target compounds. There are still key steps in the biosynthetic pathways of cycloastragenol and astragaloside IV that have not been resolved, which also thwarts the biosynthesis process of astragaloside. It can be seen that the metabolic pathways of secondary metabolites are still in need of a breakthrough in biotechnology research. Therefore, it is urgent to conduct more in-depth studies on these active ingredients and study new biotechnology. Clarifying medicinal value and supporting the further development of new products will be the focus of future research. This paper is helpful for the further study of the production and biotechnology of effective components in *A. membranaceus*.

Author Contributions: W.G. and J.W. conceived and designed the paper, B.J. and L.X. did the literature reading and analysis, Y.Z. made corrections to the manuscript, G.Z. and J.M. designed pictures and tables assistance; W.M. and J.L. reviewed the references, B.J., L.X. and Y.Z. supervised and completed the writing; K.-Y.P. and S.-Y.P. the manuscript and gave valuable opinions. All authors have read and agreed to the published version of the manuscript.

Funding: The sources of funding are National Key Research and Development Program (2020YFA0907903), the ability establishment of the sustainable use for valuable Chinese medicine resources (2060302) and the Scientific Research Transformation Foundation of Wenzhou Safety (Emergency) Institute of Tianjin University.

Data Availability Statement: No new data were created or analyzed in this study. Data sharing is not applicable to this article.

Acknowledgments: The authors gratefully acknowledge support from the School of Pharmaceutical Science and Technology, Tianjin University. And the financial support from and the open project of state key laboratory of innovative natural medicine and TCM injections.

Conflicts of Interest: The authors declare no conflict of interest for the manuscript submission. All authors have seen and approved the final version of the manuscript being submitted. We warrant that the article is the authors' original work, hasn't received prior publication and isn't under consideration for publication elsewhere.

References

1. Li, H.; Wang, Y.; Wang, R.; Chi, X.; Zhang, X. Research and utilization of astragalus germplasm resources. *Anhui Agric. Sci. Bull.* **2020**, *26*, 34–57.
2. Ivancheva, S.; Nikolova, M.; Tsvetkova, R. Pharmacological activities and biologically active compounds of bulgarian medicinal plants. *Phytochem. Adv. Res.* **2006**, *37661*, 87–103.
3. Yang, Q.Y.; Chen, K.J.; Lu, S.; Sun, H.R. Research progress on mechanism of action of radix astragalus in the treatment of heart failure. *Chin. J. Integr. Med.* **2012**, *18*, 235–240. [CrossRef] [PubMed]
4. Lysiuk, R.; Darmohray, R. Pharmacology and ethnomedicine of the genus astragalus. *Int. J. Pharmacol. Phytochem. Ethnomed.* **2016**, *3*, 46–53. [CrossRef]
5. Zhao, L.; Lv, X.; Yi, L.; Xia, Q.; Li, X. Research progress of saponins in *Astragalus membranaceus*. *J. Food Saf. Qual.* **2021**, *12*, 4937–4946.
6. Liu, Y.T.; Lv, W.L. Research progress in *Astragalus membranaceus* and its active components on immune responses in liver fibrosis. *Chin. J. Integr. Med.* **2020**, *26*, 794–800. [CrossRef]
7. Xu, N.; Wu, X.J. Research advance of pharmacological effects of astragalosides on nervous system diseases. *Zhongguo Zhong Yao Za Zhi* **2021**, *46*, 4674–4682. [PubMed]
8. Zhang, X.P.; Li, Y.D.; Luo, L.L.; Liu, Y.Q.; Li, Y.; Guo, C.; Li, Z.D.; Xie, X.R.; Song, H.X.; Yang, L.P.; et al. Astragalus saponins and liposome constitute an efficacious adjuvant formulation for cancer vaccines. *Cancer Biother. Radiopharm.* **2018**, *33*, 25–31. [CrossRef] [PubMed]
9. Lekmine, S.; Bendjedid, S.; Benslama, O.; Martin-Garcia, A.I.; Boussekine, S.; Kadi, K.; Akkal, S.; Nieto, G.; Sami, R.; Al-Mushhin, A.; et al. Ultrasound-assisted extraction, lc-ms/ms analysis, anticholinesterase, and antioxidant activities of valuablen metabolites from astragalus armatus willd.: In silico molecular docking and in vitro enzymatic studies. *Antioxidants* **2022**, *11*, 2000. [CrossRef]
10. Durazzo, A.; Nazhand, A.; Lucarini, M.; Silva, A.M.; Souto, S.B.; Guerra, F.; Severino, P.; Zaccardelli, M.; Souto, E.B.; Santini, A. Astragalus (*Astragalus membranaceus* bunge): Botanical, geographical, and historical aspects to pharmaceutical components and beneficial role. *Rend. Lincei Sci. Fis. Nat.* **2021**, *32*, 625–642. [CrossRef]
11. Yang, Z.; Yang, Y.; E, X. Research progress on differences between *Astragalus membranaceus* and *Astragalus membranaceus*. *J. Chin. Med. Mater.* **2020**, *43*, 1261–1265.
12. Wu, S.Q.; Lian, M.L.; Gao, R.; Park, S.Y.; Piao, X.C. Bioreactor application on adventitious root culture of *Astragalus membranaceus*. *Vitr. Cell Dev. Biol. Plant* **2011**, *47*, 719–724. [CrossRef]
13. Zhang, H.; Yasmin, F.; Song, B.H. Neglected treasures in the wild-legume wild relatives in food security and human health. *Curr. Opin. Plant Biol.* **2019**, *49*, 17–26. [CrossRef] [PubMed]
14. Catarino, S.; Duarte, M.C.; Costa, E.; Carrero, P.G.; Romeiras, M.M. Conservation and sustainable use of the medicinal leguminosae plants from angola. *PeerJ* **2019**, *7*, e6736. [CrossRef] [PubMed]
15. Gao, W.; Xiao, P. Application of bioengineering technology to protection of endangered medicinal plant resources. *Chin. Tradit. Herb. Drugs* **1994**, *39*, 961–964.
16. Zhang, H.C.; Liu, J.M.; Chen, H.M.; Gao, C.C.; Lu, H.Y.; Zhou, H.; Li, Y.; Gao, S.L. Up-regulation of licochalcone a biosynthesis and secretion by tween 80 in hairy root cultures of glycyrrhiza uralensis fisch. *Mol. Biotechnol.* **2011**, *47*, 50–56. [CrossRef]
17. Turgut-Kara, N.; Ari, U. In vitro plant regeneration from embryogenic cell suspension culture of astragalus chrysochlorus (leguminoseae). *Afr. J. Biotechnol.* **2008**, *7*, 1250–1255.
18. Feng, Y.C.; Zhao, Y.; Ha, Y.; Li, J.J.; Su, Z.Y.; Quan, X.L.; Wu, S.Q.; Wu, W.L. Drought stress-induced methyl jasmonate accumulation promotes calycosin-7-o-beta-d-glucoside production in *Astragalus membranaceus* adventitious roots. *Plant Cell Tissue Organ. Cult.* **2021**, *147*, 561–568. [CrossRef]
19. Zhao, X.; Zhang, H. Study on tissue culture and radiation mutation of *Astragalus membranaceus* bge. Var. Mongholicus (bge.) Hsiao. *Agric. Sci. Technol.-Hunan* **2009**, *10*, 37–40.

20. Ma, L.; Wang, X.; Gao, W. Callus initiation of *Astragalus membranaceus* var.mongholicus. *Chin. J. Pharm. Biotechnol.* **2008**, *15*, 355–358.
21. Jiao, J. Optimization of *Astragalus membranaceus* Hairy Root Cultures and Elicitation of Its Main Active Constituent Biosynthesis. Ph.D. Thesis, Northeast Forestry University, Harbin, China, 2016.
22. Du, M.; Wu, X.J.; Ding, J.; Hu, Z.B.; White, K.N.; Branford-White, C.J. Astragaloside iv and polysaccharide production by hairy roots of *Astragalus membranaceus* in bioreactors. *Biotechnol. Lett.* **2003**, *25*, 1853–1856. [CrossRef]
23. Halder, M.; Roychowdhury, D.; Jha, S. A critical review on biotechnological interventions for production and yield enhancement of secondary metabolites in hairy root cultures. In *Hairy Roots: An Effective Tool of Plant Biotechnology*; Springer: Berlin/Heidelberg, Germany, 2018; pp. 21–44.
24. Bu, Y. Effects of Culture Conditions on the Contents of Several Secondary Components in Astragalus Cells. Master's Thesis, Dalian Polytechnic University, Dalian, China, 2008.
25. Liu, Q.; Zhong, Z.; Cong, L.; Zhang, L. Lsolation of protoplast from the *Astragalus membranaceus*. *Guihaia* **2008**, *3*, 411–413.
26. Patel, H.; Krishnamurthy, R. Elicitors in plant tissue culture. *J. Pharmacogn. Phytochem.* **2013**, *2*, 60–65.
27. Naik, P.M.; Al-Khayri, J.M. Abiotic and biotic elicitors-role in secondary metabolites production through in vitro culture of medicinal plants. In *Abiotic and Biotic Stress in Plants—Recent Advances and Future Perspectives*; InTech: Rijeka, Croatia, 2016; pp. 247–277.
28. Zhang, K.; Liu, Z.; Yan, S.; Ren, W.; Liu, X.; Ma, W. Research progress on the secondary metabolism in astragalus. *World Sci. Technol.—Mod. Tradit. Chin. Med. Mater. Med.* **2016**, *18*, 875–882.
29. Al-Oubaidi, H.K.M.; Kasid, N.M. Increasing phenolic and flavonoids compounds of cicer arietinum l. From embryo explant using titanium dioxide nanoparticle in vitro. *World J. Pharm. Res* **2015**, *4*, 1791–1799.
30. Cakir, O.; Ari, S. Defensive and secondary metabolism in astragalus chrysochlorus cell cultures, in response to yeast extract stressor. *J. Env. Biol.* **2009**, *30*, 51–55.
31. Wei, H.; Cheng, L.; Wu, P.; Han, M.; Yang, L. Short-term water changes response of saponin biosynthesis process in astragalus membranceus. *Zhongguo Zhong Yao Za Zhi= China J. Chin. Mater. Med.* **2019**, *44*, 441–447.
32. Tuan, P.A.; Chung, E.; Thwe, A.A.; Li, X.; Kim, Y.B.; Mariadhas, V.A.; Al-Dhabi, N.A.; Lee, J.H.; Park, S.U. Transcriptional profiling and molecular characterization of astragalosides, calycosin, and calycosin-7-o-beta-d-glucoside biosynthesis in the hairy roots of *Astragalus membranaceus* in response to methyl jasmonate. *J. Agric. Food Chem.* **2015**, *63*, 6231–6240. [CrossRef] [PubMed]
33. Pan, H.Y.; Wang, Y.G.; Zhang, Y.F.; Zhou, T.S.; Fang, C.M.; Nan, P.; Wang, X.Q.; Li, X.B.; Wei, Y.L.; Chen, J.K. Phenylalanine ammonia lyase functions as a switch directly controlling the accumulation of calycosin and calycosin-7-o-beta-d-glucoside in *Astragalus membranaceus* var. Mongholicus plants. *J. Exp. Bot.* **2008**, *59*, 3027–3037. [CrossRef]
34. Jiao, J.; Gai, Q.Y.; Niu, L.L.; Wang, X.Q.; Guo, N.; Zang, Y.P.; Fu, Y.J. Enhanced production of two bioactive isoflavone aglycones in *Astragalus membranaceus* hairy root cultures by combining deglycosylation and elicitation of immobilized edible aspergillus niger. *J. Agric. Food Chem.* **2017**, *65*, 9078–9086. [CrossRef] [PubMed]
35. Liu, Y.; Liu, J.; Wang, Y.; Abozeid, A.; Tian, D.M.; Zhang, X.N.; Tang, Z.H. The different resistance of two astragalus plants to uv-b stress is tightly associated with the organ-specific isoflavone metabolism. *Photochem. Photobiol.* **2018**, *94*, 115–125. [CrossRef] [PubMed]
36. Jiao, J.; Gai, Q.Y.; Wang, W.; Luo, M.; Gu, C.B.; Fu, Y.J.; Ma, W. Ultraviolet radiation-elicited enhancement of isoflavonoid accumulation, biosynthetic gene expression, and antioxidant activity in *Astragalus membranaceus* hairy root cultures. *J. Agric. Food Chem.* **2015**, *63*, 8216–8224. [CrossRef]
37. Jin, H.Y.; Yu, Y.; Quan, X.L.; Wu, S.Q. Promising strategy for improving calycosin-7-o-beta-d-glucoside production in *Astragalus membranaceus* adventitious root cultures. *Ind. Crops Prod.* **2019**, *141*, 111792. [CrossRef]
38. Hwang, S. High-yield production of astragalosides from transgenic hairy root cultures of *Astragalus membranaceus*. *KSBB J.* **2006**, *21*, 123–128.
39. Gai, Q.Y.; Jiao, J.; Luo, M.; Wang, W.; Yao, L.P.; Fu, Y.J. Deacetylation biocatalysis and elicitation by immobilized penicillium canescens in *Astragalus membranaceus* hairy root cultures: Towards the enhanced and sustainable production of astragaloside iv. *Plant Biotechnol. J.* **2017**, *15*, 297–305. [CrossRef] [PubMed]
40. Gao, Y.; Sun, H.; Zhang, C.; Cao, Q.; Qin, X. Effects of hexanal on growth and accumulation of secondary metabolites in *Astragalus membranaceus* var. Mongholicus adventitious roots. *Plant Physiol. J.* **2015**, *51*, 476–480.
41. Li, X.H.; Kim, J.K.; Park, S.U. Heterologous expression of three transcription factors differently regulated astragalosides metabolic biosynthesis in *Astragalus membranaceus* hairy roots. *Plants* **2022**, *11*, 1897. [CrossRef] [PubMed]
42. Park, Y.J.; Thwe, A.A.; Li, X.H.; Kim, Y.J.; Kim, J.K.; Arasu, M.V.; Al-Dhabi, N.A.; Park, S.U. Triterpene and flavonoid biosynthesis and metabolic profiling of hairy roots, adventitious roots, and seedling roots of *Astragalus membranaceus*. *J. Agric. Food Chem.* **2015**, *63*, 8862–8869. [CrossRef] [PubMed]
43. Liu, R.R.; Xu, S.H.; Li, J.L.; Hu, Y.L.; Lin, Z.P. Expression profile of a pal gene from *Astragalus membranaceus* var. Mongholicus and its crucial role in flux into flavonoid biosynthesis. *Plant Cell Rep.* **2006**, *25*, 705–710. [CrossRef]
44. Bao, R.; Liu, J.; Yuan, Y.; Yu, Y.; Quan, X.; Wu, S. Cloning and expression analysis of chalcone reductase gene from *Astragalus membranaceus*. *Nor. Horticul.* **2019**, 141–146.
45. Zheng, H.; Yuan, Y.; Jin, H.; Yu, Y.; Quan, X.; Wu, S. Cloning and expression analysis of chs gene in *Astragalus membranaceus*. *J. Agric. Sci. Yanbian Univ.* **2019**, *41*, 9–14.

46. Liu, X. Responses of Artemisia Annua to Polyploidy Induction and the Regualtion of Agnps on Salvia Miltiorrhiza Hairy Roots. Ph.D Thesis, Northeast Forestry University, Xianyang, China, 2013.
47. Wang, W. Study on the Nutrition Absorption Characteristics and Second Ary Metabolish Regulation of *Astragalus membranaceus* (fisch.) bge. Ph.D Thesis, Northwest A&F University, Xianyang, China, 2008.
48. Xiao, Y.; Wang, R.; Zhao, S. Research progress on biosynthesis of saponins of astragali radix. *Acta Univ. Tradit. Med. Sin. Pharmacol. Shanghai* **2021**, *35*, 80–88.
49. Chen, J.; Wu, X.T.; Xu, Y.Q.; Zhong, Y.; Li, Y.X.; Chen, J.K.; Li, X.; Nan, P. Global transcriptome analysis profiles metabolic pathways in traditional herb *Astragalus membranaceus* bge. Var. Mongolicus (bge.) Hsiao. *BMC Genom.* **2015**, *16*, S15. [CrossRef]
50. Chen, K.; Zhang, M.; Xu, L.; Yi, Y.; Wang, L.; Wang, H.; Wang, Z.; Xing, J.; Li, P.; Zhang, X.; et al. Identification of oxidosqualene cyclases associated with saponin biosynthesis from *Astragalus membranaceus* reveals a conserved motif important for catalytic function. *J. Adv. Res.* **2023**, *43*, 247–257. [CrossRef]
51. Yao, M.; Liu, J.; Jin, S.; Jiao, J.; Gai, Q.; Wei, Z.; Fu, Y.; Zhao, J. A novel biotransformation of astragalosides to astragaloside iv with the deacetylation of fungal endophyte penicillium canescens. *Process Biochem.* **2014**, *49*, 807–812. [CrossRef]
52. Duan, Y.; Du, W.; Song, Z.; Chen, R.; Xie, K.; Liu, J.; Chen, D.; Dai, J. Functional characterization of a cycloartenol synthase and four glycosyltransferases in the biosynthesis of cycloastragenol-type astragalosides from *Astragalus membranaceus*. *Acta Pharm. Sin. B* **2023**, *13*, 271–283.58. [CrossRef]
53. Meng, X.T.; Yue, S.J.; Yang, Z.R.; Liu, J.; Feng, W.W.; Su, J.; Zhang, Y. Investigation on Biotransformation Characteristics of Astragalosides in Human Intestinal Microbiota. *Food Drug* **2018**, *20*, 161–167. (In Chinese)
54. Zhang, J.; Wang, S. Expression of *Astragalus membranaceus* phenylalanine ammonia-lyase gene in pichia pastoris. *J. Zhejiang Univ. Agric. Life Sci.* **2014**, *40*, 1–8.
55. Chen, C.; Fu, Y.; Zu, Y.; Wang, W.; Mu, F.; Luo, M.; Li, J.Y.; Gu, C.; Zhao, C. Biotransformation of saponins to astragaloside iv from radix astragali by immobilized aspergillus niger. *Biocatal. Agric. Biotechnol.* **2013**, *2*, 196–203. [CrossRef]
56. Ye, L.; Liu, X.H.; Zhou, W.; Feng, M.Q.; Shi, X.L.; Li, J.Y.; Chen, D.F.; Zhou, P. Microbial transformation of astragalosides to astragaloside iv by absidia corymbifera as2. *Process Biochem.* **2011**, *46*, 1724–1730. [CrossRef]
57. Bedir, E.; Kula, C.; Oner, O.; Altas, M.; Tag, O.; Ongen, G. Microbial transformation of astragalus sapogenins using cunninghamella blakesleeana nrrl 1369 and glomerella fusarioides atcc 9552. *J. Mol. Catal. B-Enzym.* **2015**, *115*, 29–34. [CrossRef]
58. Li, Q.; Wu, T.; Zhao, L.; Pei, J.; Wang, Z.; Xiao, W. Highly efficient biotransformation of astragaloside iv to cycloastragenol by sugar-stimulated beta-glucosidase and beta-xylosidase from dictyoglomus thermophilum. *J. Microbiol. Biotechnol.* **2019**, *29*, 1882–1893. [CrossRef] [PubMed]
59. Zhou, W.; Liu, X.; Ye, L.; Feng, M.; Zhou, P.; Shi, X. The biotransformation of astragalosides by a novel acetyl esterase from absidia corymbifera as2. *Process Biochem.* **2014**, *49*, 1464–1471. [CrossRef]
60. Cheng, L.; Zhang, H.; Liang, H.; Sun, X.; Shen, X.; Wang, J.; Wang, W.; Yuan, Q.; Ri, H.; Kim, T. Enzymatic bioconversion of cycloastragenol-6-o-β-d-glucoside into cycloastragenol by a novel recombinant β-glucosidase from *phycicoccus* sp. Soil748. *Process Biochem.* **2020**, *90*, 81–88. [CrossRef]

Disclaimer/Publisher's Note: The statements, opinions and data contained in all publications are solely those of the individual author(s) and contributor(s) and not of MDPI and/or the editor(s). MDPI and/or the editor(s) disclaim responsibility for any injury to people or property resulting from any ideas, methods, instructions or products referred to in the content.

Review

In Vitro Cultivation and Ginsenosides Accumulation in *Panax ginseng*: A Review

Fengjiao Xu [1,†], Anjali Kariyarath Valappil [2,†], Ramya Mathiyalagan [2], Thi Ngoc Anh Tran [1], Zelika Mega Ramadhania [1], Muhammad Awais [1] and Deok Chun Yang [1,2,*]

1. Graduate School of Biotechnology, College of Life Sciences, Kyung Hee University, Yongin-si 17104, Gyeonggi-do, Republic of Korea; fengjiaoxu96@gmail.com (F.X.); tranngocanh@khu.ac.kr (T.N.A.T.); zelika.mega@unpad.ac.id (Z.M.R.); awaiskazmi@khu.ac.kr (M.A.)
2. Department of Biopharmaceutical Biotechnology, College of Life Science, Kyung Hee University, Yongin-si 17104, Gyeonggi-do, Republic of Korea; anjalikv111@khu.ac.kr (A.K.V.); ramyabinfo@gmail.com (R.M.)
* Correspondence: dcyang@khu.ac.kr
† These authors contributed equally to this work.

Abstract: The use of in vitro tissue culture for herbal medicines has been recognized as a valuable source of botanical secondary metabolites. The tissue culture of ginseng species is used in the production of bioactive compounds such as phenolics, polysaccharides, and especially ginsenosides, which are utilized in the food, cosmetics, and pharmaceutical industries. This review paper focuses on the in vitro culture of *Panax ginseng* and accumulation of ginsenosides. In vitro culture has been applied to study organogenesis and biomass culture, and is involved in direct organogenesis for rooting and shooting from explants and in indirect morphogenesis for somatic embryogenesis via the callus, which is a mass of disorganized cells. Biomass production was conducted with different types of tissue cultures, such as adventitious roots, cell suspension, and hairy roots, and subsequently on a large scale in a bioreactor. This review provides the cumulative knowledge of biotechnological methods to increase the ginsenoside resources of *P. ginseng*. In addition, ginsenosides are summarized at enhanced levels of activity and content with elicitor treatment, together with perspectives of new breeding tools which can be developed in *P. ginseng* in the future.

Keywords: *P. ginseng*; in vitro tissue culture; ginsenosides accumulation; ginseng breeding

1. Introduction

Panax species, commonly referred as ginseng, which belong to the Araliaceae family, are slow-growing perennial herbal medicines with adaptive properties [1]. The word 'Panax' comes from the Greek word 'pan' (all) and 'zxos' (treatment of medicine), which means cure-all [2]. There are 15 species in the Panax genus, and they are listed in Table 1 [3,4]. Among these, there are three commonly used commercial ginseng species, including *Panax. ginseng*, *P. quinquefolius*, and *P. notoginseng* [5]. Most of the secondary compounds, especially ginsenosides, have been recorded in the roots. They act as tonic agents and stimulants that have been used in Asian countries for thousands of years, and they are becoming increasingly popular all over the world [6]. Pharmacological studies have demonstrated that ginseng species are rich in bioactive compounds such as ginsenosides, polysaccharides, flavonoids, phenolics, and volatile oils [7]. Among them, ginsenosides are known as the main bioactive ingredients responsible for the pharmaceutical efficacy of ginseng species [8], such as their anti-cancer [9], anti-fatigue [10], anti-inflammatory [11] activity and their prevention of cardiovascular disease [12], obesity [13], and cerebrovascular diseases [14], etc.

Table 1. Ginseng species.

No.	Scientific Name	Common Name	Rank	Cultivation Area
1	P. ginseng C. A. Meyer	Korean ginseng, Ginseng	Species	China, Republic of Korea, Russia
2	P. notoginseng (Burkill) F. H. Chen	Chinese ginseng, Sanchi ginseng	Species	China
3	P. quinquefolius	American ginseng	Species	China, America, Canada
4	P. japonicus C.A. Meyer	Japanese ginseng	Species	China, Japan
5	P. pseudoginseng Wallich	Himalayan ginseng	Species	China, Nepal
6	P. vietnamensis Ha & Grushv	Vietnamese ginseng	Species	China, Vietnam
7	P. stipuleanatus H.T. Tsai & K.M. Feng	Not mentioned	Species	China, Vietnam
8	P. trifolius L.	Dwarf ginseng	Species	America, Canada, Germany
9	P. zingiberensis C.Y. Wu & K.M. Feng	Not mentioned	Species	China, Nepal, Bhutan, Myanmar
10	P. wangianum S.C. Sun	Not mentioned	Species	China
11	P. assamocus R.N. Banerjee	Not mentioned	Species	India
12	P. variabilis J. Wen	Not mentioned	Species	China, India
13	P. omeiensis J. Wen	Not mentioned	Species	Not mentioned
14	P. sinensis J. Wen	Not mentioned	Species	East Himalaya
15	P. shangianus	Not mentioned	Species	Not mentioned

The increasing demand for herbal remedies has escalated the market value of ginseng species, but it has also created huge challenges for industries and governments to standardize and regulate plant-derived natural products to ensure consumer safety [15]. To address these issues, good standardized guidelines of agricultural cultivation should be established. For example, the Government of Canada established the Natural Health Products Directorate (NHPD) to enact the new legislation (JUS-601727) to govern the manufacture and marketing of natural health products [16].

However, the prolonged cultivation period, susceptibility to pathogens and replant diseases, limited availability of arable land, and labor-intensive cultivation practices have impeded farmers from meeting the growing market demand [15]. Moreover, the use of pesticides and the fluctuating environmental conditions resulting from global warming have compelled researchers and plant scientists to explore alternative methods to meet the demands of a rapidly increasing population [17]. Traditionally, there are two sources for obtaining ginseng species, one of which involves harvesting wild ginseng species. However, due to the over-exploitation of wild ginseng species and the destruction of arable land for growing ginseng species, the amount of wild ginseng is decreasing [18]. Another origin of ginseng species supply is to grow it in fields or forests, which is a time-consuming and labor-intensive process [19]. Furthermore, replanting disease will also result from intensive replanting, where replanting a second time in the same place will often lead to failure [20]. For these reasons, ginseng is becoming increasingly difficult to obtain and more expensive.

To address the above problems, tissue culture approaches have developed rapidly in recent years to produce bioactive compounds with high content and activities that not only have health-promoting properties but also significantly alter natural sources. The first attempt at plant cell cultivation was by the Austrian botanist Haberlandt in 1902, who isolated plant cells and cultivated them outside the whole plant [21]. The successful development of a nutrient medium by Murashige and Skoog in 1962, commonly known as MS medium, has remained in use, with minor adjustments [22]. The introduction of this specific nutrient medium, along with a range of plant growth regulators (PGRs), has significantly revolutionized the field of plant tissue culture research, leading to its successful integration as a viable commercial venture offering numerous advantages and possibilities. Multiple investigations have subsequently demonstrated that undifferentiated plant cells, such as calluses and cell suspensions, can be a valuable resource for producing identical secondary metabolites found in naturally occurring plants. It represents a significant advancement in plant research, over a century after Haberlandt's initial attempts in the field [23]. Plant tissue culture technology is helpful for plant transformation, clonal propagation, breeding, and protection of pharmaceutical plants and crops. Figure 1 describes the history and establishment of ginseng species' in vitro plant tissue culture [24–30].

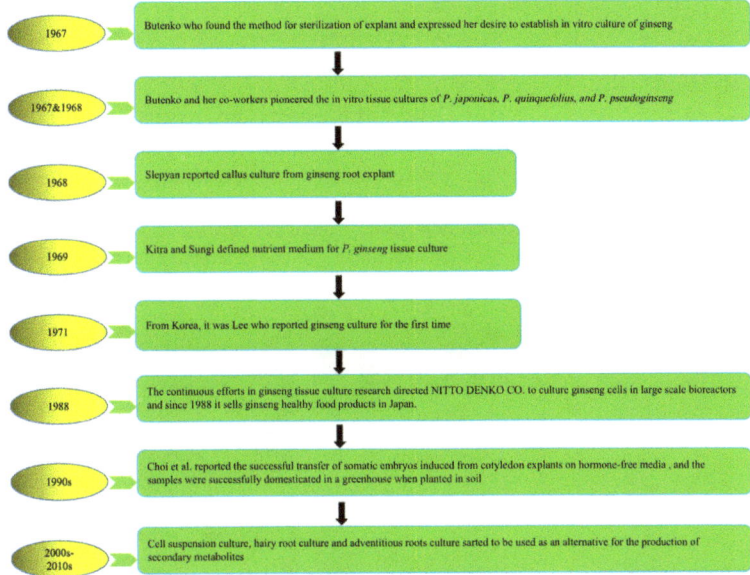

Figure 1. History of in vitro plant tissue culture of ginseng species. Note: This figure shows the in vitro cultivation of ginseng species from 1967 to present.

Recent research shows that in vitro tissue culture methods produce ginsenosides successfully from *P. ginseng* [31]. Therefore, this review summarizes the in vitro tissue culture methods on *P. ginseng*, which include direct root and shoot induction organogenesis without the formation of callus and indirect organogenesis from callus for further embryogenesis. In addition, in vitro biotechnological methods for ginsenosides accumulation include adventitious roots culture, cell suspension culture, hairy root culture, and bioreactor culture for large-scale propagation, which are also discussed, as depicted in Figure 2. Furthermore, this review discusses specific plant breeding methods that open new opportunities for ginseng breeding activities.

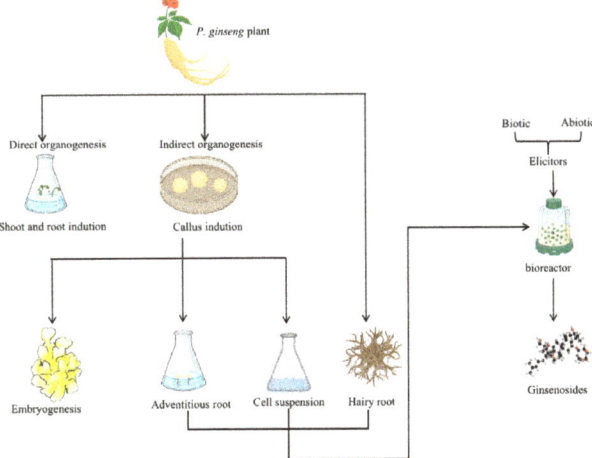

Figure 2. A summary of the standard procedures of *P. ginseng*'s in vitro culture methods and ways to increase ginsenosides accumulation. Note: The direct and indirect organogenesis methods can obtain

the in vitro culture of *P. ginseng*. Likewise, the accumulation of ginsenosides can be acquired through the adventitious roots, cell suspension, and hairy root cultures. In addition, the large-scale culture in bioreactors treated with biotic and abiotic elicitors also increases the biomass and the ginsenosides accumulation.

2. In Vitro Culture of *P. ginseng* Technologies

Tissue culture is classified based on the purpose of the culture and the source of materials. Several processes were established in *P. ginseng* based on the organization of the cells and organs (Figure 3), producing a consistent quality of *P. ginseng* and promoting the sustainable application of the species. In addition, under controlled culture conditions, numerous factors influence the quantity and quality of ginsenosides, such as medium constituents, pH, light conditions, culture temperature, explants, and abiotic factors.

Figure 3. Types of *P. ginseng* tissue culture. Note: (**a**) Callus (bar 1 cm) [32], (**b**) somatic embryos (bar 1 mm) [33], (**c**) shoots (bar 1 cm) [32], (**d**) hairy roots (bar 820 µm) [34], (**e**) cell suspension [35], and (**f**) adventitious roots [36].

2.1. Direct Organogenesis of P. ginseng

Direct organogenesis is the induction of roots and shoots directly from explants without forming a callus. Shoot culture demonstrated genetic stability and the potential to produce secondary metabolites. However, the research on the direct organogenesis of ginsenoside production is limited. Among the limited research available, it is vital to discuss the work of Hee-Young Lee et al., who studied the regeneration of *P. ginseng* from embryos obtained from the cultures of anthers. The results from the study indicated the optimum conditions required for the regeneration of *P. ginseng*—for example, cold treatment matters. The highest callus induction rate was achieved when the explants were cultured post-pretreatment at 4 °C.

On the other hand, the findings also report that the application of PGRs also affects shoot and root production. The shoots and roots can be induced on a medium supplemented with Gibberellin A3 (GA3) and 3-Indolebutyric acid (IBA) at the concentration of 28.9 µM and 14.7 µM, respectively [32]. Another study suggested that supplementing naphthaleneacetic acid (NAA) and IBA enhances the organogenic potential. Though IBA attained the highest shoot and root production rates, the roots induced by NAA showed better growth and were thicker than those of IBA. In addition, the roots induced by NAA also attained the highest ginsenosides production rates [37].

2.2. Indirect Organogenesis of P. ginseng

Indirect organogenesis, called callogenesis, is regenerating plantlets from the callus. The morphology and characteristics of calluses also influence organogenesis and biomass production. Friable and compact calluses are the two types of callus used in suspension culture and regeneration research, respectively [38].

2.2.1. Callus Culture

To date, explants, such as roots, stems, seeds, leaves, buds, petioles, anthers, and hypocotyls, have been used to induce ginseng callus. Among them, the leaves and roots are the most common ones. Typical callus induction and culture are carried out using Murashige and Skoog's (MS) basic medium or Gamborg medium (B5) with 3% sucrose and various PGRs at different concentrations. Researchers have investigated the effects of PGRs, among which 2,4-Dichlorophenoxyacetic acid (2,4-D) is the most potent one for the induction of the callus of many plant species. A summary of callus cultures is given in Table 2. Generally, the ginseng explants are cultured in the dark at 23 ± 2 °C. Wang et al. successfully induced callus from *P. ginseng* roots using MS medium supplemented with 2 mg/L of 2,4-D and 0.5 mg/L of Kinetin (KT) [39]. Similarly, Chang et al. [40] induced callus from ginseng roots using MS medium enriched with 1 mg/L of 2,4-D. However, the growth of the callus was initially slow, with only 1 cm of elongation in diameter after ten weeks. Nevertheless, it grew vigorously when the callus was subcultured on a new medium at 6–8-week intervals. In another study, Liu et al. used 3-year-old fresh ginseng roots as explants to induce callus on a modified MS medium enriched with 2 ppm of 2,4-D, 0.5 ppm of thidiazuron (TDZ), and 1 g/L of peptone [41]. They also induced another callus from 2-year-old ginseng roots on MS medium supplemented with 1 mg/L of 2,4-D and 0.1 mg/L of KT, sub-culturing every 15 days. As a result, after six months, they obtained three types of calluses [42].

Table 2. Callus induction and culture of *P. ginseng*.

Explants	Medium	PGRs		Other Factors	Ref.
		2,4-D	KT		
roots	MS	2 mg/L	0.5 mg/L		[39]
roots	MS	1 mg/L			[40]
roots	MS	2 mg/L		1 g/L peptone, 0.5 mg/L TDZ	[41]
roots	MS	1 mg/L	0.1 mg/L		[42]

2.2.2. Somatic Embryogenesis of P. ginseng

Using somatic embryogenesis for propagation allows a quick propagation of the superior ginseng lines while decreasing the variability commonly associated with seed propagation. The first and most crucial step in this process is the transition of somatic cells to embryonic cells. Somatic embryogenesis involves de-differentiating somatic cells into totipotent embryonic stem cells, which can produce embryos under appropriate in vitro conditions, ultimately developing into a whole plant [43,44]. Several studies have been conducted in ginseng to explore and optimize the conditions necessary for successful somatic embryogenesis [45].

In the next stage, they will develop into a whole plant after somatic embryogenesis. In *P. ginseng*, the first observation of somatic embryogenesis was reported in the callus derived from the roots by Butenko [25]. Since then, the regeneration of plants has been achieved through somatic embryogenesis using ginseng calluses derived from roots [40,46], zygotic embryos [47], somatic embryos [48], and protoplasts [49] isolated from somatic embryos (Table 3). The basic medium provides the nutritional composition and necessary elements for the growth and development of the explants. Most studies have employed MS medium for callus formation, proliferation, and somatic embryogenesis. In addition, Schenk and Hildebrandt medium (SH) and B5 have seldom been used for shoot regeneration and embryoid formation [50,51]. The in vitro propagation of somatic embryos in *P. ginseng* has been achieved using 2,4-D, KT, and NAA, but their concentrations and combinations vary depending on the type of explants.

Somatic embryo germination requires either chilling treatment for 8 weeks or GA3 hormone treatment at concentrations over 1.0 mg/L. Ultrastructural observation of cotyledon cells showed that without the treatment of chilling or GA3, somatic embryos contained large amounts of lipid reserves, dense cytoplasm, proplastids, and inactive mitochondria. Conversely, after chilling or GA3 treatment, the well-developed chloroplasts and functioning mitochondria with multiple cristae were seen in somatic embryos, indicating they may enter dormancy after maturation, similar to zygotic embryos. Recent studies have reported over 80% plant survival in hybrid ginseng, achieved by culturing embryos on GA3-supplemented medium, transferring them to hormone-free 1/2 SH medium, treating developed taproots with GA3 to break shoot dormancy, and transferring them to the soil. Therefore, GA3 pretreatment is crucial for successful transplantation [52].

Other factors, such as the salt content of the medium, also play a crucial role in somatic embryo induction. Choi et al. studied the effects of macrosalt stress on the embryogenesis of *P. ginseng*. The results showed that the highest frequency of somatic embryogenesis was observed on a medium containing 61.8 mM NH_4NO_3 with a ratio of $NH_4^+:NO_3^-$ at 21:39. Among the test media, including MS, B5, and SH, the maximum formation rate of the somatic embryo was observed when cotyledon explants were cultured on 1% agar MS medium with the supplementation of sucrose at 5% [29,53].

Different attempts have been made to regenerate ginseng through tissue culture using somatic embryogenesis techniques [54,55]. However, most regenerated plants cannot survive when transferred to soil. Shoot or multiple shoot formation has been successful from somatic embryos. However, taproots cannot be obtained, as the reproductive capacity of the multi-shoot complex gradually decreases and eventually disappears during long-term subculture (over 12–18 months). This phenomenon was observed in ginseng, where a single somatic embryo can regenerate into a plant with well-developed roots and shoots.

In contrast, multiple fused somatic embryos result in multiple shoots [50]. Only the study by Choi et al. [29,30] reported the successful transfer of somatic embryos induced from cotyledon explants on hormone-free media at 12–66% frequency in the regenerated plant. In his study, ginseng plants with well-developed shoots and roots regenerated from single embryos were successfully domesticated in a greenhouse when planted in soil (Figure 4) [56]. This regeneration protocol is very effective in the induction of whole plants.

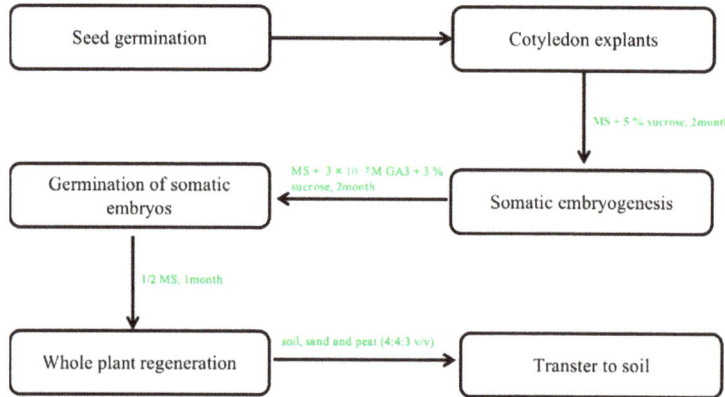

Figure 4. The protocol of somatic embryogenesis from callus to whole plants. Note: This figure shows the indirect organogenesis. First, the optimal callus was selected to form the embryogenic callus. In the next step of somatic embryogenesis, a regenerated whole plant can be obtained under optimal culture conditions.

Furthermore, optimal physical conditions, including light, temperature, and relative humidity, were identified for the in vitro propagation of ginseng species. A photoperiod of 14–16 h per day, with a cold white fluorescent lamp providing a light intensity of 24–80 µmol m^{-2} s^{-1}, and a temperature of 23 ± 2 °C were found to be appropriate for the incubation and maintenance of cultures [33,53].

In addition, there is a desperate need for a reliable and fast method to propagate the superior chemotype of ginseng species. For this purpose, it is crucial to establish a fully controlled in vitro micro-saline environment using shakers, temporary soaking, or bioreactors, which will enable the production of healthy and uniform seedlings. Additionally, molecular-marker-assisted protocols are highly recommended for verifying clonal fidelity and ensuring the production of identical clones. A well-known barrier to the effectiveness of plant production is the deterioration of culture vigor and regenerability over time. Various phenotypes, such as changes in plant height, biomass, grain yield, resistance to disease and pests, acid and salt tolerance, and agronomic performance, have all been linked to somaclonal variation. Over the 20 years of ginseng cell subculture, ginsenosides comprised just around 0.024% of the dry weight [57]. Raul Sanchez-Muñoz et al. indicated that the fundamental obstacle in creating commercially viable plant biofactories appears to be the alterations in methylation patterns, the primary mechanism predicted to be implicated in yield loss over time [58]. Therefore, clone maintenance should be investigated further for stable biomass and secondary metabolites production.

Table 3. List of the common conditions for somatic embryogenesis of ginseng.

Explants	Medium	PGRs	Embryogenesis Rate	Other Factors	Ref.
seeds	MS	2,4-D+ kinetin/ hormone free	45%/32.5%	Most of the single embryos were formed on a hormone-free medium, but multiple embryos were formed on a hormone-containing medium.	[50]
cotyledons	MS	2,4D+BA+ lactalbumin hydrolysate	87%	The use of glucose can enhance somatic embryo formation.	[59]
cotyledon	MS	61.8 mM of NH$_4$NO$_3$	56.3%	The highest frequency of somatic embryo formation occurred in the following order: NH$_4$NO$_3$ > KNO$_3$ > KH$_2$PO$_4$ > MgSO$_4$ > CaCl$_2$.	[29]
zygotic embryos	MS	2,4-D+ kinetin	NM	NM	[60]

NM: not mentioned.

3. Ginsenoside Biosynthesis and Biotechnological Production

3.1. Biosynthetic Pathways of Ginsenosides

Ginsenosides are triterpenoids or saponins, secondary metabolites with significant medical value, particularly in the pharmaceutical industry. Because they resemble steroidal hormones, these secondary metabolites have a variety of pharmacological characteristics. According to the aglycone structure, ginsenosides are classified into dammarane or oleanane types. Ocotillol-type ginsenosides are derived from oleanolic acid precursors. In contrast, the dammarane-type ginsenosides can further be classified into protopanaxadiol (PPD) and protopanaxatriol (PPT) ginsenosides [61], which are the major ginsenosides in *P. ginseng* and the main bioactive constituents for its biological activities.

The biosynthetic pathways of ginsenosides have been demonstrated, as shown in Figure 5. Generally, two pathways and three stages are involved in the biosynthesis of ginsenosides. The production of ginsenosides occurs in the cytosol and plastids through the mevalonic acid (MVA) pathway and the methylerythritol (MEP) pathway. The three stages are as follows: (1) firstly, isopentenyl diphosphate (IPP) and its isomer dimethylallyl diphosphate (DMAPP) are produced via the MVA and MEP pathways; (2) IPP and DMAPP are then transformed into 2,3-oxidosqualene; and (3) ginsenosides (such as Rh1, Rh2, Rg1, Rg3, Rd, C-K, F2, and Ro) are created via three reaction steps from 2,3-oxidosqualene, which include cyclization, hydroxylation, and glycosylation. The biosynthetic pathways of ginsenosides consist of more than 20 steps of consecutive enzymatic reactions, including enzymes, for example: 3-hydroxy-3-methylglutaryl coenzyme A reductase (HMGR), farnesyl pyrophosphate synthase (FPS), squalene synthase (SS), squalene epoxidase (SQE), dammarenediol-II synthase (DS), β-amyrin synthase (AS), cytochrome P450 (CYP450), and UDP-glycosyltransferase (UGT) [62–64].

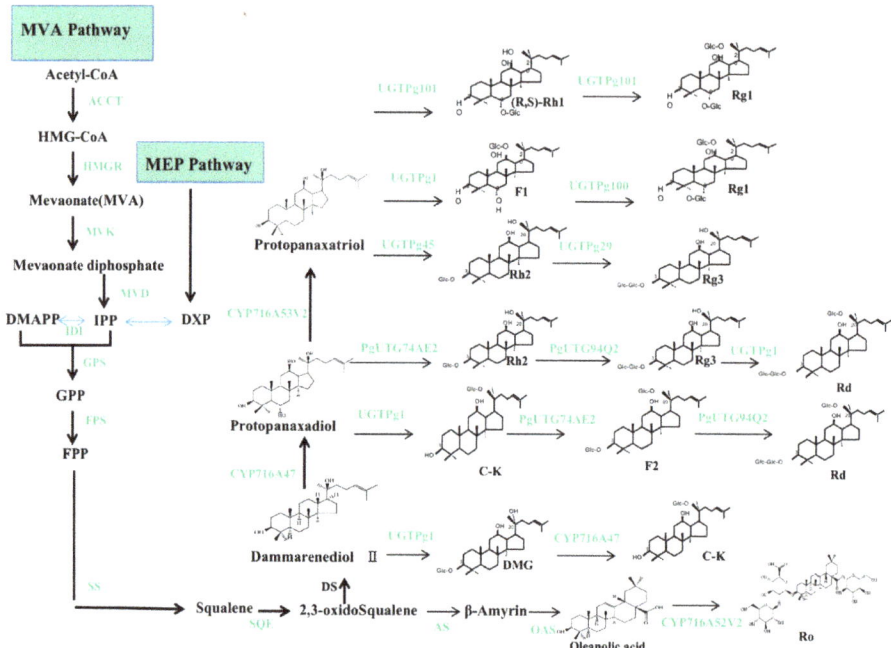

Figure 5. The biosynthetic pathways of ginsenosides. Note: the two pathways are MVA and MEP, respectively; the three stages are the formation of IPP and DMAPP; IPP and DMAPP are converted into 2,3-oxidoSqualene; ginsenosides of dammarenediol type; PPD and PPT types are synthesized from 3 steps. The green color represents the related genes and enzymes.

3.2. Ginsenosides Accumulation in In Vitro Cultivation of P. ginseng

Based on the purpose and types of tissues applied for in vitro culture, *P. ginseng* culture, which has proven successful in producing secondary metabolites, was reported in adventitious roots, cell suspension culture, and hairy roots.

3.2.1. Ginsenosides Accumulation via Adventitious Roots Culture

Adventitious roots culture is considered an alternative and prospective method for cell culture because of its higher biomass, production, stability in different physical and chemical environments, and higher ginsenosides production in large-scale bioreactors. The commercial-scale production of ginseng roots has been realized only in recent years in the Republic of Korea despite the first patent on ginseng root tissue culture being invented by Metz and Lang in 1966 [35]. The adventitious roots form from unusual parts such as calluses, stems, roots, and leaves. There are four discrete stages: the induction of callus, somatic embryogenesis, the formation of adventitious roots, and root elongation [65]. Optimal conditions, such as the types of explants, types and concentrations PGRs, and medium constituents, are needed in these four stages.

NAA and IBA are the two most used exogenous hormones for adventitious root induction from the callus. IBA was found to be more effective than NAA in the induction and elongation of lateral roots. Roots that were greatly elongated and slender formed on the IBA-containing medium compared to the NAA-containing medium when cultured under dark conditions [66,67]. A study reported the effect of NH_4NO_3 in the medium on the adventitious root induction, and the results showed that NH_4NO_3 free medium was better for the adventitious root formation. At the same time, it was shown to be necessary for the further elongation of post-induced adventitious roots [68].

Plant cells' defense mechanisms can be activated to respond to the attack of pathogens and biotic and abiotic stresses, which includes the biosynthesis of secondary metabolites [69]. In order to increase the contents of the secondary metabolite "ginsenosides," some compounds have been used as elicitors to increase the expression and critical enzymes relevant to its biosynthesis [70]. Methyl jasmonate (MJ) is the vital signaling compound involved in the biosynthetic pathways responsible for accumulating secondary metabolites [71]. Though MJ inhibits the fresh weight, dry weight, and growth ratio of ginseng roots in the in vitro culture, the production contents of ginsenosides were up to 5.5–9.7 times as high as in the untreated roots [72]. Another kind of elicitor, organic germanium, a food supplement, was used to work as an elicitor to increase biomass accumulation. When treated on the cultured roots with organic germanium at 60 mg/L, the accumulation of ginsenosides Rb and Rg and the dry biomass of adventitious roots was enhanced [73].

Owing to the high price of MJ, which limits the mass production of ginsenosides in large-scale bioreactors, the scientific community has started exploring other approaches to increase ginsenosides production in in vitro root cultures. *Endophytes* are bacterial or fungal microbes that can colonize healthy plant tissues without showing any apparent symptoms and protect their host by producing a variety of substances. Some reports have illustrated that endophytes can stimulate secondary metabolite accumulation when serving as elicitors. For example, a remarkable enhancement effect on ginsenosides accumulation was found when treating a 28-day-old adventitious root in a suspension culture of *P. ginseng* with dried mycelium of *Aspergillus niger*. Similarly, the application of this elicitor decreased the growth of the adventitious root, and the dry weight was reduced with the increasing concentrations of the elicitor [74]. An Endophyte bacterium, strain LB 5-3, from ginseng roots cultivated in the ginseng field showed the capacity to increase biomass and ginsenosides content by four times in ginseng adventitious root cultures [75]. A fungal suspension homogenate of pathogenic fungi (*Alternaria panax* Whetz) isolated from ginseng grown on the field was processed to be utilized as an elicitor. When the 30-day-old ginseng adventitious roots were treated with this fungal elicitor at the concentration of 200 mg/L for 8 days, the maximum ginsenosides accumulation content (29.6 mg/g dry weight) was obtained, and the biomass of the adventitious roots was not significantly inhibited [76].

Some other biotechnological methods, such as the induction of polyploidy and mutagenesis, can be used as alternative technologies for the enhancement of ginsenosides accumulation. A study reported that the mutagenesis induced via γ-irradiation enhanced the ginsenosides production content up to 16-fold compared with regularly cultured ginseng roots, and this study also indicated that the ginsenosides accumulation in the mutated adventitious roots is almost 1.6-fold that in of the normal roots cultivated in the ginseng field. However, it should be noted that the results of irradiation-based mutant breeding can vary. The growth of plants is impacted differently by each spectrum of γ-irradiation [77]. In another study, different concentrations of colchicine were used to treat the adventitious roots for different lengths of time to induce octoploid roots. The results showed that the total ginsenosides and Rb-group ginsenosides contents in octoploid roots were lower than that in untreated roots. However, the treated roots with colchicine contained more Rg-group ginsenosides, especially Rg1. These results indicated that polyploid adventitious roots can enhance secondary metabolite production in ginseng. Compared to the naturally tetraploid root, the fresh and dry biomasses of the octoploid adventitious roots were significantly higher [78].

Further studies are also highly desired to promote ginsenoside yields of at least the same level as those in 6-year-old ginseng roots cultivated in the field.

3.2.2. Ginsenosides Accumulation via Cell Suspension

Adventitious root cultures are an alternative method for producing stable secondary metabolites; however, root cultures of some advanced plants exhibit difficulties in harvesting bioactive ingredients and slower growth [79]. Besides adventitious roots culture, cell suspension culture has been used to produce secondary metabolites in many plant species for decades, especially in pharmaceutical botany [80]. Plant cells are regarded as green factories for synthesizing medicinal components in bulk. These cells in intact tissues, such as leaf, stem, root, and callus, are difficult to cultivate in ideal production methods and limit labor-intensive culture practices on a commercial scale. However, cell culture systems offer staggering opportunities to establish batches and continuous cultures in bioreactors of commercial scale. There have been many reports on producing pharmaceutical compounds in bulk via cell suspension culture [17]. The most commonly used method, "micro-propagation," is related to the proliferation of shoots through a semi-solid medium. While this semi-solid system has gained moderate or high success in increasing productivity and reducing the time required to propagate commercially essential materials, it is becoming increasingly important. Micropropagation via conventional techniques is usually a time-consuming method of clonal propagation. To overcome this, shaken culture methods using a liquid medium have been promoted. A liquid medium permits close contact with plant tissues to stimulate and promote the absorption of nutrients and phytohormones, thereby promoting the growth of branches and roots [81].

Research on ginseng suspension culture has mainly focused on various factors that influence cell growth and the synthesis of secondary metabolites. These factors include selecting optimal cell lines, using elicitors, and the impact of environmental and chemical factors such as light, pH, temperature, plant growth regulators (Table 4), nitrogen, carbon, and inorganic ions. Recent findings indicate that the rate of ginsenoside synthesis does not necessarily correlate with the growth rate of ginseng callus cells. As a result, a two-stage culture approach has become increasingly popular for producing secondary metabolites in ginseng cell suspension culture. This approach involves a cell growth stage followed by a ginsenoside production stage.

As mentioned above, many researchers have indicated that many physical and chemical factors can affect the production of secondary metabolites [82]. For example, optimizing the concentrations and combinations of various hormones and nutrients is frequently effective. In one study [83], the authors established a cell suspension culture system of mountain ginseng (*P. ginseng* C.A. Meyer) in an attempt to increase the production yield of ginsenosides via manipulating their culture methods and related factors, such as media strength, the concentrations and combinations of PGRs, the presence of sucrose, and the ratio of NO_3^+/NH_4^-. The maximum biomass content was obtained in the medium containing 2,4-D. However, the ginsenosides yield was much higher in the medium containing either NAA or IBA. IBA at the concentration of 7 mg/L was optimal for accelerating cell growth and saponin productivity. The level of ginsenosides, especially of the Rb group, was enhanced by adding cytokinins (benzylaminopurine (BA) at 0.5 mg/L and KT at 0.5 mg/L) despite having no effect on cell growth. The treatment of an initial nitrogen at 30 mM showed maximum cell growth and ginsenosides production. The amount of saponins will increase when the test medium has a high NO_3^+/NH_4^- ratio. The production of saponins was highest when nitrate was the only nitrogen source; however, when ammonium was used as a sole source, it was not beneficial for saponin biosynthesis [84]. The effect of another inorganic ion, phosphate, on cell growth and saponin accumulation was tested in the suspension culture of *P. ginseng* [85]. The results showed that the vital phosphate concentration for cell growth and the optimal concentration for simultaneous production of ginsenosides were 1.04 mM and 0.42 mM, respectively.

Elicitation has proven to be a successful method for increasing the production yield of various secondary metabolites. In a study conducted by Lu et al. [86], the effects of elicitor concentration and the time of elicitor addition on ginsenoside synthesis and cell growth in *P. ginseng* cell suspension cultures were investigated. The yeast extract and the MJ were tested, and both elicitors significantly increased saponin production. The highest level of supplemental ginsenosides measured by dry weight was 2.07%, 28 times greater than the control. The optimal time to add any elicitor was found to be on the day of inoculation. The results also showed that when MJ was used as an elicitor, removing 2,4-D from the medium was recommended, as MJ interacts antagonistically with 2,4-D. In another study [87], the impacts of MJ on cell growth and saponin accumulation in 5 L bioreactor cell suspension cultures were also studied. This study reveals that when the amount of MJ was between 50 and 400 µM, the ginsenosides accumulation was enhanced; however, with the increased concentration of MJ, the growth ratios and the fresh and dry weights of cells were strongly inhibited. The highest ginsenosides yield was obtained when MJ was used at 200 µM.

Table 4. Cell suspension culture of *P. ginseng*.

Medium	PGRs					Cell Growth Rate	Total Ginsenosides Content	Other Factors	Ref.
	2,4-D	6-BA	NAA	IBA	KT				
MS	1 mg/L	0.5 mg/L	1, 3, 5, 7, 9 mg/L	1, 3, 5, 7, 9 mg/L	0.5 mg/L	10 g/L	7.29 mg/g	Nitrite of 30 nM can increase both cell growth and total saponins	[83]
MS	0.4 m/L			2.5 mg/L	0.1 mg/L	11 g/L	21.4	Inorganic phosphate can promote cell growth and enhance saponin accumulation	[85]
MS	1 mg/L						28-fold higher than control	The MS medium was supplemented with inorganic salts: nicotinic acid, pyridoxine-HCl, etc.	[86]
MS			2 mg/L	7 mg/L	0.1 mg/L	8.82 mg/g	2.9 times higher than control	The highest ginsenosides yield were obtained when 200 µM MJ was added on day 15 during incubation	[87]

However, one of the omnipresent domain obstacles is the metabolic diversity in plant cell cultures, leading to the physically and genetically unstable production of secondary metabolites. From the perspective of biological process operation, any commercial attempt would be inhibited if there is no solution to this instability before it is scaled up [88].

3.2.3. Ginsenosides Accumulation via Hairy Roots

Agrobacterium rhizogenes, a bacterium found in soil, can induce hairy roots through genetic transformation. This process involves the genetic modification of plant cells via the plasmid T-DNA of *A. rhizogenes*, resulting in the formation of hairy roots during auxin metabolism. Research has demonstrated that hairy root cultures of ginseng have great potential for producing large amounts of biomass and ginsenosides. According to the literature, the mother plant's ability to manufacture secondary metabolites is on par with or surpasses the hairy root cells [89–91]. Hairy root culture has advantages over cell suspension culture, such as inherent genetic stability. Hairy roots have a genetic stability that is one of their hallmarks.

Moreover, using the hairy root system holds immense potential for incorporating other genes besides the Ri T-DNA genes, which can alter the metabolic pathways and generate valuable metabolites or compounds [92]. In addition, the growth rate of hairy roots is usually similar to or faster than that of cell culture, and they do not necessitate the use of PGRs in the culture medium [93]. The technique of hairy root cultivation can be traced back to the 1980s and is still undergoing refinement and standardization. In the case of ginseng, Inomata et al. [94] and Yoshikawa et al. [95] reported that the hairy roots of ginseng showed a higher content of ginsenosides and swift growth, as compared to its suspension cells and adventitious root cultures.

The light conditions are crucial in hairy root culture for the higher production of ginsenosides. For example, the effect of light on the growth and ginsenosides accumulation of *P. ginseng* hairy roots induced by *A. rhizogene* A4 was studied, and the results showed that the growth and ginsenosides accumulation was higher when cultured in the dark for 1 week and then transferred into the light condition (3500 lux) for 4 weeks. The yields of ginsenosides Rg1 and Rf increased by 3.3- and 2.4-fold, respectively [96]. The effects of electronic inhibitors on ginseng's root growth and ginsenoside content were tested by Yang et al. Ginsenoside production was higher when hairy roots were cultured in MS medium for 4 weeks and then transferred to 1/2MS medium containing ascorbic acid or 2,5-dimethylfuran for 1 week under light conditions. When investigating the effects of culture conditions on the growth and accumulation of ginsenosides, the research demonstrated that the accumulation of ginsenosides in ginseng hair roots cultured in a 20 L bioreactor was 34% higher than that in dark culture. During culture, ginseng hair roots were irradiated with ultraviolet light. Therefore, the growth of ginseng hairy roots was decreased following UV irradiation for a long time, but the accumulation of ginsenosides increased with the extension of UV irradiation time [97].

As discussed in the adventitious root culture, the method of induction of root mutagenesis can contribute to biomass and ginsenosides accumulation and hairy root culture. Studies were conducted to assess the role of *P. ginseng* hairy roots caused by ^{60}Co γ-ray irradiation. After removing the apical meristem of hairy roots irradiated below 2 Krad, lateral roots were used as cell lines. Furthermore, 206 hairy root cell lines were selected with various growth rates and forms and cultured in 1/2 MS medium without hormones. Then, 10 out of the 206 samples which showed excellent growth were chosen. Among them, y-GHR 70 and y-GHR 94 showed higher growth.

Different elicitors could be used for the high production of growth and ginsenosides. Hairy roots of *P. ginseng* established after induction with *A. rhizogenes* KCTC 2703 were cultured in liquid MS medium free of plant hormones supplemented with different concentrations of MJ and other inducers to promote ginsenoside accumulation. The results indicated that MJ significantly increased the total ginsenoside production, especially in the Rb group [98]. Seung-Yong Oh et al. studied the effects of chitin and chitosan on the production and growth of ginsenosides, and the results showed that when ginseng hair roots were cultured on 40 mg/L chitin and applied in the third week of culture, ginsenoside content and yield were the highest. The growth of ginseng hair root culture with 1 mg/L chitosan was the best, but the ginsenoside content with 30 mg/L chitosan was the highest [99]. Additional techniques, such as including Tween 80 in hairy root cultures, have significantly increased total ginsenoside production up to three-fold [100].

Various types and concentrations of salt affect the biomass and ginsenosides accumulation. A study investigated these effects on ginseng hair root growth and ginsenoside accumulation by adding different concentrations of potassium phosphate to 1/2 MS medium. The results showed that 1.25 mM potassium phosphate supplementation increased biomass and ginsenosides accumulation [101]. Another study determined the growth rate and yield of ginsenosides against NaCl in the hairy roots of *P. ginseng*. In MS liquid culture, the highest ginsenoside content and yield appeared 4 weeks after the onset of 0.1 M NaCl treatment [102]. To study the effects of inducers on the growth and biosynthesis of ginseng hairy roots, the hairy roots were treated with different concentrations of *Haliotidis concha* according to different time processes. *Haliotidis concha* supplementation increased the biomass and ginsenoside accumulation at the concentration of 10 mg/L [103].

3.2.4. Large-Scale Production of *P. ginseng* via Bioreactors

Elements like sluggish growth rates, constrained planting areas, climatic dependence, and labor scarcity primarily constrain the large-scale generation of biochemical compounds with economic value using field-grown plants. Advancements in plant cell and tissue culture techniques have facilitated the creation of significant phytochemicals. These plant cell and tissue culture procedures should reduce these adverse effects.

In plant biotechnology, significant progress has been made in utilizing bioreactor cultivation as a viable and appealing approach for biomass and bioactive compound production [104]. Compared to traditional tissue culture methods, the bioreactor system offers enhanced sophistication, allowing for individualized optimization of culture conditions. Factors such as temperature, pH, oxygen, carbon dioxide concentrations, and nutrient levels in the medium can be controlled precisely. In addition, the continuous circulation of the medium further improves nutrient availability. It is also possible to speed up cell regeneration and proliferation. Thus, production time and cost could be greatly decreased, product quality could be regulated and standardized, products could be free of contamination by pesticides, and production could be carried out throughout the year without being limited by geography [105]. The engineering of ginseng adventitious roots, cell suspension, and hairy root cultures has become a leading food biotechnology in Korea, China, and Japan. There are numerous varieties of bioreactors on the market. The stirred-tank bioreactor is the type that is used most frequently. It enables simple cell collection at various phases due to its larger size and capacity to boost the amount of nutrients. Even airlift and balloon-type bubble bioreactors produce ginsenosides in large quantities because they are more effective at transporting oxygen and have accurate flow predictions, reducing the shearing of cells [62].

Different bioreactors possess various advantages, such as ginsenoside accumulation in *P. ginseng* adventitious roots, cell suspension, and hairy root culture. In one study, the effects of organic nutrients on growth, the development of biomass, and ginsenosides production from the adventitious roots of *P. ginseng* in a balloon-type bioreactor were investigated. The results showed that a maximum ginsenosides yield of up to 12.42 mg/g dry weight extract under appropriate conditions can be obtained after 5 weeks of culture [105]. Another study compared the properties of *P. ginseng* hairy roots between a flask and aerated column or stirred bioreactor, and the results showed that it was almost three times as high as the flask culture of both bioreactors [104]. Another essential factor for the bioreactors' cell and root suspension cultures is the inoculum size, which can affect cell growth and secondary metabolite production [106–108]. Differences in the cell inoculum size can cause a significant difference in cell density during culture. Thus, it can lead to changes in culture conditions, such as the concentration of dissolved oxygen and gas metabolites, as well as the related enzyme activities, depending on the number of accumulated cells. These changes could affect cellular metabolism both directly and indirectly. There have been reports about the effect of inoculum size on cell growth and secondary metabolite accumulation [35,109–111], and the effect of cell density varies depending on the type of vessels and period of culture [112].

Faster biomass production increases the efficiency of producing secondary metabolites of economic interest. Therefore, using bioreactors to manufacture biomass in vitro is a novel strategy, frequently used to meet needs that are challenging to meet in the field because of pesticide use, climate change, and water restrictions. However, due to some limitations of plant tissue culture, it cannot be viewed as a replacement for actions to prevent or combat climate change. According to previous reports, the plantlets obtained from in vitro culture were initially small and had unfavorable traits. In vitro, plants must go through a transitional stage before independent growth because they cannot function autotrophically when cultured in vitro. The potential for creating plants with genetic aberrancy may increase. Plants are more vulnerable to contamination and water loss since they are grown in an atmosphere with high relative humidity [113].

Moreover, different cultivation conditions result in significant variations in quantitative and qualitative material characteristics originating from plants. Hence, the relationship between the propagation method and the quality of secondary metabolites should be made clear when using these technologies to obtain the most balanced cost–benefit ratio.

4. Perspectives on the Breeding of *P. ginseng* and Conclusions

Medicinal plants offer significant advantages for both people and the environment, and their functional and therapeutic values are higher than any other crop. However, the study and creation of therapeutic plants have been largely overlooked, and little is known about their genetic makeup, heterozygosity, growth patterns, and self-incompatibility [114]. This limitation further impedes the progress of medicinal plant breeding. Given the diversity of therapeutic plants and the environmental conditions in which they grow, breeding them is often an exceedingly intricate process. Therefore, collaborative efforts are necessary to address these challenges. The following sections will discuss current breeding strategies and future directions to overcome these challenges in *P. ginseng*.

4.1. Molecular Breeding

Molecular breeding is an essential content of molecular pharmacognosy. The process of breeding at the molecular level using molecular biology techniques is known as "molecular breeding", which is a new development of conventional breeding. Conventional breeding emphasizes phenotypic selection, while molecular breeding focuses on genotype selection. Molecular breeding is inseparable from conventional breeding. It also takes excellent phenotypes as the breeding goal, establishes the connection between genotype and phenotype, and selects phenotypes by genotype. Medicinal plant breeding is a crucial method for improving the quality of medicinal materials. In the past 20 years, through systematic breeding, crossbreeding, polyploid breeding, and other conventional breeding methods, many medicinal materials have been cultivated, such as *P. ginseng*, *P. quinquefolium*, *Rehmannia glutinosa*, *Salvia miltiorrhiza*, *Platycodon grandiflorum*, *Magnolia officinalis*, etc. However, due to the variety of medicinal plants, long growth cycles, high heterozygosity, unique breeding objectives, and other reasons, the overall level of medicinal plant breeding and breeding efficiency is not high [115].

While some progress has been made in using molecular markers for studies of therapeutic plants, most research has concentrated on identifying species and genetic polymorphism. There have been relatively few reports on marker-assisted breeding for these plants [116]. For instance, the successful application of DNA-marker-assisted selection and systematic breeding is developing a new variety of *P. notoginseng* called "Miao Xiang Kang qi" [117]. In this case, specific single-nucleotide polymorphisms (SNPs) identified in resistant varieties associated with root rot resistance act as genetic markers to assist systematic breeding. The incidence of root rot and rust was reduced by 83.6% and 71.8%, respectively, compared with conventional varieties. There are few reports of DNA-marker-assisted methods for selecting new ginseng cultivars [118–121]. Molecular breeding is fast, efficient, and accurate. Therefore, it can be used as a new reference for breeding and to direct the breeding of new variations of *P. ginseng*. More DNA molecular markers should be exploited to direct the future development of *P. ginseng* breeding [122].

4.2. Transgenic Breeding

Transgenic breeding is a molecular method in which one or more foreign genes are transferred to a plant through genetic engineering so that the plant can effectively express the corresponding products. The basic principle of genetic modification (GM) is like that of conventional crossbreeding: crossbreeding involves the transfer of whole gene chains (chromosomes). In contrast, gene transfer involves the selection of the most practical small gene segments, so GM is more selective than crossbreeding. Since the establishment of the *Agrobacterium*-mediated method [123], gene gun-mediated method [124], and pollen tube channel method [125], significant achievements have been made in the transgenic breeding of crops and horticultural crops for disease resistance, insect resistance, and stress resistance.

However, the uncertain genetic background, high heterozygosity, and repetitive sequences of medicinal plants make transgenic breeding more challenging than in other crops. Nevertheless, there have been successful examples of genetically modified medicinal plants, such as *Artemisia annua*. Despite this, the research on transgenic breeding of medicinal plants remains limited [126]. The agrobacterium-mediated method has been studied predominantly for the synthesis of secondary metabolites. There is also limited research on the transgenic breeding of *P. ginseng*. An attempt has been made using the Agrobacterium-mediated method to produce herbicide-resistant transgenic *P. ginseng* plants via the introduction of the phosphinothricin acetyl transferase (PAT) gene that confers resistance to the herbicide Basta. The results showed that transgenic *P. ginseng* grown in soil exhibited high Basta resistance [127]. In another report, thermotolerant transgenic *P. ginseng* was produced by introducing the isoprene synthase gene through Agrobacterium-mediated transformation. The transgenic plant appeared healthy when exposed to a high temperature of 46 °C for 1 h. In contrast, the non-transformed ones were wilted from heat shock, which suggested that the exogenous isoprene synthase gene can be added as an alternative technique for producing thermotolerant ginseng [128]. There is no available research report on transgenic breeding of medicinal plants via the pollen tube method. More transgenic genes must be exploited to produce either high tolerance or secondary metabolites production of *P. ginseng* plants.

Genetic modification is a promising approach that enables us to understand regulatory mechanisms through genetic alterations of single or few genes. It is worth noting that plants obtained from transgenic performances are only sometimes uniquely responsive, especially when performing classical genetic improvement activities in the field. Some studies reported that the increased yield of transgenic plants was carried out in controlled greenhouse conditions, and the response to a particular transgene can be reversed in the field [129]. However, transgenic lines were unable to maintain the advantages observed under control settings in field testing [130].

4.3. Digital Breeding

Due to advancements in DNA sequencing technologies and bioinformatics, many crop genomes are now publicly available. While having a reference genome sequence (the size of *P. ginseng* species is 3.4 Gb according to GenBank accession number GCA_020205605.1 in NCBI) is valuable, it does not fully represent the genetic diversity within a particular species. Therefore, information on DNA polymorphisms is essential for crop breeding. Hence, techniques such as whole-genome resequencing [131], sequence capture, target enrichment, resequencing methods [132], partial genome sequencing strategies [133–135], and high-density genotyping arrays [135] are highly beneficial. Genetic diversity studies have recently been conducted on staple and "orphan" crops [135–138].

Bioinformatics is a rapidly growing research area due to the crucial importance of extracting knowledge from diverse data, known as data mining. Analyzing a large amount of SNP and phenotypic data requires sufficient computing infrastructure and bioinformatics and shell scripting expertise, which is not commonly available in laboratories. Furthermore, there is a rising need to combine various "omics" data, such as genomics and phenomics, with mathematical and statistical models.

Therefore, developing bioinformatics skills among plant researchers and breeders is critical to ensuring they can analyze and interpret their data [139]. However, finding individuals with bioinformatics and plant breeding skills is challenging. The best solution is to form an interdisciplinary team where researchers can share knowledge and skills to advance crop improvements. The first open-access platform to offer extensive genetic resources of *P. ginseng* was created by Murukarthick Jayakodi et al. The most up-to-date draft genome sequence is available in the current version of this database, along with 59,352 gene structural and functional annotations and digital expression of genes based on transcriptome data from various tissues, growth stages, and treatments [140]. In another report, Woojong Jang et al. revealed the plastome diversity and established a standard haplotype grouping system according to various genotypes of ginseng plastomes. Eighteen polymorphic sites were identified, among which 11 SNPs and 7 INDels are included, with the help of a comparative investigation of the plastomes of 44 cultivated and wild ginseng accessions from Northeast Asian nations. Based on the SNP variants, 10 KASP markers were created to identify various haplotypes and their cultivation histories in various genetic resources. The understanding of ginseng evolution is intensified by establishing a digital haplotyping approach based on plastome diversity, which also acts as a powerful molecular breeding tool [141].

In contemporary times, the skill sets of breeders are evolving rapidly, which are rich enough that it is time we start thinking about breeding with different tools than in the past. Because of technological advances in phenotypic and genotypic analysis, as well as in biotechnology and the digital revolution, breeding cycles will be shortened cost-effectively [142]. Therefore, we can consider these new tools for breeding *P. ginseng* in the future.

4.4. Conclusions

This review has summarized various in vitro cultivation methods via direct and indirect organogenesis technologies and the ginsenosides' biosynthetic pathways. In addition, biotechnological approaches for ginsenosides accumulation, including adventitious root culture, cell suspension culture, hairy root culture, biotic and abiotic factors, and large scale-up culture for the high production of ginsenosides, have been explored. Despite significant advancements in ginseng in vitro culture, there is still more room for the research community to identify superior chemotypes of ginseng species for propagation. The perception of new techniques like transplanting seedlings and aeroponics culture methods are proposed as a significant requirement to grow high-quality ginseng rapidly. Finally, we have also introduced some breeding technologies, which may provide new insights as better options for prior cell lines of *P. ginseng* for better application and highly stable production of its secondary metabolites.

Author Contributions: Conceptualization: F.X. and D.C.Y. Software: F.X. and A.K.V. Validation: F.X. and D.C.Y. Resources: D.C.Y. Data curation: R.M. and A.K.V. Writing—original draft preparation: F.X. Writing—review and editing: F.X., A.K.V., R.M., T.N.A.T., Z.M.R. and M.A. Funding acquisition: D.C.Y. All authors have read and agreed to the published version of the manuscript.

Funding: This work was funded by Korea Institute of Planning and Evaluation for Technology in Food, Agriculture and Forestry (IPET) through the Agri-Food Export Business Model Development Program, funded by the Ministry of Agriculture, Food and Rural Affairs (MAFRA) (Project No: 320104-3).

Data Availability Statement: The data are contained within the article.

Acknowledgments: We would like to thank Hangbang Bio Inc. (South Korea) for supporting the relevant resources.

Conflicts of Interest: The authors declare no conflict of interest.

References

1. Uchendu, E.E.; Shukla, M.R.; Reed, B.M.; Brown, D.C.W.; Saxena, P.K. Improvement of Ginseng by In Vitro Culture. In *Comprehensive Biotechnology*; Academic Press: Burlington, MA, USA, 2011; pp. 317–329.
2. Qiang, B.; Miao, J.; Phillips, N.; Wei, K.; Gao, Y. Recent Advances in the Tissue Culture of American Ginseng (*Panax quinquefolius*). *Chem. Biodivers.* **2020**, *17*, e2000366. [CrossRef]
3. Cho, I.H.; Lee, H.J.; Kim, Y.-S. Differences in the Volatile Compositions of Ginseng Species (*Panax* sp.). *J. Agric. Food Chem.* **2012**, *60*, 7616–7622. [CrossRef]
4. Yue, J.; Zuo, Z.; Huang, H.; Wang, Y. Application of Identification and Evaluation Techniques for Ethnobotanical Medicinal Plant of Genus Panax: A Review. *Crit. Rev. Anal. Chem.* **2021**, *51*, 373–398. [CrossRef]
5. Szczuka, D.; Nowak, A.; Zakłos-Szyda, M.; Kochan, E.; Szymańska, G.; Motyl, I.; Blasiak, J. American Ginseng (*Panax quinquefolium* L.) as a Source of Bioactive Phytochemicals with Pro-Health Properties. *Nutrients* **2019**, *11*, 1041. [CrossRef]
6. Park, M.-J.; Kim, M.K.; In, J.-G.; Yang, D.-C. Molecular identification of Korean ginseng by amplification refractory mutation system-PCR. *Food Res. Int.* **2006**, *39*, 568–574. [CrossRef]
7. Li, P.; Liu, J. *Ginseng Nutritional Components and Functional Factors [Electronic Resource]*, 1st ed.; Li, P., Liu, J., Eds.; Springer: Singapore, 2020.
8. Liu, X.Y.; Xiao, Y.K.; Hwang, E.; Haeng, J.J.; Yi, T.H. Antiphotoaging and Antimelanogenesis Properties of Ginsenoside C-Y, a Ginsenoside Rb2 Metabolite from American Ginseng PDD-ginsenoside. *Photochem. Photobiol.* **2019**, *95*, 1412–1423. [CrossRef]
9. Indra, B. Assessment of Ginsenosides Efficacy on Three Dimensional HepG2 Liver Cancer Spheroids/Indra Batjikh. Master's Thesis, Graduate School of Kyung Hee University, Yongin, Republic of Korea, 2020.

10. Jiadan, Y.; Rongfeng, X.; Qing, D.A.I.; Xiaodan, L.A.I. Effects of ginsenoside Rg3 on fatigue resistance and skeletal muscle mitochondrial function in rats exposed to a simulated altitude of 5000 m. *Di 3 Jun Yi Da Xue Xue Bao* **2019**, *41*, 110–115. [CrossRef]
11. Li, J.; Yang, C.; Zhang, S.; Liu, S.; Zhao, L.; Luo, H.; Chen, Y.; Huang, W. Ginsenoside Rg1 inhibits inflammatory responses via modulation of the nuclear factor-κB pathway and inhibition of inflammasome activation in alcoholic hepatitis. *Int. J. Mol. Med.* **2018**, *41*, 899–907. [CrossRef]
12. Fu, C.; Yin, D.; Nie, H.; Sun, D. Notoginsenoside R1 protects HUVEC against oxidized low density lipoprotein (Ox-LDL)-Induced atherogenic response via down-regulating miR-132. *Cell Physiol. Biochem.* **2018**, *51*, 1739–1750. [CrossRef]
13. Pu, J.; Akter, R.; Rupa, E.J.; Awais, M.; Mathiyalagan, R.; Han, Y.; Kang, J.; Yang, D.C.; Kang, S.C. Role of ginsenosides in browning of white adipose tissue to combat obesity: A narrative review on molecular mechanism. *Arch. Med. Res.* **2022**, *53*, 231–239. [CrossRef]
14. Wu, T.; Jia, Z.; Dong, S.; Han, B.; Zhang, R.; Liang, Y.; Zhang, S.; Sun, J. Panax notoginseng Saponins Ameliorate Leukocyte Adherence and Cerebrovascular Endothelial Barrier Breakdown upon Ischemia-Reperfusion in Mice. *J. Vasc. Res.* **2019**, *56*, 1–10. [CrossRef] [PubMed]
15. Adil, M.; Ren, X.; Kang, D.I.; Thi, L.T.; Jeong, B.R. Effect of explant type and plant growth regulators on callus induction, growth and secondary metabolites production in Cnidium officinale Makino. *Mol. Biol. Rep.* **2018**, *45*, 1919–1927. [CrossRef] [PubMed]
16. Murch, S.J.; Saxena, P.K. St. John's wort (*Hypericum perforatum* L.): Challenges and strategies for production of chemically-consistent plants. *Can. J. Plant Sci.* **2006**, *86*, 765–771. [CrossRef]
17. Adil, M.; Jeong, B.R. In vitro cultivation of Panax ginseng C.A. Meyer. *Ind. Crops Prod.* **2018**, *122*, 239–251. [CrossRef]
18. Nantel, P.; Gagnon, D.; Nault, A. Population Viability Analysis of American Ginseng and Wild Leek Harvested in Stochastic Environments. *Conserv. Biol.* **1996**, *10*, 608–621. [CrossRef]
19. Wu, C.H.; Popova, E.V.; Hahn, E.J.; Paek, K.Y. Linoleic and α-linolenic fatty acids affect biomass and secondary metabolite production and nutritive properties of Panax ginseng adventitious roots cultured in bioreactors. *Biochem. Eng. J.* **2009**, *47*, 109–115. [CrossRef]
20. Gao, W.-Y.; Jia, W.; Duan, H.-Q.; Xiao, P.-G. Industrialization of medicinal plant tissue culture. *Zhongguo Zhong Yao Za Zhi* **2003**, *28*, 385–390.
21. Georgiev, M.I.; Weber, J.; Maciuk, A. Bioprocessing of plant cell cultures for mass production of targeted compounds. *Appl. Microbiol. Biotechnol.* **2009**, *83*, 809–823. [CrossRef] [PubMed]
22. Murashige, T.; Skoog, F. A revised medium for rapid growth and bio assays with tobacco tissue cultures. *Physiol. Plant.* **1962**, *15*, 473–497. [CrossRef]
23. Jamwal, K.; Bhattacharya, S.; Puri, S. Plant growth regulator mediated consequences of secondary metabolites in medicinal plants. *J. Appl. Res. Med. Aromat.* **2018**, *9*, 26–38. [CrossRef]
24. Butenko, R. Tissue culture of medicinal plants and prospective of its usage in medicine. *Probl. Pharmacog* **1967**, *21*, 184–191.
25. Butenko, R.; Brushwitzky, I.; Slepyan, L. Organogenesis and somatic embryogenesis in the tissue culture of Panax ginseng CA Meyer. *Bot Zh.* **1968**, *7*, 906–913.
26. Slepyan, L. Pharmacological activity of callus tissues of Ginseng grown under in vitro conditions. *Trans. Leningr. Khim-Farm. Inst.* **1968**, *26*, 236–244.
27. Lee, K.-D.; Huemer, R.P. Antitumor Al Activity of Panax Ginseng Extracts. *Jpn. J. Pharmacol.* **1971**, *21*, 299–302. [CrossRef] [PubMed]
28. Li, P.; Hang, B.; Ding, J. Comparison of immune function between polysaccharides from tissue culture of Panax ginseng and from ginseng root. *J. China Pharm. Univ.* **1989**, *20*, 216–218.
29. Choi, Y.-E.; Yang, D.-C.; Choi, K.-T. Induction of somatic embryos by macrosalt stress from mature zygotic embryos of Panax ginseng. *Plant Cell Tissue Organ Cult.* **1998**, *52*, 177–181. [CrossRef]
30. Choi, Y.; Yang, D.; Kim, H.; Choi, K. Distribution and changes of reserve materials in cotyledon cells of Panax ginseng related to direct somatic embryogenesis and germination. *Plant Cell Rep.* **1997**, *16*, 841–846. [CrossRef]
31. Zhang, T.; Chen, C.; Chen, Y.; Zhang, Q.; Li, Q.; Qi, W. Changes in the Leaf Physiological Characteristics and Tissue-Specific Distribution of Ginsenosides in Panax ginseng During Flowering Stage Under Cold Stress. *Front. Bioeng. Biotechnol.* **2021**, *9*, 637324. [CrossRef]
32. Lee, H.-Y.; Khorolragchaa, A.; Sun, M.-S.; Kim, Y.-J.; Kim, Y.-J.; Kwon, W.-S.; Yang, D.-C. Plant regeneration from anther culture of Panax ginseng. *Korean J. Plant Resour.* **2013**, *26*, 383–388. [CrossRef]
33. Choi, Y.-E.; Yang, D.-C.; Yoon, E.-S.; Choi, K.-T. High-efficiency plant production via direct somatic single embryogenesis from preplasmolysed cotyledons of Panax ginseng and possible dormancy of somatic embryos. *Plant Cell Rep.* **1999**, *18*, 493–499. [CrossRef]
34. Yang, D.-C.; Choi, Y.-E. Production of transgenic plants via Agrobacterium rhizogenes-mediated transformation of Panax ginseng. *Plant Cell Rep.* **2000**, *19*, 491–496. [CrossRef] [PubMed]
35. Thanh, N.T.; Murthy, H.N.; Paek, K.Y. Optimization of ginseng cell culture in airlift bioreactors and developing the large-scale production system. *Ind. Crops Prod.* **2014**, *60*, 343–348. [CrossRef]
36. Natalie, K.; Chandra, S.P.; Christanti, P.; Hak, K.J.; Yang, D.-C.; Sukweenadhi, J. Influence of volume medium on growth and ginsenoside level in adventitious root culture of Panax ginseng CA Meyer. In *IOP Conference Series: Earth and Environmental Science*; IOP Publishing: Bristol, UK, 2022.

37. Bonfill, M.; Cusidó, R.M.; Palazón, J.; Piñol, M.T.; Morales, C. Influence of auxins on organogenesis and ginsenoside production in Panax ginseng calluses. *Plant Cell Tissue Organ Cult.* **2002**, *68*, 73–78. [CrossRef]
38. Martin, K. Rapid propagation of Holostemma ada-kodien Schult., a rare medicinal plant, through axillary bud multiplication and indirect organogenesis. *Plant Cell Rep.* **2002**, *21*, 112–117.
39. Wang, J.; Man, S.; Gao, W.; Zhang, L.; Huang, L. Cluster analysis of ginseng tissue cultures, dynamic change of growth, total saponins, specific oxygen uptake rate in bioreactor and immuno-regulative effect of ginseng adventitious root. *Ind. Crops Prod.* **2013**, *41*, 57–63. [CrossRef]
40. Chang, W.-C.; Hsing, Y.-I. Plant regeneration through somatic embryogenesis in root-derived callus of ginseng (Panax ginseng CA Meyer). *Theor. Appl. Genet.* **1980**, *57*, 133–135. [CrossRef]
41. Liu, K.-H.; Lin, H.-Y.; Thomas, J.L.; Shih, Y.-P.; Chen, J.-T.; Lee, M.-H. Magnetic analogue-imprinted polymers for the extraction of ginsenosides from the Panax ginseng callus. *Ind. Crops Prod.* **2021**, *163*, 113291. [CrossRef]
42. Bonfill, M.; Cusido, R.M.; Palazon, J.; Canut, E.; Pinol, M.T.; Morales, C. Relationship between peroxidase activity and organogenesis in Panax ginseng calluses. *Plant Cell Tissue Organ Cult.* **2003**, *73*, 37–41. [CrossRef]
43. Ikeuchi, M.; Iwase, A.; Rymen, B.; Harashima, H.; Shibata, M.; Ohnuma, M.; Breuer, C.; Morao, A.K.; de Lucas, M.; De Veylder, L. PRC2 represses dedifferentiation of mature somatic cells in Arabidopsis. *Nat. Plants* **2015**, *1*, 1–7. [CrossRef]
44. Verdeil, J.-L.; Alemanno, L.; Niemenak, N.; Tranbarger, T.J. Pluripotent versus totipotent plant stem cells: Dependence versus autonomy? *Trends Plant Sci.* **2007**, *12*, 245–252. [CrossRef]
45. Guan, Y.; Li, S.-G.; Fan, X.-F.; Su, Z.-H. Application of somatic embryogenesis in woody plants. *Front. Plant Sci.* **2016**, *7*, 938. [CrossRef]
46. Ahn, I.; Choi, K.; Kim, B. Relationship between somatic embryogenesis and anthocyanin synthesis in callus cultures of Panax ginseng. *Korean J. Plant Tissue Cult.* **1991**, *18*, 227–232.
47. Lee, H.S.; Liu, J.R.; Yang, S.G.; Lee, Y.H.; Lee, K.-W. In vitro flowering of plantlets regenerated from zygotic embryo-derived somatic embryos of ginseng. *HortScience* **1990**, *25*, 1652–1654. [CrossRef]
48. Arya, S.; Arya, I.D.; Eriksson, T. Rapid multiplication of adventitious somatic embryos of Panax ginseng. *Plant Cell Tissue Organ Cult.* **1993**, *34*, 157–162. [CrossRef]
49. Arya, S.; Liu, J.R.; Eriksson, T. Plant regeneration from protoplasts of Panax ginseng (CA Meyer) through somatic embryogenesis. *Plant Cell Rep.* **1991**, *10*, 277–281. [CrossRef]
50. Kim, Y.-J.; Lee, O.R.; Kim, K.-T.; Yang, D.-C. High Frequency of Plant Regeneration through Cyclic Secondary Somatic Embryogenesis in Panax ginseng. *J. Ginseng Res.* **2012**, *36*, 442–448. [CrossRef]
51. Uchendu, E.E.; Paliyath, G.; Brown, D.C.W.; Saxena, P.K. In vitro propagation of North American ginseng (*Panax quinquefolius* L.). *Vitr. Cell. Dev. Biol.—Plant* **2011**, *47*, 710–718. [CrossRef]
52. Kim, J.Y.; Adhikari, P.B.; Ahn, C.H.; Kim, D.H.; Chang Kim, Y.; Han, J.Y.; Kondeti, S.; Choi, Y.E. High frequency somatic embryogenesis and plant regeneration of interspecific ginseng hybrid between Panax ginseng and Panax quinquefolius. *J. Ginseng Res.* **2019**, *43*, 38–48. [CrossRef]
53. Kim, O.T.; Kim, T.S.; In, D.S.; Bang, K.H.; Kim, Y.C.; Choi, Y.E.; Cha, S.W.; Seong, N.S. Optimization of direct somatic embryogenesis from mature zygotic embryos ofPanax ginseng CA Meyer. *J. Plant Biol.* **2006**, *49*, 348–352. [CrossRef]
54. Shoyama, Y.; Zhu, X.; Matsushita, H.; Kishira, H. Somatic embryogenesis in ginseng (*Panax* species). In *Somatic Embryogenesis and Synthetic Seed II*; Springer: Berlin/Heidelberg, Germany, 1995; pp. 343–356.
55. Kim, Y.; Kim, M.; Shim, J.; Pulla, R.; Yang, D. Somatic embryogenesis of two new Panax ginseng cultivars, Yun-Poong and Chun-Poong. *Russ. J. Plant Physiol.* **2010**, *57*, 283–289. [CrossRef]
56. Choi, Y.; Yang, D.; Park, J.; Soh, W.; Choi, K. Regenerative ability of somatic single and multiple embryos from cotyledons of Korean ginseng on hormone-free medium. *Plant Cell Rep.* **1998**, *17*, 544–551. [CrossRef]
57. Kiselev, K.V.; Dubrovina, A.S.; Shumakova, O.A. DNA mutagenesis in 2-and 20-yr-old Panax ginseng cell cultures. *Vitr. Cell. Dev. Biol.-Plant* **2013**, *49*, 175–182. [CrossRef]
58. Sanchez-Muñoz, R.; Moyano, E.; Khojasteh, A.; Bonfill, M.; Cusido, R.M.; Palazon, J. Genomic methylation in plant cell cultures: A barrier to the development of commercial long-term biofactories. *Eng. Life Sci.* **2019**, *19*, 872–879. [CrossRef] [PubMed]
59. Tang, W. High-frequency plant regeneration via somatic embryogenesis and organogenesis and in vitro flowering of regenerated plantlets in Panax ginseng. *Plant Cell Rep.* **2000**, *19*, 727–732. [CrossRef]
60. Langhansová, L.; Konrádová, H.; Vaněk, T. Polyethylene glycol and abscisic acid improve maturation and regeneration of Panax ginseng somatic embryos. *Plant Cell Rep.* **2004**, *22*, 725–730. [CrossRef] [PubMed]
61. Mohanan, P.; Yang, T.-J.; Song, Y.H. Genes and regulatory mechanisms for ginsenoside biosynthesis. *J. Plant Biol.* **2023**, *66*, 87–97. [CrossRef] [PubMed]
62. Luthra, R.; Roy, A.; Pandit, S.; Prasad, R. Biotechnological methods for the production of ginsenosides. *S. Afr. J. Bot.* **2021**, *141*, 25–36. [CrossRef]
63. Yang, J.-L.; Hu, Z.-F.; Zhang, T.-T.; Gu, A.-D.; Gong, T.; Zhu, P. Progress on the studies of the key enzymes of ginsenoside biosynthesis. *Molecules* **2018**, *23*, 589. [CrossRef]
64. Gantait, S.; Mitra, M.; Chen, J.-T. Biotechnological interventions for ginsenosides production. *Biomolecules* **2020**, *10*, 538. [CrossRef]
65. Praveen, N.; Manohar, S.; Naik, P.; Nayeem, A.; Jeong, J.; Murthy, H. Production of andrographolide from adventitious root cultures of Andrographis paniculata. *Curr. Sci.* **2009**, *96*, 694–697.

66. Kim, Y.-S.; Hahn, E.-J.; Yeung, E.C.; Paek, K.-Y. Lateral root development and saponin accumulation as affected by IBA or NAA in adventitious root cultures of Panax ginseng CA Meyer. *Vitr. Cell. Dev. Biol.-Plant* **2003**, *39*, 245–249. [CrossRef]
67. Jeong, C.; Murthy, H.; Hahn, E.; Lee, H.; Paek, K. Inoculum size and auxin concentration influence the growth of adventitious roots and accumulation of ginsenosides in suspension cultures of ginseng (Panax ginseng CA Meyer). *Acta Physiol. Plant.* **2009**, *31*, 219–222. [CrossRef]
68. Han, J.-Y.; Jung, S.-J.; Kim, S.-W.; Kwon, Y.-S.; Yi, M.-J.; Yi, J.-S.; Choi, Y.-E. Induction of adventitious roots and analysis of ginsenoside content and the genes involved in triterpene biosynthesis in Panax ginseng. *J. Plant Biol.* **2006**, *49*, 26–33. [CrossRef]
69. Vasconsuelo, A.; Boland, R. Molecular aspects of the early stages of elicitation of secondary metabolites in plants. *Plant Sci.* **2007**, *172*, 861–875. [CrossRef]
70. Hu, X.; Neill, S.; Cai, W.; Tang, Z. Hydrogen peroxide and jasmonic acid mediate oligogalacturonic acid-induced saponin accumulation in suspension-cultured cells of Panax ginseng. *Physiol. Plant.* **2003**, *118*, 414–421. [CrossRef]
71. Ali, M.B.; Yu, K.-W.; Hahn, E.-J.; Paek, K.-Y. Methyl jasmonate and salicylic acid elicitation induces ginsenosides accumulation, enzymatic and non-enzymatic antioxidant in suspension culture Panax ginseng roots in bioreactors. *Plant Cell Rep.* **2006**, *25*, 613–620. [CrossRef]
72. Kim, O.T.; Bang, K.H.; Kim, Y.C.; Hyun, D.Y.; Kim, M.Y.; Cha, S.W. Upregulation of ginsenoside and gene expression related to triterpene biosynthesis in ginseng hairy root cultures elicited by methyl jasmonate. *Plant Cell Tissue Organ Cult. (PCTOC)* **2009**, *98*, 25–33. [CrossRef]
73. Yu, K.; Murthy, H.; Jeong, C.; Hahn, E.; Paek, K. Organic germanium stimulates the growth of ginseng adventitious roots and ginsenoside production. *Process Biochem.* **2005**, *40*, 2959–2961. [CrossRef]
74. Li, J.; Liu, S.; Wang, J.; Li, J.; Liu, D.; Li, J.; Gao, W. Fungal elicitors enhance ginsenosides biosynthesis, expression of functional genes as well as signal molecules accumulation in adventitious roots of Panax ginseng CA Mey. *J. Biotechnol.* **2016**, *239*, 106–114. [CrossRef]
75. Song, X.; Wu, H.; Yin, Z.; Lian, M.; Yin, C. Endophytic bacteria isolated from Panax ginseng improves ginsenoside accumulation in adventitious ginseng root culture. *Molecules* **2017**, *22*, 837. [CrossRef]
76. Hao, Y.-J.; An, X.-L.; Sun, H.-D.; Piao, X.-C.; Gao, R.; Lian, M.-L. Ginsenoside synthesis of adventitious roots in Panax ginseng is promoted by fungal suspension homogenate of Alternaria panax and regulated by several signaling molecules. *Ind. Crops Prod.* **2020**, *150*, 112414. [CrossRef]
77. Kim, D.S.; Song, M.; Kim, S.-H.; Jang, D.-S.; Kim, J.-B.; Ha, B.-K.; Kim, S.H.; Lee, K.J.; Kang, S.-Y.; Jeong, I.Y. The improvement of ginsenoside accumulation in Panax ginseng as a result of γ-irradiation. *J. Ginseng Res.* **2013**, *37*, 332. [CrossRef]
78. Kim, Y.-S.; Hahn, E.-J.; Murthy, H.N.; Paek, K.-Y. Effect of polyploidy induction on biomass and ginsenoside accumulations in adventitious roots of ginseng. *J. Plant Biol.* **2004**, *47*, 356–360. [CrossRef]
79. Bhatia, S.; Bera, T.; Dahiya, R.; Bera, T.; Bhatia, S.; Bera, T. Classical and nonclassical techniques for secondary metabolite production in plant cell culture. In *Modern Applications of Plant Biotechnology in Pharmaceutical Sciences*; Elsevier: Amsterdam, The Netherlands, 2015; pp. 231–291.
80. Le, K.-C.; Jeong, C.-S.; Lee, H.; Paek, K.-Y.; Park, S.-Y. Ginsenoside accumulation profiles in long-and short-term cell suspension and adventitious root cultures in Panax ginseng. *Hortic. Environ. Biotechnol.* **2019**, *60*, 125–134. [CrossRef]
81. Mehrotra, S.; Goel, M.K.; Kukreja, A.K.; Mishra, B.N. Efficiency of liquid culture systems over conventional micropropagation: A progress towards commercialization. *Afr. J. Biotechnol.* **2007**, *6*, 1484–1492. [CrossRef]
82. Bourgaud, F.; Gravot, A.; Milesi, S.; Gontier, E. Production of plant secondary metabolites: A historical perspective. *Plant Sci.* **2001**, *161*, 839–851. [CrossRef]
83. Lian, M.-L.; Chakrabarty, D.; Paek, K.-Y. Effect of plant growth regulators and medium composition on cell growth and saponin production during cell-suspension culture of mountain ginseng (Panax ginseng CA Mayer). *J. Plant Biol.* **2002**, *45*, 201–206. [CrossRef]
84. Zhong, J.-J.; Wang, S.-J. Effects of nitrogen source on the production of ginseng saponin and polysaccharide by cell cultures of Panax quinquefolium. *Process Biochem.* **1998**, *33*, 671–675. [CrossRef]
85. Liu, S.; Zhong, J. Phosphate effect on production of ginseng saponin and polysaccharide by cell suspension cultures of Panax ginseng and Panax quinquefolium. *Process Biochem.* **1998**, *33*, 69–74. [CrossRef]
86. Lu, M.; Wong, H.; Teng, W. Effects of elicitation on the production of saponin in cell culture of Panax ginseng. *Plant Cell Rep.* **2001**, *20*, 674–677. [CrossRef]
87. Thanh, N.; Murthy, H.; Yu, K.; Hahn, E.; Paek, K. Methyl jasmonate elicitation enhanced synthesis of ginsenoside by cell suspension cultures of Panax ginseng in 5-l balloon type bubble bioreactors. *Appl. Microbiol. Biotechnol.* **2005**, *67*, 197–201. [CrossRef]
88. Qu, J.; Zhang, W.; Yu, X.; Jin, M. Instability of anthocyanin accumulation in Vitis vinifera L. var. Gamay Fréaux suspension cultures. *Biotechnol. Bioprocess Eng.* **2005**, *10*, 155. [CrossRef]
89. Kim, Y.; Wyslouzil, B.E.; Weathers, P.J. Secondary metabolism of hairy root cultures in bioreactors. *In Vitro Cell. Dev. Biol. -Plant* **2002**, *38*, 1–10. [CrossRef]
90. Srivastava, S.; Srivastava, A.K. Hairy root culture for mass-production of high-value secondary metabolites. *Crit. Rev. Biotechnol.* **2007**, *27*, 29–43. [CrossRef]
91. Mishra, B.N.; Ranjan, R. Growth of hairy-root cultures in various bioreactors for the production of secondary metabolites. *Biotechnol. Appl. Biochem.* **2008**, *49*, 1–10. [CrossRef]

92. Giri, A.; Narasu, M.L. Transgenic hairy roots: Recent trends and applications. *Biotechnol. Adv.* **2000**, *18*, 1–22. [CrossRef] [PubMed]
93. Flores, H.E.; Vivanco, J.M.; Loyola-Vargas, V.M. 'Radicle'biochemistry: The biology of root-specific metabolism. *Trends Plant Sci.* **1999**, *4*, 220–226. [CrossRef]
94. Inomata, S.; Yokoyama, M.; Gozu, Y.; Shimizu, T.; Yanagi, M. Growth pattern and ginsenoside production of Agrobacterium-transformed Panax ginseng roots. *Plant Cell Rep.* **1993**, *12*, 681–686. [CrossRef] [PubMed]
95. Yoshikawa, T.; Furuya, T. Saponin production by cultures of Panax ginseng transformed with Agrobacterium rhizogenes. *Plant Cell Rep.* **1987**, *6*, 449–453. [CrossRef]
96. Yang, D.; Choi, H.; Kim, Y.; Yun, K. Growth and ginsenosides production of hairy root (Panax ginseng CA Meyer) via light energy. *Korean J. Ginseng Sci.* **1996**, *20*, 318–324.
97. In, J.-G.; Park, D.-S.; Lee, B.-S.; Kim, S.-Y.; Rho, Y.-D.; Cho, D.-H.; Kim, S.-M.; Yang, D.-C. Effects of white light and UV irradiation on growth and saponin production from ginseng hairy root. *Korean J. Med. Crop Sci.* **2006**, *14*, 360–366.
98. Yu, K.-W.; Gao, W.-Y.; Son, S.-H.; Paek, K.-Y. Improvement of ginsenoside production by jasmonic acid and some other elicitors in hairy root culture of ginseng (Panax ginseng CA Meyer). *In Vitro Cell. Dev. Biol. -Plant* **2000**, *36*, 424–428. [CrossRef]
99. Oh, S.-Y.; Park, H.-J.; Choi, K.-H.; Meang, S.-J.; Yang, K.-J.; Yang, D.-C. The Production of ginsenosides from ginseng hairy root by treatment of the chitin and chitosan. *J. Ginseng Res.* **2000**, *24*, 68–73.
100. Liang, Y.; Wu, J.; Li, Y.; Li, J.; Ouyang, Y.; He, Z.; Zhao, S. Enhancement of ginsenoside biosynthesis and secretion by Tween 80 in Panax ginseng hairy roots. *Biotechnol. Appl. Biochem.* **2015**, *62*, 193–199. [CrossRef] [PubMed]
101. In, J.-G.; Park, D.-S.; Lee, B.-S.; Lee, T.-H.; Kim, S.-Y.; Rho, Y.-D.; Cho, D.-H.; Kim, S.-M.; Yang, D.-C. Effect of potassium phosphate on growth and ginsenosides biosynthesis from ginseng hairy root. *Korean J. Med. Crop Sci.* **2006**, *14*, 371–375.
102. Kim, Y.-J.; Sim, J.-S.; Lee, C.-H.; In, J.-G.; Lee, B.-S.; Yang, D.-C. The Effect of NaCl on the Growth and Ginsenoside Production from Ginseng Hairy Root. *Korean J. Med. Crop Sci.* **2008**, *16*, 94–99. [CrossRef]
103. Jeong, D.-Y.; Kim, Y.-J.; Shim, J.-S.; Lee, J.-H.; Jung, S.-K.; Kim, S.-Y.; In, J.-G.; Lee, B.-S.; Yang, D.-C. The Effect of Haliotidis Concha on the Growth and Ginsenoside Biosynthesis of Korean Ginseng Hairy Root. *J. Ginseng Res.* **2009**, *33*, 206–211.
104. Jeong, G.-T.; Park, D.-H.; Hwang, B.; Park, K.; Kim, S.-W.; Woo, J.-C. Studies on mass production of transformed Panax ginseng hairy roots in bioreactor. *Appl. Biochem. Biotechnol.* **2002**, *98*, 1115–1127. [CrossRef]
105. Sivakumar, G.; Yu, K.; Paek, K. Production of biomass and ginsenosides from adventitious roots of Panax ginseng in bioreactor cultures. *Eng. Life Sci.* **2005**, *5*, 333–342. [CrossRef]
106. Sakurai, M.; Mori, T.; Seki, M.; Furusaki, S. Changes of anthocyanin composition by conditioned medium and cell inoculum size using strawberry suspension culture. *Biotechnol. Lett.* **1996**, *18*, 1149–1154. [CrossRef]
107. Zhang, Y.-H.; Zhong, J.-J. Hyperproduction of ginseng saponin and polysaccharide by high density cultivation of Panax notoginseng cells. *Enzym. Microb. Technol.* **1997**, *21*, 59–63. [CrossRef]
108. Mavituna, F.; Buyukalaca, S. Somatic embryogenesis of pepper in bioreactors: A study of bioreactor type and oxygen-uptake rates. *Appl. Microbiol. Biotechnol.* **1996**, *46*, 327–333. [CrossRef]
109. Akalezi, C.; Liu, S.; Li, Q.; Yu, J.; Zhong, J. Combined effects of initial sucrose concentration and inoculum size on cell growth and ginseng saponin production by suspension cultures of Panax ginseng. *Process Biochem.* **1999**, *34*, 639–642. [CrossRef]
110. Thanh, N.T.; Murthy, H.N.; Paek, K.-Y. Ginseng cell culture for production of ginsenosides. In *Production of Biomass and Bioactive Compounds Using Bioreactor Technology*; Springer: Berlin/Heidelberg, Germany, 2014; pp. 121–142.
111. Thanh, N.T.; Murthy, H.N.; Yu, K.-W.; Jeong, C.S.; Hahn, E.-J.; Paek, K.-Y. Effect of inoculum size on biomass accumulation and ginsenoside production by large-scale cell suspension cultures of Panax ginseng. *J. Plant Biotechnol.* **2004**, *6*, 265–268.
112. Paek, K.-Y.; Hahn, E.-J.; Son, S.-H. Application of bioreactors for large-scale micropropagation systems of plants. *Vitro Cell. Dev. Biol.-Plant* **2001**, *37*, 149–157. [CrossRef]
113. Bhatia, S.; Sharma, K.; Dahiya, R.; Bera, T. *Modern Applications of Plant Biotechnology in Pharmaceutical Sciences*; Academic Press: Cambridge, MA, USA, 2015.
114. Dong, L.; Wei-Wen, L.I.; Ning, Z.Y.; Liao, H.J.; Jiang, Q.; Yao, Q.S. Status and Prospect of Medicinal Plant Breeding in China. *Res. Pract. Chin. Med.* **2014**, *28*, 3–6.
115. Ma, X.J.; Mo, C.M. Prospects of molecular breeding in medical plants. *Zhongguo Zhong Yao Za Zhi* **2017**, *42*, 2021–2031. [CrossRef]
116. Ying, Z.; Awais, M.; Akter, R.; Xu, F.; Baik, S.; Jung, D.; Yang, D.C.; Kwak, G.-Y.; Wenying, Y. Discrimination of Panax ginseng from counterfeits using single nucleotide polymorphism: A focused review. *Front. Plant Sci.* **2022**, *13*, 903306. [CrossRef]
117. Dong, L.-L.; Chen, Z.-J.; Wang, Y.; Wei, F.-G.; Zhang, L.-J.; Xu, J.; Wei, G.-F.; Wang, R.; Yang, J.; Liu, W.-L.; et al. DNA marker-assisted selection of medicinal plants (I). Breeding research of disease-resistant cultivars of Panax notoginseng. *Zhongguo Zhong Yao Za Zhi=China J. Chin. Mater. Medica* **2017**, *42*, 56–62. [CrossRef]
118. Wang, H.; Sun, H.; Kwon, W.-S.; Jin, H.; Yang, D.-C. Molecular identification of the Korean ginseng cultivar "Chunpoong" using the mitochondrial nad7 intron 4 region. *Mitochondrial Dna* **2009**, *20*, 41–45. [CrossRef]
119. Wang, H.; Xu, F.; Wang, X.; Kwon, W.-S.; Yang, D.-C. Molecular discrimination of Panax ginseng cultivar K-1 using pathogenesis-related protein 5 gene. *J. Ginseng Res.* **2019**, *43*, 482–487. [CrossRef]
120. Lee, J.-W.; Kim, Y.-C.; Jo, I.-H.; Seo, A.-Y.; Lee, J.-H.; Kim, O.-T.; Hyun, D.-Y.; Cha, S.-W.; Bang, K.H.; Cho, J.-H. Development of an ISSR-derived SCAR marker in Korean ginseng cultivars (Panax ginseng CA Meyer). *J. Ginseng Res.* **2011**, *35*, 52–59. [CrossRef]

121. Choi, H.-I.; Kim, N.H.; Kim, J.H.; Choi, B.S.; Ahn, I.-O.; Lee, J.-S.; Yang, T.-J. Development of reproducible EST-derived SSR markers and assessment of genetic diversity in Panax ginseng cultivars and related species. *J. Ginseng Res.* **2011**, *35*, 399. [CrossRef]
122. Wang, W.; Xu, J.; Fang, H.; Li, Z.; Li, M. Advances and challenges in medicinal plant breeding. *Plant Sci.* **2020**, *298*, 110573. [CrossRef]
123. Shaw, C.H.; Leemans, J.; Shaw, C.H.; Van Montagu, M.; Schell, J. A general method for the transfer of cloned genes to plant cells. *Gene* **1983**, *23*, 315–330. [CrossRef]
124. Klein, T.M.; Wolf, E.D.; Wu, R.; Sanford, J.C. High-velocity microprojectiles for delivering nucleic acids into living cells. *Nature* **1987**, *327*, 70–73. [CrossRef]
125. Hess, D. Investigations on the intra-and interspecific transfer of anthocyanin genes using pollen as vectors. *Z. Für Pflanzenphysiol.* **1980**, *98*, 321–337. [CrossRef]
126. Wang, Y.X.; Long, S.P.; Zeng, L.X.; Xiang, L.E.; Lin, Z.; Chen, M.; Liao, Z.H. Enhancement of artemisinin biosynthesis in transgenic Artemisia annua L. by overexpressed HDR and ADS genes. *Yao Xue Xue Bao* **2014**, *49*, 1346–1352.
127. Choi, Y.; Jeong, J.; In, J.; Yang, D. Production of herbicide-resistant transgenic Panax ginseng through the introduction of the phosphinothricin acetyl transferase gene and successful soil transfer. *Plant Cell Rep.* **2003**, *21*, 563–568. [CrossRef] [PubMed]
128. Kim, O.T.; Hyun, D.Y.; Bang, K.H.; Jung, S.J.; Kim, Y.C.; Shin, Y.S.; Kim, D.H.; Kim, S.W.; Seong, N.S.; Cha, S.W. Thermotolerant transgenic ginseng (Panax ginseng CA Meyer) by introducing isoprene synthase gene through Agrobacterium tumefaciens-mediated transformation. *J. Korean Med. Crops* **2007**, *15*, 95–99.
129. Passioura, J.B. Phenotyping for drought tolerance in grain crops: When is it useful to breeders? *Funct. Plant Biol.* **2012**, *39*, 851–859. [CrossRef] [PubMed]
130. Khan, S.; Anwar, S.; Yu, S.; Sun, M.; Yang, Z.; Gao, Z.-Q. Development of drought-tolerant transgenic wheat: Achievements and limitations. *Int. J. Mol. Sci.* **2019**, *20*, 3350. [CrossRef] [PubMed]
131. Huang, X.; Feng, Q.; Qian, Q.; Zhao, Q.; Wang, L.; Wang, A.; Guan, J.; Fan, D.; Weng, Q.; Huang, T.; et al. High-throughput genotyping by whole-genome resequencing. *Genome Res.* **2009**, *19*, 1068–1076. [CrossRef] [PubMed]
132. Terracciano, I.; Cantarella, C.; Fasano, C.; Cardi, T.; Mennella, G.; D'Agostino, N. Liquid-phase sequence capture and targeted re-sequencing revealed novel polymorphisms in tomato genes belonging to the MEP carotenoid pathway. *Sci. Rep.* **2017**, *7*, 5616. [CrossRef]
133. Elshire, R.J.; Glaubitz, J.C.; Sun, Q.; Poland, J.A.; Kawamoto, K.; Buckler, E.S.; Mitchell, S.E. A Robust, Simple Genotyping-by-Sequencing (GBS) Approach for High Diversity Species. *PLoS ONE* **2011**, *6*, e19379. [CrossRef]
134. Miller, M.R.; Dunham, J.P.; Amores, A.; Cresko, W.A.; Johnson, E.A. Rapid and cost-effective polymorphism identification and genotyping using restriction site associated DNA (RAD) markers. *Genome Res.* **2007**, *17*, 240–248. [CrossRef]
135. Davey, J.W.; Blaxter, M.L. RADSeq: Next-generation population genetics. *Brief. Funct. Genom.* **2010**, *9*, 416–423. [CrossRef]
136. Pavan, S.; Marcotrigiano, A.R.; Ciani, E.; Mazzeo, R.; Zonno, V.; Ruggieri, V.; Lotti, C.; Ricciardi, L. Genotyping-by-sequencing of a melon (Cucumis melo L.) germplasm collection from a secondary center of diversity highlights patterns of genetic variation and genomic features of different gene pools. *BMC Genom.* **2017**, *18*, 59. [CrossRef]
137. Pavan, S.; Curci, P.L.; Zuluaga, D.L.; Blanco, E.; Sonnante, G. Genotyping-by-sequencing highlights patterns of genetic structure and domestication in artichoke and cardoon. *PLoS ONE* **2018**, *13*, e0205988. [CrossRef]
138. D'Agostino, N.; Taranto, F.; Camposeo, S.; Mangini, G.; Fanelli, V.; Gadaleta, S.; Miazzi, M.; Pavan, S.; di Rienzo, V.; Sabetta, W.; et al. GBS-derived SNP catalogue unveiled wide genetic variability and geographical relationships of Italian olive cultivars OPEN. *Sci. Rep.* **2018**, *8*, 15877. [CrossRef]
139. Brazas, M.D.; Blackford, S.; Attwood, T.K. Plug gap in essential bioinformatics skills. *Nature* **2017**, *544*, 161. [CrossRef] [PubMed]
140. Jayakodi, M.; Choi, B.-S.; Lee, S.-C.; Kim, N.-H.; Park, J.Y.; Jang, W.; Lakshmanan, M.; Mohan, S.V.; Lee, D.-Y.; Yang, T.-J. Ginseng Genome Database: An open-access platform for genomics of Panax ginseng. *BMC Plant Biol.* **2018**, *18*, 1–7. [CrossRef] [PubMed]
141. Jang, W.; Jang, Y.; Cho, W.; Lee, S.H.; Shim, H.; Park, J.Y.; Xu, J.; Shen, X.; Liao, B.; Jo, I.-H. High-throughput digital genotyping tools for Panax ginseng based on diversity among 44 complete plastid genomes. *Plant Breed. Biotechnol.* **2022**, *10*, 174–185. [CrossRef]
142. Varshney, R.K.; Singh, V.K.; Kumar, A.; Powell, W.; Sorrells, M.E. Can genomics deliver climate-change ready crops? *Curr. Opin. Plant Biol.* **2018**, *45*, 205–211. [CrossRef]

Disclaimer/Publisher's Note: The statements, opinions and data contained in all publications are solely those of the individual author(s) and contributor(s) and not of MDPI and/or the editor(s). MDPI and/or the editor(s) disclaim responsibility for any injury to people or property resulting from any ideas, methods, instructions or products referred to in the content.

Review

Bioreactor Systems for Plant Cell Cultivation at the Institute of Plant Physiology of the Russian Academy of Sciences: 50 Years of Technology Evolution from Laboratory to Industrial Implications

Maria Titova [1,*], Elena Popova [1] and Alexander Nosov [1,2]

[1] K.A. Timiryazev Institute of Plant Physiology, Russian Academy of Sciences, 127276 Moscow, Russia; elena_aygol@hotmail.com (E.P.); al_nosov@mail.ru (A.N.)
[2] Department of Biology, M.V. Lomonosov Moscow State University, 119234 Moscow, Russia
* Correspondence: titomirez@mail.ru

Citation: Titova, M.; Popova, E.; Nosov, A. Bioreactor Systems for Plant Cell Cultivation at the Institute of Plant Physiology of the Russian Academy of Sciences: 50 Years of Technology Evolution from Laboratory to Industrial Implications. *Plants* 2024, *13*, 430. https://doi.org/10.3390/plants13030430

Academic Editor: Mikihisa Umehara

Received: 30 December 2023
Revised: 29 January 2024
Accepted: 29 January 2024
Published: 1 February 2024

Copyright: © 2024 by the authors. Licensee MDPI, Basel, Switzerland. This article is an open access article distributed under the terms and conditions of the Creative Commons Attribution (CC BY) license (https:// creativecommons.org/licenses/by/ 4.0/).

Abstract: The cultivation of plant cells in large-scale bioreactor systems has long been considered a promising alternative for the overexploitation of wild plants as a source of bioactive phytochemicals. This idea, however, faced multiple constraints upon realization, resulting in very few examples of technologically feasible and economically effective biotechnological companies. The bioreactor cultivation of plant cells is challenging. Even well-growing and highly biosynthetically potent cell lines require a thorough optimization of cultivation parameters when upscaling the cultivation process from laboratory to industrial volumes. The optimization includes, but is not limited to, the bioreactor's shape and design, cultivation regime (batch, fed-batch, continuous, semi-continuous), aeration, homogenization, anti-foaming measures, etc., while maintaining a high biomass and metabolite production. Based on the literature data and our experience, the cell cultures often demonstrate cell line- or species-specific responses to parameter changes, with the dissolved oxygen concentration (pO_2) and shear stress caused by stirring being frequent growth-limiting factors. The mass transfer coefficient also plays a vital role in upscaling the cultivation process from smaller to larger volumes. The Experimental Biotechnological Facility at the K.A. Timiryazev Institute of Plant Physiology has operated since the 1970s and currently hosts a cascade of bioreactors from the laboratory (20 L) to the pilot (75 L) and a semi-industrial volume (630 L) adapted for the cultivation of plant cells. In this review, we discuss the most appealing cases of the cell cultivation process's adaptation to bioreactor conditions featuring the cell cultures of medicinal plants *Dioscorea deltoidea* Wall. ex Griseb., *Taxus wallichiana* Zucc., *Stephania glabra* (Roxb.) Miers, *Panax japonicus* (T. Nees) C.A.Mey., *Polyscias filicifolia* (C. Moore ex E. Fourn.) L.H. Bailey, and *P. fruticosa* L. Harms. The results of cell cultivation in bioreactors of different types and designs using various cultivation regimes are covered and compared with the literature data. We also discuss the role of the critical factors affecting cell behavior in bioreactors with large volumes.

Keywords: plant cell culture; cell suspension; bioreactors; biotechnology; periodic cultivation; continuous cultivation; phytochemicals; plant secondary metabolites

1. Introduction

Over 100,000 plant-produced secondary metabolites have been identified to date, and the discovery of new compounds continues on a daily basis. Plant secondary metabolites possess extremely diverse chemical structures. The alkaloids, isoprenoids (terpenoids), and phenolic compounds are the most well-studied metabolite classes so far, each of them being subdivided into numerous subgroups composed of thousands of chemicals. Other bioactive molecules belong to plant amines, non-protein amino acids, cyanogenic glycosides, glucosinolates, polyacetylenes, betalaines, alkylamides, thiophenes, etc. [1–4].

Many of the identified plant secondary metabolites are economically important products and are widely used in the pharmacological, cosmetic, and food industries, veterinary medicine and agriculture. Anticancer drugs, adaptogens, immunostimulants, and antimicrobials remain in the greatest demand [1,5–10]. The development of new synthetic drugs may cost about 100 million USD and take about 10 years; hence, the practical interest in plants as natural raw materials for drug production is economically justified [11–13].

The ever-increasing demand for bioactive compounds of plant origin leads to the overexploitation of wild plant diversity and, as a consequence, to the search for alternative renewable sources of valuable secondary metabolites. Plantations may partially solve the problem but have other issues such as significant costs, the use of herbicides and insecticides, land occupation, certain climatic conditions requirements, etc. Moreover, both the composition and the number of secondary metabolites produced by plantation plants are subject to change depending on the plant's age and growth environments. Plant cell cultures provide fundamentally different opportunities for the production of plant bioactive compounds [4,14–17].

Cell biotechnology for the production of bioactive phytochemicals has numerous advantages over traditional plant raw materials as highlighted in earlier reviews [15,16,18–20]. The main advantages are independence from seasonal, climatic, and soil conditions, eco-friendly production process, a variety of strategies available for maximizing the yield of cell biomass and the compounds of interest, unlimited application to plant species with rare/endangered/at-risk status, and the use of standard equipment designed for microbiological productions including bioreactors, post-fermentation systems, etc., with only minor modifications. Recent advances in bioengineering and in vitro selection technologies may prompt the development of superior cell lines with intensive growth and the production of the desired metabolites at similar or higher levels compared to wild plants [21–23].

Recent advances in in vitro cultivation methods underpinned the development of both practical and fundamental aspects of plant cell and tissue culture [13,16,24]. In addition to biotechnological applications, plant cell cultures have been used to investigate the biosynthetic pathways and regulatory processes in plants and successfully utilized as natural systems for biotransformation, both at the final and intermediate production stages [25,26].

Numerous studies conducted since the 1940s have demonstrated that suspension cultures of plant cells are capable of synthesizing the entire range of secondary metabolites, often in amounts exceeding their concentration in plants [13,15,19,23,27–30]. Furthermore, plant cell cultures possess a high potential to synthesize phytochemicals that are found in a minority in donor plants [23]. Suspension cultures of plant cells can also be used to produce therapeutic proteins, including monoclonal antibodies, human serum albumin, human hemoglobin, interferon, immunostimulatory allergenic proteins, and others [13,25,31–33].

The list of plant cell cultures that have been tested for biotechnological application is quite wide and is constantly updated with new species. The most known and described in the literature are cell cultures of *Panax* spp. producing ginsenosides, *Taxus* spp. synthesizing paclitaxel, *Dioscorea* spp. producing steroidal glycosides, *Coleus blumei* Benth. producing rosmarinic acid, *Aralia cordata* Thunb. producing anthocyanins, *Lithospermum erythrorhizon* var. erythrorhizon Siebold and Zucc. producing shikonin derivatives, etc. [15,19,22,28,34–37]. Until the 2000s, plant cell-based biotechnologies have been mainly considered most relevant for the products that are unprofitable or unfeasible to manufacture using traditional methods of wild plant collections or plantation cultivation, for example, for bioactive metabolites produced by rare, endemic, or slowly growing plants [38,39]. To date, only a few effective biotechnological productions based on the large-scale cultivation of plant cell suspensions have been described [17,32]. The reasons for the limited industrial application of plant cell cultures are the high cost and the complexity of the hardware design, resulting in high production costs and the uncompetitively high prices of the final product as well as difficulties in developing productive cell strains [17,40,41].

The K.A. Timiryazev Institute of Plant Physiology of the Russian Academy of Sciences (IPPRAS) is a research institute that performed pioneer studies in different aspects of plant cell cultures [42–45]. The institute hosts the Experimental Biotechnological Facility, a large department with unique bioreactor systems specifically designed for the cultivation of plant suspension cell cultures from the laboratory (2–20 L) to the pilot (75 L) and the industrial (630 L) scale (Figures 1 and 2). The facility serves both scientific and commercial applications, with the main goals of obtaining and selecting cell lines with enhanced production of bioactive metabolites, optimization of the cell cultivation conditions by adapting nutrient media and cultivation regimes, the upscaling of cultivation to industrial volumes, and the design and modification of equipment for cell cultivation. The research team conducts comprehensive studies aiming at developing the large-scale bioreactor cultivation of cell culture producers of biologically active compounds, taking into account the productivity and individual physiological characteristics of cell strains and the technological characteristics of the equipment used.

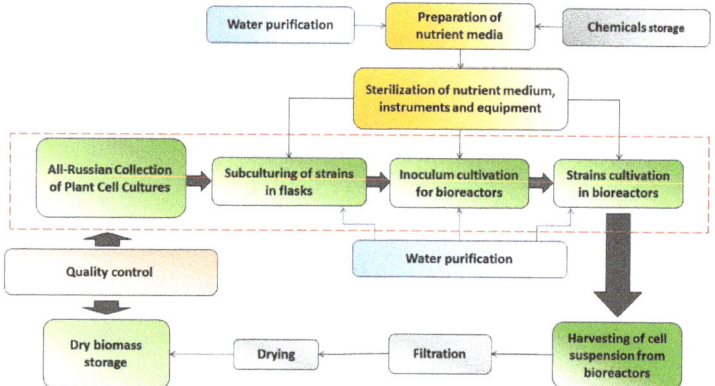

Figure 1. A simplified process flow chart of the Experimental Biotechnological Facility of the IPPRAS.

Figure 2. Experimental Biotechnological Facility of the IPPRAS: (**a–d,f**) bioreactors of different volumes for cultivation of plant cell suspensions; (**e**) cell biomass harvested from 630 L bioreactors, dried and packed for shipment; (**g**) example of suspension cells under microscope: cells of *Panax japonicus* strain 62 adapted for cultivation in industrial bioreactors.

Here, we provide a historical perspective and a comprehensive review of our experience gained through several decades of plant cell cultivation using bioreactor systems of different types, volumes, designs, and operation regimes. Most studies were performed using cell culture strains from the All-Russian Collection of Plant Cell Cultures of IPPRAS featured in the previous review [22]. Each section gives a brief introduction to the different aspects of bioreactor cultivation and discusses their implications for plant cell cultures based on the literature and our experience.

2. Bioreactor Types for Plant Cell Cultivation and Their Specifics

The bioreactor system for plant cell cultivation had evolved from microbial production and is mainly based on the same principles but accomplished its own specifics. From the technological perspective, the process of living organisms' cultivation in a bioreactor involves "in" and "out" flows: inoculum, air or gas mixtures, nutrient components, defoamers, etc. are supplied into bioreactors, constantly or periodically, while heat, exhaust air, culture medium, and cell biomass are removed from the system. The process is controlled by measuring the main physical and chemical parameters and their stabilization at the optimum level for maximizing the yield of the desired product (biomass or compound of interest). In the process of cultivation, a complex mixture is formed composed of cells and cell aggregates, extracellular metabolites, and residual concentrations of the initial substrate, while target products are usually found in small concentrations and may be easily destroyed [46–50]. These general principles are followed in the bioreactor cultivation of plant cells with certain specifics of regime optimization and bioreactor design and construction [15,51,52]. Most authors acknowledged that the major differences between microbial and plant cell production systems are due to the following specifics [15,19,20,51,53]:

- The large size and vacuoles make plant cells particularly sensitive to physical and mechanical stresses;
- The specifically high requirements for maintaining aseptic conditions during long cultivation due to the relatively low growth rate and long cultivation cycle of plant cell cultures compared to microbial and animal cell cultures;
- The high requirements of uniform mixing due to the high sedimentation rate of cell aggregates and the increasing viscosity of cell suspensions at the high concentrations of cell biomass;
- The intensive foaming and adhesion of cell biomass to the walls of a bioreactor;
- The complex mechanisms of regulating the cell growth and biosynthesis of target metabolites.

To meet these specifics, bioreactors for plant cell cultivation should be both structurally and operationally complex. The main engineering challenges are to maintain a high intensity of mass and energy exchange between the cells and culture medium, to minimize cell damage during mixing, to aseptically monitor technological parameters, in particular, the concentrations of cell biomass and target metabolites, and to minimize the cost of the process [15,52,54].

2.1. Bioreactor Classification Based on Their Design

Different types of bioreactors varying in design, operating regime, and size, from several liters to several thousands of tons, have been tested for plant cell cultivation. The large spectrum of cell cultures led to a huge variety of engineering solutions based on cell strain characteristics, medium used, production scale, specifics of the product isolation, etc. [15,52,54–56]. Several attempts have been made to classify bioreactors for plant cell cultivation. In the literature, bioreactors are most often grouped based on their constructions [20,52,54–56] into the following types:

- Bioreactors where mixing is performed by compressed air supply;
- Bioreactors with mechanical stirring;
- Wave bioreactors.

These types of bioreactors have been reviewed in detail [15,55]. Here, we provide only a brief description of their main characteristics that are important for further reading.

Bioreactors with air mixing. In this type of bioreactor, the aeration and mixing of the cell suspension are performed by compressed air. The most well-known are the bubble type, airlift bioreactors that are usually shaped as a vertical tank equipped with a gas distribution device or spargers installed at the bottom. Such a construction is relatively simple, with no rubbing or moving parts, but highly functional and reliable [57,58]. This type of bioreactor was used in the first experiments on the scaling-up cultivation of plant cell suspension cultures [59], followed by decades of effective exploitation. Bubble-type bioreactors were employed for experimental cultivation of *Taxus cuspidata* Siebold. and Zucc. and *Eurycoma longifolia* Jack cells [60,61]. Several cases of commercial cultivation based on suspension cell cultures of *Panax ginseng* C.A. Mey., *Digitalis lanata* Ehrh., *Lithospermum erythrorhizon*, *Taxus baccata* Thunb., and *Taxus wallichiana* Zucc. in airlift bioreactors were also reported [36,40,62–64]. However, a number of studies demonstrated that bubble-type and airlift bioreactors have relatively low mass transfer characteristics, and they are, therefore, not recommended for cell suspensions with a high viscosity or high final cell biomass concentration [57,58].

Bioreactors with mechanical stirring. In these bioreactors, aeration is performed by compressed air, while a mechanical stirrer is used for mixing. Usually, these bioreactors are made as cylindrical vessels equipped with mechanical stirring devices and a sparger, which is normally installed under the bottom tier of the stirring device. The oxygen mass transfer coefficient values in these bioreactors may vary within a very wide range [55,58,65,66]. According to the literature, bioreactors with mechanical stirring are most widely used for upscaling the cultivation process from the laboratory to industrial volumes [67,68]. One of the first large-scale commercial bioreactor systems of this type was developed by the Diversa (later Phyton Biotech) company (Germany) and employed a cascade of five mechanically stirred bioreactors (75, 750, 7500, 15,000, and 75,000 L). This system was used for the cultivation of *Echinacea purpurea* (L.) Moench, *Rauvolfia serpentina* Benth. ex Kurz cell suspensions [69], followed by *Taxus chinensis* Roxb. for the production of paclitaxel [17]. Other examples of the successful cultivation of plant cells in bioreactors with mechanical stirring include *Panax* spp., *Catharanthus roseus* (L.) G. Don, *Podophyllum hexandrum* Royle, *Azadirachta indica* A. Juss., and some others [70–74].

Wave bioreactors are rather complex in design, consume a substantial amount of energy, and are rarely used for the large-scale commercial cultivation of plant cells. They are characterized by wave-induced motion, where the mass and energy transfer are manually adjusted via the rocking angle, agitation rate, medium level, and culture vessel geometry. The disadvantages of these bioreactors are high energy losses during liquid agitation, engineering difficulties due to the lack of reliable methods for the calculation of optimum parameters, and the relatively high cost of additional equipment [55,75,76].

The given examples cover the most widely used industrial and laboratory types of bioreactors. However, new experimental bioreactors are constantly being developed, which are quite difficult to classify and do not fall into the usual categories [76–78].

2.2. Bioreactors of the Experimental Biotechnological Facility of the IPPRAS

The Experimental Biotechnological Facility of the IPPRAS has been operating since the 1970s and was the first Russian facility focused specifically on the bioreactor cultivation of plant cells. During its operation history, bioreactors of different types and volumes were tested and adapted for the cultivation of plant cell suspensions (Table 1). The very first experiments were performed in 1972–1979 using laboratory glass barbotage V-shape bioreactors (no. 1 in Table 1, total volume 1.5–3.0 L) and laboratory bioreactors MF-107 with a modified mechanical stirrer (no. 5 in Table 1, total volume 7.0 L). These bioreactors provided satisfactory conditions for both growth and the triterpene glycoside biosynthesis of the suspension cell cultures of *Dioscorea deltoidea* Wall. ex Griseb. in a series of physiological and biochemical studies [42]. The same systems were used in later experiments for culturing other strains of *Dioscorea deltoidea* [45,79–82], as well as for the suspension cell cultures of *Polyscias filicifolia* (C. Moore ex E. Fourn.) L.H. Bailey [83], *Panax ginseng* [80], and *Taxus baccata* [84]. Mechanically stirred laboratory bioreactors Fermus-apparatus and AK-210 (no. 3 и 4 in Table 1) were also actively operated in the 1990s, in particular for developing and optimizing cultivation regimes for different strains of *Dioscorea deltoidea* cell suspensions [85]. However, these bioreactors had various design flaws, and their current use is restricted to experimental purposes with a very limited number of suspension cell strains (Table 1).

Table 1. Bioreactors of the Experimental Biotechnological Facility of the IPPRAS.

No.	Bioreactor	Material	Total/Working Volume, L	Mixing	Sparger Type	Impeller Type	Manufacturer	Advantages (A), Disadvantages (D)	Exploitation Period
					Laboratory (bench-top) bioreactors				
1.	Bubble-type bioreactors	Glass	1.5/1.0, 3.0/2.5	Aeration	Single point sparger, ⌀~2 mm *	n/a	IPPRAS, Moscow, Russia	A: Easy upscaling, simple construction, low cost D: Small volume, intense foaming, non-optimal mass transfer, application is limited to fine-aggregated, non-foaming cell lines	Until 2014 [84]
2.		Glass	10/7, 20/15		Single point sparger, ⌀~6 mm				Currently in use [86]
3.	Fermus-apparatus	Glass + stainless steel	8/6	Magnetic stirrer + aeration	Single point sparger, ⌀~4 mm	Open turbine impeller	R&D Center "Bioavtomatika", N. Novgorod, Russia	A: Highly efficient mass transfer D: Intense shear stress, foaming, non-optimal magnetic drive configuration, limited options for modification, higher chances for contamination due to construction specifics	Until 1995 [85]
4.	AK-210	Glass + stainless steel	10/8	Magnetic stirrer + aeration	Single point sparger, ⌀~4 mm	Open turbine impeller	R&D Bureau, Pushchino, Russia		Until 1995 [85]
5.	MF-107	Glass + stainless steel	7/5	Magnetic stirrer + aeration	Single point sparger, ⌀~4 mm	Three-impeller stirrer (two open turbine impellers and one marine-type impeller)	New Brunswick, USA		Until 2000 [83]
					Pilot-scale bioreactors				
6.	Tank bioreactor	Stainless steel	75/50	Magnetic stirrer for media sterilization, aeration for cell cultivation	Single point sparger, ⌀~6 mm or ring-type gas distributor ⌀~200 mm with multiple holes ⌀~1 mm	Marine-type impeller	Electrolux, Sweden	A: Highly efficient mass transfer, suitable for viscous cell suspensions D: Intense shear stress, high energy cost due to mechanical agitation	Currently in use [87–89]
					Industrial-scale bioreactors				
7.	Tank bioreactor	Stainless steel	630/550	Magnetic stirrer for media sterilization, aeration for cell cultivation	Ring-type gas distributor ⌀~750 mm with multiple holes ⌀~1 mm	Marine-type impeller	1T series, "EBEE" Research & Manufacturing facility, Yoshkar-Ola, Russia	A: Highly efficient mass transfer, suitable for viscous cell suspensions D: Intense shear stress, high energy cost due to mechanical agitation	Currently in use [21,87,89]

* here and further in the table, the inner diameter is specified.

During the past 20 years, the Biotechnological Facility of the IPPRAS primarily used laboratory glass bubble-type 10 L and 20 L bioreactors (No. 2, Table 1) or stainless steel 75 L and 630 L tank bioreactors (No. 6 and 7, Table 1) for plant cell suspension cultivation. Our studies confirmed that these bioreactors were the most favorable for the cultivation of undifferentiated plant cells and had fewer disadvantages compared to bioreactors of other types [21,80,87–90].

3. Cultivation Regimes

3.1. Cultivation Regimes Suitable for Plant Cell Cultures

The choice of bioreactor design and cultivation regime for maximizing the yield of biomass and/or a product of interest is mainly determined by the growth and biosynthetic characteristics of individual cell culture strains [15,32]. The most common cultivation regimes and their implications for culturing cells of different plant species are discussed below.

The batch, or periodic, cultivation is a type of "closed" cultivation when the system undergoes dynamic changes that might be difficult to control. The concentrations of cells and nutrient medium components, products of cell metabolism, and other factors are constantly changing in the course of the cultivation process following the development of the cell population inside the bioreactor [32,91]. The main advantages of batch systems are the following:

- A reduced risk of contamination and cell mutations due to the relatively short cultivation cycle compared to other regimes;
- The high degree of substrate utilization;
- The relatively low cost (compared to the cost of continuous cultivation).

Because of its simplicity and universality, this method has been widely used in plant cell cultivation experiments and industrial-scale productions, for example, for the cell cultivation of *Azadirachta indica* (azadirachtin), *Catharanthus roseus* (ajmalicine, catharanthine, serpentine alkaloids), *Panax notoginseng* (Burkill) F.H. Chen (ginsenosides), and *Taxus cuspidata* (taxane production) [29,60,92–94]. Hibino et al. [95] reported the cultivation of the suspension culture of *Panax ginseng* cells in 20,000 and 25,000 L bioreactors with mechanical stirring (periodic regime) with biomass productivity reaching 20 g dry weight (DW) $(L \cdot day)^{-1}$ in four weeks.

However, some authors noted the reduction in the cell biomass and secondary metabolite accumulation during cell cultivation as a result of batch cultivation compared to other regimes. This may be due to a gradual accumulation of the inhibitory products of cell metabolism in the medium and depletion of substrate during cultivation. In addition, the efficiency of the process is corrupted by the cultivation pauses for equipment preparation including cleaning, refilling, and sterilization of the bioreactor, as well as for preparing the required amount of cell culture inoculum for every new cultivation cycle. Upon upscaling to industrial volumes, there is an additional risk of contamination from inoculating bioreactors with substantial volumes of cell suspension [15,32,91].

The continuous regime for plant cell cultivation are mainly open systems; they can be organized according to the principle of complete mixing [91,96]. During continuous cultivation, fresh medium is continuously fed into the system at a constant rate under mixing, whereas the total volume of the cell suspension is kept stable by continuously removing a portion of the suspension culture at the same rate. A continuous regime creates uniform stationary conditions in the whole volume of the bioreactor and stabilizes the cell strain in the required state ('steady state"), e.g., in the phase of exponential growth [91,96]. Compared to batch cultivation, continuous systems demonstrate a number of important advantages:

- The production of the cell biomass or compound of interest with predetermined and reproducible characteristics due to the stable and thoroughly controlled cultivation conditions;
- The possibility to shift the composition of the cell population and their metabolic activity by manipulating the oxygen supply and nutrient components;
- The possibility to regulate the growth rate of the culture and the concentration of cell biomass within a wide range by changing the flow rate of the nutrient medium.

In addition, the time for equipment preparation is reduced since there is no need for multiple re-sterilizations of tanks.

There are also a number of problems associated with the continuous cultivation [96,97]:

- Difficulties to control the production of secondary metabolites that are not directly correlated with the growth of the cell population;
- Difficulties in providing stable cultivation conditions for cell cultures with high aggregation level and viscosity;
- The risk of losing the culture strain due to cell mutation or due to the auto-selection of cells with a high proliferation rate;
- The high cost and complexity of controlling and automation systems;
- The increased risk of contamination due to long cultivation cycles and the use of additional equipment.

In continuous cultivation, the process is controlled in several ways. The **chemostat** is based on measuring and regulating the parameters of the flow entering the system. In this case, the concentration of oxygen or one of the components of the nutrient medium supplied to the bioreactor is fixed at a level at which other nutrient components are in excess. The rate of cell multiplication in the culture is thus limited by a concentration of a regulated component. Wilson et al. [98] and Kurz et al. [99] were among the first to test the chemostat for suspension cultures of *Acer pseudoplatanus* L., *Glycine max* L. Merr., and *Triticum monococcum* L. cells [98,99]. Later, a chemostat was used, to give some examples, for *Petunia hybrida* E.Vilm., *Catharanthus roseus*, and *Nicotiana tabacum* L. cell cultures [100–102].

The turbidostat is based on measuring the turbidity of the outlet flow. In this case, the change in the optical density of the cell suspension is used to regulate the rate of fresh nutrient medium entering the bioreactor. The turbidostat is rarely used in plant cell culture because of the narrow range of correlation between the optical density of the suspension and the actual cell concentration [15,91,103,104].

The continuous regime is most often used to obtain the growth-associated products of primary and secondary metabolism, to study cell populations in the phase of intensive proliferation, and to culture cell suspensions whose growth is inhibited by cellular metabolic products [15,96,97].

The continuous cultivation regime may be also used to study the metabolic profiles of cell populations in the growth retardation and stationary phases; in this case, cell biomass is not removed from the bioreactor. Such systems are called **closed continuous cultivation systems** [15,91,105]. An important characteristic of the closed continuous mode is the need for a constant supply of nutrient medium and the withdrawal of cell-free culture fluid [91]. In technical terms, this is realized with the help of peristaltic pumps, simple flow rate meters, and tanks for nutrient medium and draining cell-free culture fluid. The technically challenging task here is the continuous separation of cell biomass from the liquid phase, for example, by sedimentation and/or using membranes, while maintaining cell viability and aseptic conditions. Closed continuous systems offer a number of advantages:

- The continuous operation of the system without the problem of cell washout;
- The separated cells are protected from shear stress;
- The possibility of achieving high cell concentrations, up to 30–40 g L^{-1} medium;
- The intercellular contacts are increased in closed cultivation systems.

On the other hand, the closed system for large-scale cultivation of plant cells is limited by a number of negative factors [105]:

- The high chances of cell viability reduction caused by cell separation from the culture fluid or immobilization;
- The difficulties in controlling the growth and biosynthetic parameters of the cell population;
- The significant gradients of nutrients and oxygen within the system in case of cell immobilization or sedimentation;
- The high cost and complexity of the additional equipment.

Cell growth and metabolite production in the closed system are mainly controlled by manipulating the limiting substrate concentration and flow velocity as well as by removing or reducing the concentration of growth-inhibiting metabolites secreted by plant cells to the medium. The varying content of nutrients in the supplied medium also affects the dynamics of intracellular metabolite accumulation. In a closed cultivation regime, cells are not washed out from the system; hence, the flow rate of the nutrient medium may vary within a very wide range, allowing the multifactorial control of the cultured cell population [15,91,96,105]. Closed continuous cultivation systems were used for *Glycyrrhiza inflata* Batalin and *Anchusa officinalis* Thunb. cell culturing [106,107] and to produce recombinant proteins in plant cell suspensions [108].

Many bioreactor systems are hybrids of the batch and continuous cultivation regimes. This includes **periodic substrate-fed cultures** (the periodical addition of the nutrient medium or individual limiting components without the removal of cell biomass) [109–112], **semi-continuous systems** (the periodic addition of fresh medium, while removing part of the cell suspension) [113,114], **two-stage systems** [70,115], etc. Unlike continuous culture regimes (chemostats in particular), such hybrid systems imply periodic changes in suspension volume and velocity, periodic suspension dilution, as well as varying the specific growth rate, productivity, and other parameters [15,19,91]. Such systems are relatively simple in design and combine the advantages of continuous and periodic cultivation models:

- Multiple options to control and optimize cultivation conditions depending on the phase of the growth cycle, productivity, or culture age;
- Reduced risk of mutations, contamination, or cell washout during cultivation;
- A high degree of substrate utilization;
- The duration of subcultivations may be varied depending on the physiological requirements of the cell population;
- No time-consuming preparation of equipment and inoculum for each new subcultivation cycle.

Combined systems better suit the purpose of upscaling the cultivation process than periodic and continuous regimes. In particular, they are efficient in studies of substrate-limited cell growth, for the cultivation of highly aggregated or slow-growing cell suspensions, and for the production of metabolites whose biosynthesis is not directly associated with growth intensity [19]. For example, a semi-continuous regime was employed by Villarreal et al. [116] for the cultivation of a *Solanum chrysotrichum* C.H. Wright cell suspension in a 10 L airlift bioreactor. The authors observed a 60% increase in biomass and phytochemical productivity in the bioreactor compared to batch cultivation in flasks. Choi et al. [117] compared different regimes for a suspension cell culture of *Thalictrum rugosum* Poir. The highest cell viability, growth rate, and berberine accumulation were observed with semi-continuous cultivation. With suspension cell cultures of *Taxus chinensis* and *Panax notoginseng*, a higher biomass and target metabolite accumulation was observed in the substrate-fed mode compared to periodic cultivation [94,118].

3.2. The Use of Different Cultivation Regimes at the Experimental Biotechnological Facility of the IPPRAS

Table 2 presents examples of using different operation regimes for the bioreactor cultivation of suspension cell cultures at the Experimental Biotechnological Facility of the IPPRAS.

Transferring cell culturing from flasks to laboratory bioreactors is the first step toward upscaling the cultivation process. At the Experimental Biotechnological Facility of the IPPRAS, batch cultivation has been widely used since 1979 in preliminary experiments with laboratory and pilot bioreactors of different types to identify critical factors affecting the productivity of various cell cultures (Figure 3a). The suspension cell cultures tested for batch cultivation were *Dioscorea deltoidea*, *Polyscias filicifolia*, *Stephania glabra* (Roxb.) Miers (synonym of *S. rotunda* Lour.), *Panax japonicus* (T. Nees) C.A.Mey., *Alhagi persarum* Boiss. and Buhse (synonym of *A. pseudalhagi* subsp. persarum (Boiss. and Buhse) Takht., and some others [21,42,43,79,81,83,119]. In these studies, the batch mode was used for the optimization of aeration and agitation regimes, gas mixture compositions, inoculum density, the primary assessment of bioreactor-induced changes in cell aggregation, the growth dynamics, and synthesis of target metabolites under changing cultivation conditions.

Continuous cultivation regimes have been successfully applied for *Dioscorea deltoidea* and *Panax japonicus* cell suspensions (Table 2, Figure 3b). For example, Kandarakov et al. [82] performed a 115-day-long experiment with *Dioscorea deltoidea*, switching between batch and continuous culture regimes and testing four dilution rates. The specific growth rate of cell suspension varied from 0.12 to 0.25 day^{-1} during the exponential growth phase in batch culture and from 0.08 to 0.23 day^{-1} during continuous culture. A viability value of 52–90% was recorded during the whole cultivation cycle. The maximum total furostanol glycoside content was 3.2–4.0%DW. The continuous mode significantly changed the pattern of furostanol glycoside accumulation, likely due to the auto-selection of highly proliferating cells. For the suspension cell culture of *Panax japonicus*, continuous cell cultivation (chemostat) was performed for 70 days with the dilution rate 0.11–0.16 day^{-1}. The stable growth of the culture was demonstrated, with the maximum dry cell weight varying depending on the dilution rate within 4.9–7.8 g L^{-1}, a viability value of 77–84%, and the total ginsenoside content reaching 5%DW [120].

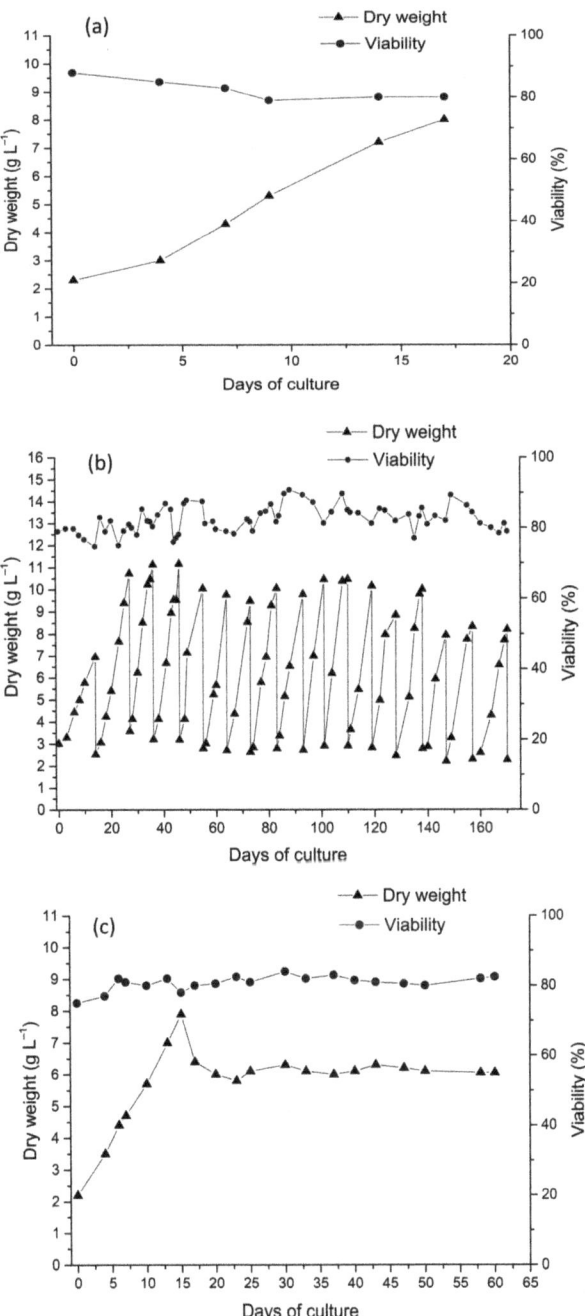

Figure 3. The representative growth curves of plant cell suspensions in bioreactors under different culture regimes: (**a**) *Dioscorea deltoidea*, periodic subculture. The process was finished upon achieving the maximum biomass accumulation on day 17 [121]; (**b**) *D. deltoidea*, semi-continuous regime (growth curve fragment) [121]; (**c**) *Panax japonicus*, continuous cultivation regime (growth curve fragment, medium flow with dilution rate D = 0.11 day^{-1} was initiated on the 15th day of cultivation) [120].

A closed continuous regime (Table 2) was used by Oreshnikov et al. [85] for a *Dioscorea deltoidea* suspension cell culture. All bioreactors were equipped with peristaltic pumps, vessels for the nutrient medium and cell-free culture medium, and a system to separate the cells from the medium by sedimentation. A maximum dry cell weight of 15.5 g L^{-1} (up to 32 g L^{-1} with an increased medium supply), viability of 60–70%, and maximum total furostanol glycoside content of 4.0–6.0%DW or up to 10%DW depending on the medium concentration were recorded. The dilution rate was maintained at 0.15 day^{-1} (Table 2). Increasing the flow rate to 0.30–0.45 day^{-1} (above the specific growth rate values) led to a reduction in all culture parameters and cell lysis. Moreover, the authors demonstrated that variations in the flow rate, the concentration of nutrient components, and the degree of mechanical stress may be used to purposefully alter and regulate the phases of cell development in the bioreactors.

Table 2. The different bioreactor operation regimes for the cultivation of plant cell suspensions used at the Experimental Bitechnological Facility of the IPPRAS.

Species	Bioreactor *	Cultivation Cycle (Days)	Maximum Biomass Accumulation (gDW L^{-1}), Cell Viability (%)	Maximum Metabolites Content Achieved	Operating Conditions	Reference
			Periodic (batch) cultivation regime			
Dioscorea deltoidea	Bubble-type bioreactors (no. 1)	21	10.0–11.5 g L^{-1}	ND	27 °C, daylight, air flow 0.5–1.0 L min^{-1}	[42]
		15–28	9.5–10.0 g L^{-1}	ND	27 °C, darkness, air flow 0.4 L min^{-1}	[45,79,81]
	MF-107 (no. 5)	21	10.0–11.5 g L^{-1}	Diosgenin 7.4–13.7 mg gDW^{-1}	27 °C, daylight, stirring rate 350–500 rpm, air flow 0.5–1.0 L min^{-1}	[42]
		14–15	9.0–9.5 g L^{-1}	Diosgenin 6.2–6.3 mg gDW^{-1}	26 °C, darkness, stirring rate 300–500 rpm, pO$_2$ 70–90% of saturation volume	[43]
Polyscias filicifolia	Bubble-type bioreactors (no. 1)	18	11.0–16.0 g L^{-1}	ND	26 °C, darkness, pO$_2$–ND	[83]
	MF-107 (no. 5)	24–30	12.8–17.4 g L^{-1}	ND	26 °C, darkness, pO$_2$–ND	
Stephania glabra	Bubble-type bioreactors (no. 2)	21	8.0–16.0 g L^{-1}, 75–90%	Stepharin 0.05–0.16%DW	26 °C, darkness, pO$_2$ 10–40% of saturation volume	[119]
	75 L tank bioreactor (no. 6)	14	7.0–9.0 g L^{-1}, 65–90%	Stepharin, traces	26 °C, darkness, stirring rate 30–65 rpm, pO$_2$ 10–40% of saturation volume (single point sparger)	
Alhagi persarum	Bubble-type bioreactors (no. 2)	16	13.71 ± 1.84 g L^{-1}, 74.1 ± 2.16%	ND	26 °C, darkness, pO$_2$ 10–40% of saturation volume	[122]
Polyscias filicifolia	75 L tank bioreactor (no. 6)	22	9.3–13.7 g L^{-1}, 77–85%	ND	26 °C, darkness, pO$_2$ 10–40% of saturation volume (ring-type gas distributor)	[89]
			Continuous cultivation regime			
Dioscorea deltoidea	MF-107 (no. 5)	115	~12.6 g L^{-1}, 52–90%	Total furostanol glycosides 3.2–4.0%DW	26 °C, darkness, stirring rate 100–360 rpm, dilution rates (D) 0.14–0.23 day^{-1}	[82]
Panax japonicus var. *repens*	Bubble-type bioreactors (no. 2)	86	4.9–7.8 g L^{-1}, 77–84%	Total ginsenosides 2.5–3.0%DW	26 °C, darkness, pO$_2$ 10–40% of saturation volume, D 0.11–0.22 day^{-1}	[120]
			Closed continuous cultivation regime			
Dioscorea deltoidea	Fermus-apparatus (no. 3)	57	~14.0 g L^{-1}, 60–70%	Total furostanol glycosides 2.0–3.0%DW	26 °C, darkness, pO$_2$ 20–60% of saturation volume, stirring rate 20–250 rpm, D 0.15 day^{-1} (days 20–30 and 46–57) **	[85]
	AK-210 (no. 4)	19	15.0–15.5 g L^{-1}, 60–80%	Total furostanol glycosides 4.0–6.0%DW	26 °C, darkness, pO$_2$ 20–60% of saturation volume, stirring rate 20–250 rpm, D 0.15 day^{-1} (days 7 to 19) **	
		20	30–32 g L^{-1}, 62–84%	Total furostanol glycosides 9.5%DW	Same as above, with ×2 medium concentration	[123]

Table 2. Cont.

Species	Bioreactor *	Cultivation Cycle (Days)	Maximum Biomass Accumulation (gDW L^{-1}), Cell Viability (%)	Maximum Metabolites Content Achieved	Operating Conditions	Reference
			Semi-continuous cultivation regime			
Stephania glabra	Bubble-type bioreactors (no. 2)	Multicycle 40–60	11.0–16.0 g L^{-1}, 78–92%	Stepharin 0.06–0.16%DW	26 °C, darkness, pO$_2$ 10–40% of saturation volume ***	[119]
Dioscorea deltoidea	Bubble-type bioreactors (no. 2)	Multicycle 182	8.50–12.50 g L^{-1}, 80–85%	Total furostanol glycosides 4.2–8.0%DW	26 °C, darkness, pO$_2$ 10–40% of saturation volume ***	[124]
	630 L tank bioreactor (no 7)	Multicycle 170	8.87–11.13 g L^{-1}, 79–86%	Total furostanol glycosides 7.7–13.9%DW		
Polyscias filicifolia	630 L tank bioreactor (no. 7)	Multicycle 112	10.8–16.2 g L^{-1}, 79–87%	ND	26 °C, darkness, pO$_2$ 10–40% of saturation volume ***	[89]
Taxus wallichiana	Bubble-type bioreactors (no. 2)	Multicycle 75	10.5–17.5 g L^{-1}, ~90%	Yunnanxane 0.08–0.36 mg gDW^{-1} taxuyunnanine C 0.09–0.34 mg gDW^{-1} paclitaxel 0.06–0.15 mg gDW^{-1}	26 °C, darkness, pO$_2$ 10–40% of saturation volume ***	[88]
	75 L tank bioreactor (no. 6)	Multicycle 140	9.5–13.0 g L^{-1}, ~90%	Synenxane C ~0.55 mg gDW^{-1} yunnanxane ~0.1 mg gDW^{-1}		
Polyscias fruticosa	Bubble-type bioreactors (no. 2)	Multicycle 204	6.31–7.31 g L^{-1}, 70–90%	Ladyginoside A 0.66–0.79 mg gDW^{-1} PFS 0.78–1.03 mg gDW^{-1}	26 °C, darkness, pO$_2$ 10–40% of saturation volume ***	[86]
Panax japonicus	630 L tank bioreactor (no. 7)	Multicycle 115	8.7–10.2 g L^{-1}, 86–90%	Total ginsenosides ~7.5%DW	26 °C, darkness, pO$_2$ 10–40% of saturation volume ***	[87]

* Numbers in parentheses correspond to the bioreactor numbers in Table 1. ** Equipment to support the closed continuous regime and separate cells from the medium was mounted into the bioreactor. *** To maintain the semi-continuous cultivation, the fresh nutrient medium was fed into bioreactors at the beginning of the stationary growth phase of each subculture cycle until the suspension was diluted to a cell concentration of X0 = ~1.4 gDW L^{-1} for S. glabra, X0 = 1.4–2.3 gDW L^{-1} for P. japonicus, X0 = 1.5–2.0 gDW L^{-1} for P. filicifolia and P. fruticosa, X0 = 2.0–2.5 gDW L^{-1} for D. deltoidea, and X0 = 2.0–3.0 gDW L^{-1} for T. wallichiana. ND—No data; µ—the specific growth rate calculated on a dry weight basis; DW—dry weight; pO$_2$—the concentration of dissolved oxygen; X0—initial dry cell biomass weight; D—dilution rate; PFS—28-O-β-D-glucopyranosyl ester of oleanolic acid 3-O-β-D-glucopyranosyl-(1→4)-β-D-glucuronopyranoside. Different strains of D. deltoidea cell culture varying in productivity and the ratio of steroidal glycoside content were used in the course of facility operation [39].

The semi-continuous cultivation of plant cell suspensions was successful in bioreactors with a total volume from 10 to 630 L equipped with different stirring devices. Since 1993, repeated experiments have been performed on long-term cell cultivation in all types of bioreactors for different cell suspension cultures. The process of suspension removal and medium refilling was initiated once the suspension reached the cell density corresponding to the beginning of the growth retardation phase. After harvesting a portion of the suspension from a bioreactor, the remaining cell culture was diluted with fresh medium until reaching the minimum cell concentration, allowing further growth without the lag phase (usually above 1.0–2.5 gDW L^{-1}). In the course of the cultivation, the physiological parameters and productivity by biomass and target metabolites were evaluated. Stirring and aeration regimes were selected experimentally for each culture considering the following requirements:

- The dissolved oxygen concentration (pO$_2$) should remain above 10–15%;
- The stirrer rotation speed should be adjusted to aeration intensity to avoid any "dead" zones in the bioreactor.

The semi-continuous cultivation was successfully developed for the suspension cell cultures of Stephania glabra, Dioscorea deltoidea, Polyscias filicifolia, Taxus wallichiana, Polyscias fruticosa, and Panax japonicus (Table 2, Figure 3c). After the optimization of cultivation conditions, all cell suspensions in bioreactors retained their growth and biosynthetic characteristics at the flask culture level.

4. Strategies for Upscaling the Cultivation Process

One of the greatest problems to solve while transferring the process of cultivation from flasks to bioreactors is the optimization of aeration and mixing conditions for each specific cell strain, including the selection of the optimal design of the bioreactor internal space [125–128].

4.1. Mixing

During the cultivation of plant cell suspensions, constant mixing serves two main purposes [19,105]:

- Providing mass transfer between the gas, liquid, and solid phases of the suspension;
- Maintaining homogeneous chemical and physical conditions in the system for a uniform distribution of the nutrients and gases, heat transfer, and dispersion of cell biomass.

In flask cultures, these principles are realized by the constant agitation of cell suspensions on rotary shakers. In bioreactors, mechanical stirring, suspension mixing by compressed air, or a combination of these approaches are applied [105,127]. However, the larger the size and more complex the configuration of the cultivation system, the more difficult it is to achieve uniform mixing. Temperature gradients, fluctuations in the concentrations of limiting substrates, the formation of so-called "dead zones", etc. are often observed in bioreactors of semi-industrial and industrial scales. The efficiency of mixing is also significantly influenced by the rheological characteristics of the cell suspensions [19,67,127].

Additional difficulties are associated with the sensitivity of plant cells to hydrodynamic and mechanical stress. Hydrodynamic (shear) stress contributes to changes in cell morphology and metabolism, the release of intracellular compounds, and a decrease in viability. For example, a sensitivity to mechanical agitation has been demonstrated for tobacco (*Nicotiana tabacum*) cells cultured in a stirred-tank bioreactor, where the maximum total biomass density decreased by 27% (from 11.8 g to 8.6 gDW L^{-1}) with the increasing of the impeller speed from 100 rpm to 325 rpm [129]. Simultaneously, rapid mixing resulted in a high number of visibly damaged and deformed cells [129]. A comparatively higher tolerance to shear stress was shown for *Morinda citrifolia* L. cell culture, where turbine stirring had no detrimental effect on growth. Nevertheless, at a high agitation speed, the accumulation of anthraquinone in the cells was lower than in the flask culture [130]. Hydrodynamic stress led to a decrease in the intracellular adenosine triphosphate (ATP) content and the respiratory activity of *Carthamus tinctorius* L. cell culture, and these changes long preceded cell lysis and membrane damage [131].

The stress effect during stirring is usually minimized by the culture-targeted optimization of bioreactor designs, in particular, by the individual selection of stirrers and gas distribution devices [67,92,127,132,133], the stirring condition (stirrer rotation speed and air supply rate) [67,127,132,133], as well as the selection or creation of cell strains resistant to shear stress, while maintaining high productivity [19,134,135]. For example, with *Coleus blumei* cells, the spiral stirrer provided a 1.5–2 times higher productivity of rosmarinic acid compared to the airlift stirring system and anchor stirrer [35]. Pavlov et al. [136] experimented with the stirrer rotation speed and its effect on the biomass and secondary metabolite productivity of the suspension cell culture of *Lavandula vera* DC, grown in a 3 L bioreactor with a propeller-type stirrer, and they recorded the maximum growth performance at 100 rpm and the maximum yield of rosmarinic acid at 300 rpm. Zhong et al. [137] cultured *Perilla frutescens* (L.) Britton cell suspension in a bioreactor with a marine-type impeller and found that the impeller tip speed of 0.5–0.8 m s^{-1} most favored the accumulation of cell biomass and anthocyanins.

In general, it should be noted once again that the response of plant cell cultures to hydrodynamic stress is individual and depends both on the nature and intensity of the stress and on the physiological state of the cell culture, the age of the population (cells are most susceptible to stress in the lag phase and stationary phase of the growth cycle), the concentration of components in the nutrient medium, the presence of inhibitory metabolites, etc. The hydrodynamic effects of mixing should be particularly considered in large-scale cultivation processes [19,128,138,139].

4.2. Aeration

Along with stirring, constant aeration is necessary to achieve the homogeneity of the cell suspension, to increase the rate of the mass exchange of nutrients and products of metabolism, as well as to supply plant cells with oxygen during cultivation [19,68,140,141].

Plant cell respiration is a complex process of thoroughly regulated redox reactions that serve as a source of convertible forms of cellular energy, such as a pH gradient ($\Delta\mu H^+$) and ATP, and intermediate metabolites involved in various biosynthetic pathways. The electron-transport respiratory chain of plant cells is known to operate two pathways of electron transfer: the main cytochrome pathway (via cyanide-sensitive cytochrome oxidase) and a cyanide-resistant alternative pathway (AP) [142]. Their contribution to the total oxygen uptake varies within a rather wide range and depends on many factors, in particular, on the content and activity of the corresponding enzymes, the degree of electron transfer inhibition, the availability of respiratory substrates, and others [142–145].

An alternative transport provides an electron transfer from ubiquinone to oxygen, bypassing two parts of the electron-transport chain (complexes III and IV), thus being energetically less efficient than the cytochrome pathway. However, studies confirm the important role of AP in the maintenance of plant cell metabolism [142,143,145]. Moreover, AP regulates the balance of reduced electron transporters, reducing the possibility of reactive oxygen species formation, and it may also promote the active growth of plant cells [146]. The activation of AP in response to negative external factors suggests its participation in signaling mechanisms of plant cell defense against different types of stresses [146–149].

The measurement of the total oxygen uptake rate is a common method for monitoring the metabolic activity of plant cells during flask or bioreactor cultivation and can adequately indicate the response of cell cultures to changing conditions, including temperature, pH, osmotic stress, nutritional deficiencies, pathogen attack, etc. [143,147,150–152]. However, the link between different respiratory metabolic pathways, cell growth, and secondary metabolite biosynthesis in plant cell culture has rarely been studied so far. High cellular respiratory activity promotes intensive growth and biosynthesis processes; therefore, the constant aeration of suspension cultures is necessary to maintain aerobic growing conditions as well as to dissipate possible excess heat generated during the cultivation process [68,145,153].

To prevent an oxygen limitation of cell suspension growth, the dissolved oxygen (dO_2) concentration in the culture medium is usually maintained at a minimum of 10–15% of saturation [51]. The general O_2 uptake rate for plant cells has been shown to vary within the range of about 5–10 mmol O_2 $(L \cdot h)^{-1}$, compared with 10–90 mmol O_2 $(L \cdot h)^{-1}$ for microbial cells and 0.05–10 mmol O_2 $(L \cdot h)^{-1}$ or 0.02–0.1×10^{-9} mmol O_2 $(cell \cdot h)^{-1}$ for mammalian cells, depending on the individual characteristics of the cell lines, cultivation conditions, phases of the growth cycle, etc. [26,51,94]. For example, for *Thalictrum minus* L. cells actively synthesizing berberine, a twofold increase in the rate of oxygen consumption was observed compared to non-producing cells [154]. Pavlov et al. [136] conducted experiments to study the effects of different concentrations of dissolved oxygen (within 10–50% of the saturation level) on growth and rosmarinic acid production in a cell suspension of *Lavandula vera*. The maximum productivity of both the biomass and rosmarinic acid was observed at dO_2 30–50% of the saturation level, while reducing dO_2 to 10% of the saturation level resulted in a significant decrease in all physiological parameters. This was consistent with the observation of other authors [18,155] that an increased dO_2 level led to the enhanced respiratory activity and an intensified synthesis of β-glucuronidase and phenolic compounds in *Nicotiana tabacum* cell culture. The effect of dO_2 on the growth and accumulation of secondary metabolites has also been shown for other cultures, including *Perilla frutescens* and *Catharanthus roseus* [156].

At present, various gas-distributing devices are used for the aeration of the suspension cultures of plant cells grown in bioreactors: single point spargers, rings, lattices, etc. The design of the sparger is selected individually for each cell line/bioreactor type to ensure the optimal air flow without extensive turbulence and promote mass transfer throughout the working volume of the bioreactor. A diameter of the sparger holes of 1–3 mm allows for the efficient dispersion of air bubbles and prevents their clumping and aggregation [127]. Similar to microbiological processes [65,128,157,158], comprehensive studies of aeration efficiency, including automatic measurements of oxygen consumption by cells, and various methods of analyzing the solubility of oxygen and carbon dioxide in the nutrient medium may be useful to evaluate the physiological state of plant cell cultures during cultivation in flasks and bioreactors of different volumes [140,159,160].

4.3. Oxygen Mass Transfer Coefficient ($K_{L\alpha}$)

The difference in the hydrodynamic conditions and mass transfer characteristics of the systems must be taken into account when transferring plant cell cultivation from flasks to industrial bioreactors [19,127,139,161,162]. The upscaling process was previously performed using the "theory of similarities" considering geometric, kinematic, and dynamic properties, each with its own criteria and differential equations describing the cultivation process. However, this approach resulted in an abundant number of criteria that should be satisfied simultaneously during upscaling, which often led to contradictory results; therefore, the principle of geometric similarity was abandoned for simplicity [127,163–165].

As mentioned above, the efficiency of bioreactor cultivation is largely determined by the interaction of the growing cell population with the environment, including the transport of nutrient components and gaseous substances from the medium to the cell surface and the removal of cell metabolic products from the cell surface to the medium. This dynamic exchange is, in turn, affected by the hydrodynamic conditions in the bioreactor. Depending on the intensity of the agitation and aeration, the ratio between the turbulent and molecular diffusion changes, causing different mass transfer rates [18,127,163–165].

It is crucial to maintain the most suitable conditions of mass transfer, i.e., an optimal hydrodynamic environment, in the process of cell growth in the bioreactor as determined by the conditions of the energy input and the type of bioreactor used [18,127,164,165]. By analogy to microbial and animal cultivation systems, the volumetric coefficient of the oxygen mass transfer ($K_{L\alpha}$) was proposed as one of the key criteria to consider when upscaling the process of plant cell cultivation. For example, the importance of oxygen transfer and its limitations have been demonstrated in scaling up the cultivation of the suspension cell cultures of *Nicotiana tabacum* [129], *Digitalis lanata* [166], *Panax ginseng* [153], and *Taxus chinensis* [167,168]. However, the use of these criteria is only effective when the same macro- and micro-mixing conditions are maintained during the transition from the laboratory to industrial bioreactors.

It is worth noting that the scaling principles for the bioreactor cultivation of plant cells are still being developed. From the technological viewpoint, plant cell cultures are challenging to work with, hence the difficulties to standardize and unambiguous specify the critical scaling parameters for each cell strain [52,54].

4.4. Scale-Up Technologies at the Experimental Biotechnological Facility of the IPPRAS

As already mentioned, plant cell cultivation in bioreactors is usually focused on scaling up the developed and optimized technological processes from smaller to greater volumes using cell strains with known growth and biosynthetic performance. However, it is usually quite difficult to accurately predict adaptive changes in the cell cultures upon their transfer to bioreactors and to precisely match the geometric and technological aspects of the equipment to the culture's needs. In our studies, this problem was approached by stepwise analysis of the critical parameters reflecting the physiological state of plant cell cultures at all stages of the upscaling process, from flasks to industrial bioreactors. The development and optimization of the technology for the industrial cultivation of suspension cell cultures was mainly focused on *Dioscorea deltoidea*, *Polyscias filicifolia*, *Panax japonicus*, and *Taxus wallichiana* [21,87–89,124,169]. For these cell cultures, changes in growth dynamics and the accumulation of secondary metabolites between the different stages of the upscaling process (Figure 4) were identified and critically analyzed. The changes in the main physiological parameters demonstrated cell strains' ability to adapt to a new cultivation system. The semi-continuous cultivation was selected for upscaling as the most flexible and productive regime both in terms of biomass and secondary metabolite yield (Table 2). Bubble-type bioreactors (total volume 20–630 L, Figure 4) were chosen for the upscaling scheme based on the highest growth and biosynthetic characteristics of the cell strains observed in this type of bioreactors [83,87,88,124]. Bioreactors with a 20 L volume were inoculated directly from flasks. The cell suspension produced in bioreactors of the smaller volume was used to inoculate the larger ones (Figure 4). Our results demonstrated that ring-type aerators were more suitable than single point spargers for maintaining optimal mass transfer conditions in bioreactors of different volumes. Cultivation in mechanically stirred bioreactors usually resulted in lower growth and biosynthetic characteristics and could only be recommended for short-term use.

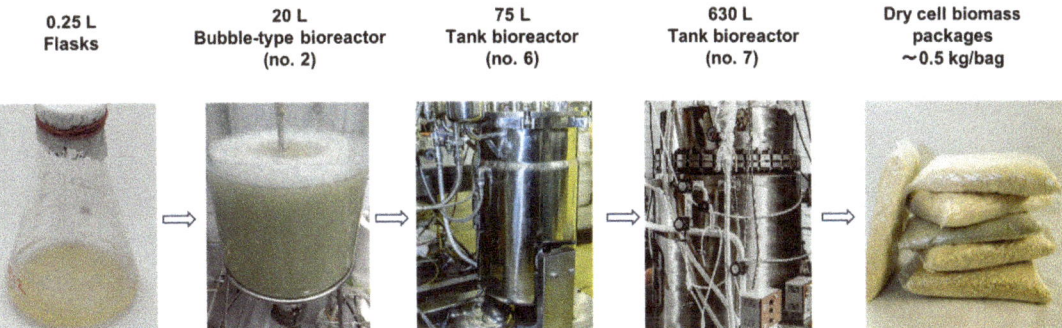

Figure 4. The scheme for upscaling the cultivation of plant cell suspensions in bioreactors at the Experimental Biotechnological Facility of the IPPRAS, from the flask culture to dry biomass product.

Important information about the physiological state of the cell population during the cultivation and upscaling process can be obtained by analyzing cells' respiration activity. In particular, a correlation between the changes in the respiration intensity and the dynamics of secondary metabolite accumulation was observed for *Dioscorea deltoidea* and *Panax japonicus* cell cultures: the maximum rate of oxygen uptake was recorded before the beginning of active metabolite synthesis, i.e., in the lag phase for *D. deltoidea* and in the exponential phase for *P. japonicus*. In other words, during plant cell cultivation, the oxygen supply rate should be set depending on the culture's biosynthetic activity [124,170]. Moreover, the activity of alternative oxidase in *Dioscorea deltoidea* cell culture was significantly affected by the cultivation conditions and, in particular, by the mass-exchange characteristics of the bioreactors. When the cell suspension was cultured in bubble-type bioreactors of different volumes with a different sparger configuration, the highest level of cyanide-resistant respiration was recorded for the 20 L bioreactors with a single point aerator, which corresponded to minimum $K_{L\alpha}$ values and the lowest production of cell biomass and furostanol glycosides. Probably, this effect was due to a non-uniform distribution of oxygen in the culture medium [124].

The reproducibility of the main growth and biosynthetic characteristics of selected strains during prolonged cultivation in bioreactors is fundamentally important for the development of industrial technologies. The scale-up cultivation of the suspension cell cultures of *Dioscorea deltoidea*, *Polyscias filicifolia*, and *Panax japonicus* using the semi-continuous regime has been repeated multiple times during the past 20 years, with similarly high growth and biosynthetic parameters successfully reproduced for all of them (Table 3).

Table 3. Examples of cell culture strains adapted for large-scale (630 L) bioreactor cultivation at the Experimental Biotechnological Facility of the IPPRAS to produce useful health products *.

Suspension Cell Culture	Metabolites Produced	Biological and Pharmacological Activities	Reference
Dioscorea deltoidea, strain DM-05-03	25(S)- and 25(R)-deltoside isomers, 25(S)- and 25(R)-protodioscin isomers, dioscin	Bioreactor-produced cell biomass was assessed for elemental composition and toxicology, and it demonstrated positive effects in rats with induced type 2 diabetes mellitus and obesity	[21,124,171,172]
Panax japonicus, strain 62	Ginsenosides: PPD: Rb1, Rc, Rb2/Rb3, Rd; PPT: Rg1, Re, Rf; OA: R0, chikusetsusaponin IVa; malonylated derivatives of ginsenosides	Bioreactor-produced cell biomass was assessed for elemental composition and toxicology and exhibited hypoglycemic and hypocholesterolemic activity in rats with diet-induced obesity	[87,169,173]
Polyscias filicifolia, strain BFT-01-95	Triterpene glycosides of the oleanane type: PFS, ladyginoside A, polysciosides A–E	Bioreactor-produced cell biomass has documented adaptogenic and anti-teratogenic activities and is currently used in commercial food supplements	[89,174,175]

*—upscaling scheme as in Figure 4. PPD—20(S)-protopanaxadiol group; PPT—20(S)-protopanaxatriol group; OA—oleanolic acid group; PFS—3-O-[β-D-glucopyranosyl-(1→4)-β-D-glucuronopyranosyl] oleanolic acid 28-O-β-D-glucopyranosyl ester.

The cell biomass of *Dioscorea deltoidea* and *Panax japonicus* produced in industrial bioreactors contained essential macro- (K, Ca, Mg, Na) and micro- (Zn, Mn, Fe, B, Al, Cu) elements in dietary safe concentrations [21,87]. Toxicology analysis on in vivo models revealed little or no effect of the cell biomass of these cultures on the animal state, organ weights, and the hematological and biochemical parameters of the blood [21,87]. Phytopreparations based on the cell culture extracts of bioreactor-produced *P. japonicus*, *D. deltoidea*, and *Tribulus terrestris* L. also demonstrated positive effects in rats with induced type 2 diabetes mellitus and diet-induced obesity [171,172].

5. Conclusions

The Experimental Biotechnological Facility of the IPPRAS was established in the 1970s as the first Russian center for biotechnological research and the production of plant cells and phytochemicals. After multi-year tests and the adaptation of bioreactors with different designs and operation regimes, a cascade of bioreactor pipelines from the laboratory (20 L) to industrial (630 L) scale were developed and optimized for cell cultures of medicinal plants, including *Polyscias filicifolia*, *Panax japonicus*, *Dioscorea deltoidea*, and *Taxus wallichiana*. The main growth and biosynthetic characteristics for all tested cell strains remained stable and were successfully reproduced during repeated long-term bioreactor cultivation. The maximum duration of semi-continuous cultivation in the industrial bioreactor reached 170 days. During the scaling up of the cultivation process to industrial volumes, all strains maintained an active synthesis of target metabolites (ginsenosides, furostanol-type glycosides, triterpene glycosides of the oleanane type, and taxanes) at a sufficiently high level, mostly corresponding to those recorded for flask cultures [21,87,171,172]. A minor decrease in the productivity of secondary metabolites and a reduction in all physiological parameters was observed only for the pilot (75 L) bioreactor with a single point sparger combined with mechanical agitation, which was likely due to cell culture response to intense shear stress. The highest productivity of secondary metabolites was recorded for the industrial bubble-type bioreactor with a ring sparger, and the hydrodynamic conditions of this model were considered the most appropriate for the cultivation of plant cell suspensions.

In general, the scale-up experiments have demonstrated the high sensitivity of plant cell suspension cultures to even minor modifications in cultivation systems. The level of biomass and secondary metabolite accumulation was notably affected by bioreactors' technical characteristics, such as the design, aeration, and mixing intensity, as well as their cultivation conditions (media composition, inoculum, etc.). This is consistent with the literature data [52,54,56,164,176–178] on the lack of uniform and comprehensive scaling criteria for geometrically and structurally dissimilar bioreactor systems. Much of the success depended on the ability of the cells to adapt to the stress caused by bioreactor cultivation. Modifications of the cultivation conditions had a stronger effect on the biosynthetic characteristics of the cell cultures than on the growth parameters. It was critically important to tailor both the operation regime and the cultivation conditions, particularly the aeration and mixing rates, depending on the bioreactor type and cell strain.

In conclusion, the research team of the Experimental Biotechnological Facility of the IPPRAS developed effective systems, methods, and criteria for scaling up the plant cell cultivation process from flasks to bioreactors of industrial volumes. These results will be helpful for the development of green biotechnological platforms and production, assessments, and the certification of plant cell biomass as a sustainable component of functional foods, food additives, and natural health products.

Author Contributions: Conceptualization, M.T.; methodology and data curation, M.T.; writing—original draft preparation, M.T.; writing—review and editing, E.P. and A.N.; funding acquisition, M.T., E.P. and A.N. All authors have read and agreed to the published version of the manuscript.

Funding: The results were obtained within the state assignments of the Ministry of Science and Higher Education of the Russian Federation, theme No. 122042700045-3 (maintenance of the collection of cell lines with high biosynthetic ability) and theme No. 122042600086-7 (bioreactor cultivation of plant cells). Bioreactor cultivation of plant cell suspensions was performed using the equipment of the large-scale research facilities "Experimental biotechnological facility" and "All-Russian Collection of cell cultures of higher plants" of the IPPRAS (EBF IPPRAS and ARCCC HP IPPRAS). These facilities were modernized with the financial support of Megagrant project no. 075-15-2019-1882.

Data Availability Statement: No new data were generated during this research.

Conflicts of Interest: The authors declare no conflict of interest.

References

1. Harnischfeger, G. Proposed Guidelines for Commercial Collection of Medicinal Plant Material. *J. Herbs. Spices Med. Plants* **2000**, *7*, 43–50. [CrossRef]
2. Misra, A. Studies on Biochemical and Physiological Aspects in Relation to Phytomedicinal Qualities and Efficacy of the Active Ingredients during the Handling, Cultivation and Harvesting of the Medicinal Plants. *J. Med. Plants Res.* **2009**, *3*, 1140–1146.
3. Verma, N.; Shukla, S. Impact of Various Factors Responsible for Fluctuation in Plant Secondary Metabolites. *J. Appl. Res. Med. Aromat. Plants* **2015**, *2*, 105–113. [CrossRef]
4. Alamgir, A.N.M. Biotechnology, In Vitro Production of Natural Bioactive Compounds, Herbal Preparation, and Disease Management (Treatment and Prevention). In *Therapeutic Use of Medicinal Plants and their Extracts*; Springer: Cham, Switzerland, 2018; Volume 2, pp. 585–664.
5. Kregiel, D.; Berlowska, J.; Witonska, I.; Antolak, H.; Proestos, C.; Babic, M.; Babic, L.; Zhang, B. Saponin-Based, Biological-Active Surfactants from Plants. In *Application and Characterization of Surfactants*; InTechOpen: London, UK, 2017.
6. Seca, A.; Pinto, D. Plant Secondary Metabolites as Anticancer Agents: Successes in Clinical Trials and Therapeutic Application. *Int. J. Mol. Sci.* **2018**, *19*, 263. [CrossRef] [PubMed]
7. Todorova, V.; Ivanov, K.; Delattre, C.; Nalbantova, V.; Karcheva-Bahchevanska, D.; Ivanova, S. Plant Adaptogens—History and Future Perspectives. *Nutrients* **2021**, *13*, 2861. [CrossRef] [PubMed]
8. Alhazmi, H.A.; Najmi, A.; Javed, S.A.; Sultana, S.; Al Bratty, M.; Makeen, H.A.; Meraya, A.M.; Ahsan, W.; Mohan, S.; Taha, M.M.E.; et al. Medicinal Plants and Isolated Molecules Demonstrating Immunomodulation Activity as Potential Alternative Therapies for Viral Diseases Including COVID-19. *Front. Immunol.* **2021**, *12*, 637553. [CrossRef] [PubMed]
9. Sellami, M.; Slimeni, O.; Pokrywka, A.; Kuvačić, G.; D Hayes, L.; Milic, M.; Padulo, J. Herbal Medicine for Sports: A Review. *J. Int. Soc. Sports Nutr.* **2018**, *15*, 14. [CrossRef] [PubMed]
10. Cragg, G.M.; Newman, D.J. Nature: A Vital Source of Leads for Anticancer Drug Development. *Phytochem. Rev.* **2009**, *8*, 313–331. [CrossRef]
11. Newman, D.J.; Cragg, G.M. Natural Products as Sources of New Drugs over the Nearly Four Decades from 01/1981 to 09/2019. *J. Nat. Prod.* **2020**, *83*, 770–803. [CrossRef] [PubMed]

12. Dias, D.A.; Urban, S.; Roessner, U. A Historical Overview of Natural Products in Drug Discovery. *Metabolites* **2012**, *2*, 303–336. [CrossRef]
13. Ochoa-Villarreal, M.; Howat, S.; Hong, S.; Jang, M.O.; Jin, Y.-W.; Lee, E.-K.; Loake, G.J. Plant Cell Culture Strategies for the Production of Natural Products. *BMB Rep.* **2016**, *49*, 149–158. [CrossRef]
14. Vasil, I.K. A History of Plant Biotechnology: From the Cell Theory of Schleiden and Schwann to Biotech Crops. *Plant Cell Rep.* **2008**, *27*, 1423–1440. [CrossRef]
15. Georgiev, M.I.; Weber, J.; Maciuk, A. Bioprocessing of Plant Cell Cultures for Mass Production of Targeted Compounds. *Appl. Microbiol. Biotechnol.* **2009**, *83*, 809–823. [CrossRef] [PubMed]
16. Wawrosch, C.; Zotchev, S.B. Production of Bioactive Plant Secondary Metabolites through in Vitro Technologies—Status and Outlook. *Appl. Microbiol. Biotechnol.* **2021**, *105*, 6649–6668. [CrossRef] [PubMed]
17. Kreis, W. Exploiting Plant Cell Culture for Natural Product Formation. *J. Appl. Bot. Food Qual.* **2019**, *92*, 216–225. [CrossRef]
18. Kieran, P.M. Bioreactor Design for Plant Cell Suspension Cultures. In *Multiphase Bioreactor Design*; Cabral, J., Mota, M., Tramper, J., Eds.; Taylor and Francis: London, UK, 2001; pp. 391–426.
19. Chattopadhyay, S.; Farkya, S.; Srivastava, A.K.; Bisaria, V.S. Bioprocess Considerations for Production of Secondary Metabolites by Plant Cell Suspension Cultures. *Biotechnol. Bioprocess Eng.* **2002**, *7*, 138–149. [CrossRef]
20. Werner, S.; Maschke, R.W.; Eibl, D.; Eibl, R. Bioreactor Technology for Sustainable Production of Plant Cell-Derived Products. In *Bioprocessing of Plant In Vitro Systems. Reference Series in Phytochemistry*; Pavlov, A., Bley, T., Eds.; Springer: Cham, Switzerland, 2018; pp. 413–432.
21. Titova, M.V.; Popova, E.V.; Konstantinova, S.V.; Kochkin, D.V.; Ivanov, I.M.; Klyushin, A.G.; Titova, E.G.; Nebera, E.A.; Vasilevskaya, E.R.; Tolmacheva, G.S.; et al. Suspension Cell Culture of *Dioscorea deltoidea*—A Renewable Source of Biomass and Furostanol Glycosides for Food and Pharmaceutical Industry. *Agronomy* **2021**, *11*, 394. [CrossRef]
22. Yuorieva, N.; Sinetova, M.; Messineva, E.; Kulichenko, I.; Fomenkov, A.; Vysotskaya, O.; Osipova, E.; Baikalova, A.; Prudnikova, O.; Titova, M.; et al. Plants, Cells, Algae, and Cyanobacteria In Vitro and Cryobank Collections at the Institute of Plant Physiology, Russian Academy of Sciences—A Platform for Research and Production Center. *Biology* **2023**, *12*, 838. [CrossRef] [PubMed]
23. Verpoorte, R.; van der Heijden, R.; ten Hoopen, H.; Memelink, J. Metabolic Engineering of Plant Secondary Metabolite Pathways for the Production of Fine Chemicals. *Biotechnol. Lett.* **1999**, *21*, 467–479. [CrossRef]
24. Smetanska, I. Production of Secondary Metabolites Using Plant Cell Cultures. In *Food Biotechnology*; Springer: Berlin/Heidelberg, Germany, 2008; pp. 187–228.
25. Doran, P.M. Foreign Protein Production in Plant Tissue Cultures. *Curr. Opin. Biotechnol.* **2000**, *11*, 199–204. [CrossRef]
26. Huang, T.-K.; McDonald, K.A. Bioreactor Engineering for Recombinant Protein Production in Plant Cell Suspension Cultures. *Biochem. Eng. J.* **2009**, *45*, 168–184. [CrossRef]
27. Gamborg, O.L. Plant Tissue Culture. Biotechnology. Milestones. *Vitr. Cell. Dev. Biol. Plant Plant* **2002**, *38*, 84–92. Available online: http://www.jstor.org/stable/20065017 (accessed on 28 November 2023). [CrossRef]
28. Wu, J.; Zhong, J.-J. Production of Ginseng and Its Bioactive Components in Plant Cell Culture: Current Technological and Applied Aspects. *J. Biotechnol.* **1999**, *68*, 89–99. [CrossRef] [PubMed]
29. Hu, W.-W.; Yao, H.; Zhong, J.-J. Improvement of *Panax notoginseng* Cell Culture for Production of Ginseng Saponin and Polysaccharide by High Density Cultivation in Pneumatically Agitated Bioreactors. *Biotechnol. Prog.* **2001**, *17*, 838–846. [CrossRef] [PubMed]
30. Baldi, A.; Dixit, V.K. Yield Enhancement Strategies for Artemisinin Production by Suspension Cultures of *Artemisia annua*. *Bioresour. Technol.* **2008**, *99*, 4609–4614. [CrossRef] [PubMed]
31. Santos, R.B.; Abranches, R.; Fischer, R.; Sack, M.; Holland, T. Putting the Spotlight Back on Plant Suspension Cultures. *Front. Plant Sci.* **2016**, *7*, 297. [CrossRef] [PubMed]
32. Xu, J.; Ge, X.; Dolan, M.C. Towards High-Yield Production of Pharmaceutical Proteins with Plant Cell Suspension Cultures. *Biotechnol. Adv.* **2011**, *29*, 278–299. [CrossRef] [PubMed]
33. Kirchhoff, J.; Raven, N.; Boes, A.; Roberts, J.L.; Russell, S.; Treffenfeldt, W.; Fischer, R.; Schinkel, H.; Schiermeyer, A.; Schillberg, S. Monoclonal Tobacco Cell Lines with Enhanced Recombinant Protein Yields Can Be Generated from Heterogeneous Cell Suspension Cultures by Flow Sorting. *Plant Biotechnol. J.* **2012**, *10*, 936–944. [CrossRef] [PubMed]
34. Eibl, R.; Meier, P.; Stutz, I.; Schildberger, D.; Hühn, T.; Eibl, D. Plant Cell Culture Technology in the Cosmetics and Food Industries: Current State and Future Trends. *Appl. Microbiol. Biotechnol.* **2018**, *102*, 8661–8675. [CrossRef] [PubMed]
35. Ulbrich, B.; Wiesner, W.; Arens, H. Large-Scale Production of Rosmarinic Acid from Plant Cell Cultures of *Coleus blumei* Benth. In *Primary and Secondary Metabolism of Plant Cell Cultures*; Springer: Berlin/Heidelberg, Germany, 1985; pp. 293–303.
36. Tabata, M.; Fujita, Y. Production of Shikonin by Plant Cell Cultures. In *Biotechnology in Plant Science*; Zaitlin, M., Day, P., Hollaender, A., Eds.; Academic Press: New York, NY, USA, 1985; pp. 207–218.
37. Kobayashi, Y.; Akita, M.; Sakamoto, K.; Liu, H.; Shigeoka, T.; Koyano, T.; Kawamura, M.; Furuya, T. Large-Scale Production of Anthocyanin by *Aralia cordata* Cell Suspension Cultures. *Appl. Microbiol. Biotechnol.* **1993**, *40*, 215–218. [CrossRef]
38. Nosov, A.M. Application of Cell Technologies for Production of Plant-Derived Bioactive Substances of Plant Origin. *Appl. Biochem. Microbiol.* **2012**, *48*, 609–624. [CrossRef]

39. Nosov, A.M.; Popova, E.V.; Kochkin, D.V. Isoprenoid Production via Plant Cell Cultures: Biosynthesis, Accumulation and Scaling-Up to Bioreactors. In *Production of Biomass and Bioactive Compounds Using Bioreactor Technology*; Springer: Dordrecht, The Netherlands, 2014; pp. 563–623.
40. Yazaki, K. *Lithospermum erythrorhizon* Cell Cultures: Present and Future Aspects. *Plant Biotechnol.* **2017**, *34*, 131–142. [CrossRef] [PubMed]
41. Popova, E.V.; Nosov, A.V.; Titova, M.V.; Kochkin, D.V.; Fomenkov, A.A.; Kulichenko, I.E.; Nosov, A.M. Advanced Biotechnologies: Collections of Plant Cell Cultures As a Basis for Development and Production of Medicinal Preparations. *Russ. J. Plant Physiol.* **2021**, *68*, 385–400. [CrossRef]
42. Butenko, R.G.; Lipsky, A.K.; Paukov, V.N.; Davidova, I.M. Applying Different Types of Bioreactors for Cultivation of Plant Cell Cultures (Kultivatoren Zur Zuchtung von Pflanzenzellen). In *Entwicklung von Laborfermentoren. VI. Reinhardsbrunner Symposium veranstaltet von der Sektion Mikrobiologie der Biologischen Gesellschaft der DDR und dem Institut für Technische Chemie der Akademie der Wissenschaften der DDR vom 21.–27. Mai 1978*; Ringpfeil, M., Dimter, L., Eds.; De Gruyter: Berlin, Germany, 1979; pp. 193–198.
43. Nosov, A.M.; Paukov, N.V.; Butenko, R. Steroid Compounds of *D. deltoidea* Wall. during Incubation of Culture in a Microbiological Fermentor. *Appl. Biochem. Microbiol.* **1984**, *20*, 119.
44. Butenko, R.G. The Tissue Culture of Medical Plants and Its Possible Use in Pharmacy. *Probl. Pharmacog.* **1967**, *21*, 184–191.
45. Lipsky, A.K.; Chernyak, N.D.; Butenko, R.G. Estimation of Biomass Amounts during the Submerged Cultivation of *Dioscorea deltoidea* Wall. by an Optical Method. *Appl. Biochem. Microbiol.* **1983**, *19*, 624–631.
46. Lara, A.R.; Galindo, E.; Ramírez, O.T.; Palomares, L.A. Living with Heterogeneities in Bioreactors: Understanding the Effects of Environmental Gradients on Cells. *Mol. Biotechnol.* **2006**, *34*, 355–382. [CrossRef] [PubMed]
47. Varley, J.; Birch, J. Reactor Design for Large Scale Suspension Animal Cell Culture. *Cytotechnology* **1999**, *29*, 177–205. [CrossRef]
48. Padilla-Córdova, C.; Mongili, B.; Contreras, P.; Fino, D.; Tommasi, T.; Díaz-Barrera, A. Productivity and Scale-up of Poly(3-hydroxybutyrate) Production under Different Oxygen Transfer Conditions in Cultures of *Azotobacter vinelandii*. *J. Chem. Technol. Biotechnol.* **2020**, *95*, 3034–3040. [CrossRef]
49. Liu, W.-C.; Inwood, S.; Gong, T.; Sharma, A.; Yu, L.-Y.; Zhu, P. Fed-Batch High-Cell-Density Fermentation Strategies for *Pichia pastoris* Growth and Production. *Crit. Rev. Biotechnol.* **2019**, *39*, 258–271. [CrossRef]
50. Ha, S.-J.; Kim, S.-Y.; Seo, J.-H.; Oh, D.-K.; Lee, J.-K. Optimization of Culture Conditions and Scale-up to Pilot and Plant Scales for Coenzyme Q10 Production by *Agrobacterium tumefaciens*. *Appl. Microbiol. Biotechnol.* **2007**, *74*, 974–980. [CrossRef]
51. Kieran, P.; MacLoughlin, P.; Malone, D. Plant Cell Suspension Cultures: Some Engineering Considerations. *J. Biotechnol.* **1997**, *59*, 39–52. [CrossRef] [PubMed]
52. Georgiev, M.I.; Weber, J. Bioreactors for Plant Cells: Hardware Configuration and Internal Environment Optimization as Tools for Wider Commercialization. *Biotechnol. Lett.* **2014**, *36*, 1359–1367. [CrossRef]
53. Scragg, A.H. The Problems Associated with High Biomass Levels in Plant Cell Suspensions. *Plant Cell. Tissue Organ Cult.* **1995**, *43*, 163–170. [CrossRef]
54. Verdú-Navarro, F.; Moreno-Cid, J.A.; Weiss, J.; Egea-Cortines, M. The Advent of Plant Cells in Bioreactors. *Front. Plant Sci.* **2023**, *14*, 1310405. [CrossRef]
55. Eibl, R.; Eibl, D. Design of Bioreactors Suitable for Plant Cell and Tissue Cultures. *Phytochem. Rev.* **2008**, *7*, 593–598. [CrossRef]
56. Motolinía-Alcántara, E.A.; Castillo-Araiza, C.O.; Rodríguez-Monroy, M.; Román-Guerrero, A.; Cruz-Sosa, F. Engineering Considerations to Produce Bioactive Compounds from Plant Cell Suspension Culture in Bioreactors. *Plants* **2021**, *10*, 2762. [CrossRef]
57. Kantarci, N.; Borak, F.; Ulgen, K.O. Bubble Column Reactors. *Process Biochem.* **2005**, *40*, 2263–2283. [CrossRef]
58. de Jesus, S.S.; Moreira Neto, J.; Maciel Filho, R. Hydrodynamics and Mass Transfer in Bubble Column, Conventional Airlift, Stirred Airlift and Stirred Tank Bioreactors, Using Viscous Fluid: A Comparative Study. *Biochem. Eng. J.* **2017**, *118*, 70–81. [CrossRef]
59. Nickell, L.G.; Tulecke, W. Submerged Growth of Cells of Higher Plants. *J. Biochem. Microbiol. Technol. Eng.* **1960**, *2*, 287–297. [CrossRef]
60. Son, S.H.; Choi, S.M.; Lee, Y.H.; Choi, K.B.; Yun, S.R.; Kim, J.K.; Park, H.J.; Kwon, O.W.; Noh, E.W.; Seon, J.H.; et al. Large-Scale Growth and Taxane Production in Cell Cultures of *Taxus cuspidata* (Japanese Yew) Using a Novel Bioreactor. *Plant Cell Rep.* **2000**, *19*, 628–633. [CrossRef]
61. Shim, K.-M.; Murthy, H.N.; Park, S.-Y.; Rusli, I.; Paek, K.-Y. Production of Biomass and Bioactive Compounds from Cell Suspension Cultures of *Eurycoma longifolia* in Balloon Type Bubble Bioreactors. *Hortic. Sci. Technol.* **2015**, *33*, 251–258. [CrossRef]
62. Thanh, N.T.; Murthy, H.N.; Paek, K.Y. Optimization of Ginseng Cell Culture in Airlift Bioreactors and Developing the Large-Scale Production System. *Ind. Crop. Prod.* **2014**, *60*, 343–348. [CrossRef]
63. Reinhard, E.; Kreis, W.; Barthlen, U.; Helmbold, U. Semicontinuous Cultivation of *Digitalis lanata* Cells: Production of B-methyldigoxin in a 300-L Airlift Bioreactor. *Biotechnol. Bioeng.* **1989**, *34*, 502–508. [CrossRef]
64. Navia-Osorio, A.; Garden, H.; Cusidó, R.M.; Palazón, J.; Alfermann, A.W.; Piñol, M.T. Taxol® and Baccatin III Production in Suspension Cultures of *Taxus baccata* and *Taxus wallichiana* in an Airlift Bioreactor. *J. Plant Physiol.* **2002**, *159*, 97–102. [CrossRef]
65. Garcia-Ochoa, F.; Gomez, E. Prediction of Gas-liquid Mass Transfer Coefficient in Sparged Stirred Tank Bioreactors. *Biotechnol. Bioeng.* **2005**, *92*, 761–772. [CrossRef] [PubMed]

66. Karimi, A.; Golbabaei, F.; Mehrnia, M.R.; Neghab, M.; Mohammad, K.; Nikpey, A.; Pourmand, M.R. Oxygen Mass Transfer in a Stirred Tank Bioreactor Using Different Impeller Configurations for Environmental Purposes. *Iranian J. Environ. Health Sci. Eng.* **2013**, *10*, 6. [CrossRef] [PubMed]
67. Doran, P.M. Design of Mixing Systems for Plant Cell Suspensions in Stirred Reactors. *Biotechnol. Prog.* **1999**, *15*, 319–335. [CrossRef] [PubMed]
68. Tanaka, H. Technological Problems in Cultivation of Plant Cells at High Density. *Biotechnol. Bioeng.* **2000**, *67*, 775–790. [CrossRef]
69. Rittershaus, E.; Ulrich, J.; Weiss, A.; Westphal, K. Large Scale Industrial Fermentation of Plant Cells: Experiences in Cultivation of Plant Cells in a Fermentation Cascade up to a Volume of 75,000 L. *BioEngineering* **1989**, *5*, 28–34.
70. Wang, J.; Gao, W.-Y.; Zhang, J.; Zuo, B.-M.; Zhang, L.-M.; Huang, L.-Q. Production of Ginsenoside and Polysaccharide by Two-Stage Cultivation of *Panax quinquefolium* L. Cells. *In Vitro Cell. Dev. Biol. Plant* **2012**, *48*, 107–112. [CrossRef]
71. Zhang, Z.-Y.; Zhong, J.-J. Scale-up of Centrifugal Impeller Bioreactor for Hyperproduction of Ginseng Saponin and Polysaccharide by High-Density Cultivation of *Panax notoginseng* Cells. *Biotechnol. Prog.* **2004**, *20*, 1076–1081. [CrossRef] [PubMed]
72. Verma, P.; Mathur, A.K.; Masood, N.; Luqman, S.; Shanker, K. Tryptophan Over-Producing Cell Suspensions of *Catharanthus roseus* (L) G. Don and Their up-Scaling in Stirred Tank Bioreactor: Detection of a Phenolic Compound with Antioxidant Potential. *Protoplasma* **2013**, *250*, 371–380. [CrossRef] [PubMed]
73. Chattopadhyay, S.; Srivastava, A.K.; Bhojwani, S.S.; Bisaria, V.S. Production of Podophyllotoxin by Plant Cell Cultures of *Podophyllum hexandrum* in Bioreactor. *J. Biosci. Bioeng.* **2002**, *93*, 215–220. [CrossRef]
74. Prakash, G.; Srivastava, A.K. Modeling of Azadirachtin Production by *Azadirachta indica* and Its Use for Feed Forward Optimization Studies. *Biochem. Eng. J.* **2006**, *29*, 62–68. [CrossRef]
75. Zhong, J.-J. Recent Advances in Bioreactor Engineering. *Korean J. Chem. Eng.* **2010**, *27*, 1035–1041. [CrossRef]
76. Eibl, R.; Eibl, D. Design and Use of the Wave Bioreactor for Plant Cell Culture. In *Plan Tissue Culture Engineering. Focus on Biotechnology*; Guptta, S.D., Ibaraki, Y., Eds.; Springer: Dordrecht, The Netherlands, 2008; Volume 6, pp. 203–227.
77. Terrier, B.; Courtois, D.; Hénault, N.; Cuvier, A.; Bastin, M.; Aknin, A.; Dubreuil, J.; Pétiard, V. Two New Disposable Bioreactors for Plant Cell Culture: The Wave and Undertow Bioreactor and the Slug Bubble Bioreactor. *Biotechnol. Bioeng.* **2007**, *96*, 914–923. [CrossRef]
78. Ducos, J.-P.; Terrier, B.; Courtois, D. Disposable Bioreactors for Plant Micropropagation and Mass Plant Cell Culture. In *Disposable Bioreactors. Advances in Biochemical Engineering/Biotechnology*; Eibl, R., Eibl, D., Eds.; Springer: Berlin/Heidelberg, Germany, 2009; Volume 115, pp. 89–115.
79. Lipsky, A.K.; Chernyak, N.D. Carbon Dioxide Requirement during Submerged Cultivation of *Dioscorea deltoidea* Wall. Cells. *Sov. Plant Physiol. Transl.* **1983**, *30*, 591–596.
80. Lipsky, A.K. Problems of Optimisation of Plant Cell Culture Processes. *J. Biotechnol.* **1992**, *26*, 83–97. [CrossRef] [PubMed]
81. Butenko, R.G.; Lipsky, A.K.; Chernyak, N.D.; Arya, H.C. Changes in Culture Medium PH by Cell Suspension Cultures of *Dioscorea deltoidea*. *Plant Sci. Lett.* **1984**, *35*, 207–212. [CrossRef]
82. Kandarakov, O.F.; Vorobev, A.S.; Nosov, A.M. Biosynthetic Characteristics of *Dioscorea deltoidea* Cell Population Grown in Continuous Culture. *Russ. J. Plant Physiol.* **1994**, *41*, 805–809.
83. Klyushin, A.G.; Oreshnikov, A.V.; Cherniak, N.D.; Nosov, A.M. The Characteristics of Growth of the Suspension Culture of *Polyscias filicifolia* Bailey Cells in Bioreactors. *Biotekhnologiya* **2000**, *3*, 67–72.
84. Orlova, L.V.; Globa, E.B.; Chernyak, N.D.; Demidova, E.V.; Titova, M.V.; Solovchenko, A.E.; Sergeev, R.V.; Nosov, A.M. Growth and Bioartificial Characteristics of *Taxus baccata* Suspension Culture (Cultivation in Flasks and Bioreactor). *Vestn. Volga State Univ. Technol. Ser. For. Ecol. Nat. Manag.* **2014**, *3*, 86–97.
85. Oreshnikov, A.V.; Nosov, A.M.; Manakov, M.N. Characterization of *Dioscorea deltoidea* Cells Grown in Closed Continuous Culture. *Russ. J. Plant Physiol.* **1994**, *41*, 810–814.
86. Titova, M.V.; Kochkin, D.V.; Sukhanova, E.S.; Gorshkova, E.N.; Tyurina, T.M.; Ivanov, I.M.; Lunkova, M.K.; Tsvetkova, E.V.; Orlova, A.; Popova, E.V.; et al. Suspension Cell Culture of *Polyscias fruticosa* (L.) Harms in Bubble-Type Bioreactors—Growth Characteristics, Triterpene Glycosides Accumulation and Biological Activity. *Plants* **2023**, *12*, 3641. [CrossRef]
87. Titova, M.V.; Popova, E.V.; Ivanov, I.M.; Fomenkov, A.A.; Nebera, E.A.; Vasilevskaya, E.R.; Tolmacheva, G.S.; Kotenkova, E.A.; Klychnikov, O.I.; Metalnikov, P.S.; et al. Toxicological Evaluation of Ginsenoside-Rich Cell Culture Biomass of *Panax japonicus* Produced in a Large-Scale Bioreactor System. *Ind. Crop. Prod.* **2024**, *208*, 117761. [CrossRef]
88. Demidova, E.; Globa, E.; Klushin, A.; Kochkin, D.; Nosov, A. Effect of Methyl Jasmonate on the Growth and Biosynthesis of C13- and C14-Hydroxylated Taxoids in the Cell Culture of Yew (*Taxus wallichiana* Zucc.) of Different Ages. *Biomolecules* **2023**, *13*, 969. [CrossRef]
89. Titova, M.V.; Popova, E.V.; Shumilo, N.A.; Kulichenko, I.E.; Chernyak, N.D.; Ivanov, I.M.; Klushin, A.G.; Nosov, A.M. Stability of Cryopreserved *Polyscias filicifolia* Suspension Cell Culture during Cultivation in Laboratory and Industrial Bioreactors. *Plant Cell Tissue Organ Cult.* **2021**, *145*, 591–600. [CrossRef]
90. Sukhanova, E.S.; Kochkin, D.V.; Titova, M.V.; Nosov, A.M. Growth and Biosynthetic Characteristics of Different *Polyscias* Plant Cell Culture Strains. *Vestn. Volga State Univ. Technol. Ser. For. Ecol. Nat. Manag.* **2012**, *2*, 57–66.
91. Pirt, S.J. *Principles of Microbe and Cell Cultivation*; Blackwell Scientific: Oxford, UK, 1975; ISBN 0632081503.
92. Prakash, G.; Srivastava, A.K. Azadirachtin Production in Stirred Tank Reactors by *Azadirachta indica* Suspension Culture. *Process Biochem.* **2007**, *42*, 93–97. [CrossRef]

93. Van Gulik, W.M.; Nuutila, A.M.; Vinke, K.L.; Ten Hoopen, H.J.G.; Heijnen, J.J. Effects of Carbon Dioxide, Air Flow Rate, and Inoculation Density on the Batch Growth of *Catharanthus roseus* Cell Suspensions in Stirred Fermentors. *Biotechnol. Prog.* **1994**, *10*, 335–339. [CrossRef]
94. Han, J.; Zhong, J. High Density Cell Culture of *Panax Notoginseng* for Production of Ginseng Saponin and Polysaccharide in an Airlift Bioreactor. *Biotechnol. Lett.* **2002**, *24*, 1927–1930. [CrossRef]
95. Hibino, K.; Ushiyama, K. Commercial Production of Ginseng by Plant Tissue Culture Technology. In *Plant Cell and Tissue Culture for the Production of Food Ingredients*; Springer: Boston, MA, USA, 1999; pp. 215–224.
96. van Gulik, W.M.; ten Hoopen, H.J.G.; Heijnen, J.J. The Application of Continuous Culture for Plant Cell Suspensions. *Enzyme Microb. Technol.* **2001**, *28*, 796–805. [CrossRef] [PubMed]
97. Hoopen, H.J.G.; Gulik, W.M.; Heijnen, J.J. Continuous Culture of Suspended Plant Cells. *In Vitro Cell. Dev. Biol. Plant* **1992**, *28*, 115–120. [CrossRef]
98. Wilson, G. A Simple and Inexpensive Design of Chemostat Enabling Steady-State Growth of *Acer pseudoplatanus* L. Cells under Phosphate-Limiting Conditions. *Ann. Bot.* **1976**, *40*, 919–932. [CrossRef]
99. Kurz, W.G.W. A Chemostat for Growing Higher Plant Cells in Single Cell Suspension Cultures. *Exp. Cell Res.* **1971**, *64*, 476–479. [CrossRef]
100. de Gucht, L.P.E.; van der Plas, L.H.W. Growth and Respiration of *Petunia hybrida* Cells in Chemostat Cultures: A Comparison of Glucose-Limited and Nitrate-Limited Cultures. *Biotechnol. Bioeng.* **2000**, *52*, 412–422. [CrossRef]
101. Meijer, J.J.; ten Hoopen, H.J.G.; van Gameren, Y.M.; Luyben, K.C.A.M.; Libbenga, K.R. Effects of Hydrodynamic Stress on the Growth of Plant Cells in Batch and Continuous Culture. *Enzyme Microb. Technol.* **1994**, *16*, 467–477. [CrossRef]
102. Sahai, O.P.; Shuler, M.L. Multistage Continuous Culture to Examine Secondary Metabolite Formation in Plant Cells: Phenolics from *Nicotiana tabacum*. *Biotechnol. Bioeng.* **1984**, *26*, 27–36. [CrossRef]
103. Wilson, S.B.; King, P.J.; Street, H.E. Studies on the Growth in Culture of Plant Cells: XII. A Versatile System for the Large Scale Batch or Continuous Culture of Plant Cell Suspensions. *J. Exp. Bot.* **1971**, *22*, 177–207. [CrossRef]
104. Peel, E. Photoautotrophic Growth of Suspension Cultures of *Asparagus officinalis* L. Cells in Turbidostats. *Plant Sci. Lett.* **1982**, *24*, 147–155. [CrossRef]
105. Georgiev, M.I.; Eibl, R.; Zhong, J.-J. Hosting the Plant Cells in Vitro: Recent Trends in Bioreactors. *Appl. Microbiol. Biotechnol.* **2013**, *97*, 3787–3800. [CrossRef] [PubMed]
106. Wang, G.R.; Qi, N.M.; Wang, Z.M. Application of a Stirrer-Tank Bioreactor for Perfusion Culture and Continuous Harvest of *Glycyrrhiza inflata* Suspension Cells. *Afr. J. Biotechnol.* **2010**, *9*, 347–351.
107. Su, W.W.; Arias, R. Continuous Plant Cell Perfusion Culture: Bioreactor Characterization and Secreted Enzyme Production. *J. Biosci. Bioeng.* **2003**, *95*, 13–20. [CrossRef]
108. De Dobbeleer, C.; Cloutier, M.; Fouilland, M.; Legros, R.; Jolicoeur, M. A High-Rate Perfusion Bioreactor for Plant Cells. *Biotechnol. Bioeng.* **2006**, *95*, 1126–1137. [CrossRef]
109. Hao, Y.-J.; Ye, W.-Q.; Wang, M.; Liu, L.-L.; Yu, S.; Piao, X.-C.; Lian, M.-L. Selection of Initial Culture Medium in Fed-Batch Bioreactor Culture of *Rhodiola sachalinensis* Cells. *J. Biotechnol.* **2022**, *346*, 15–22. [CrossRef]
110. Chastang, T.; Pozzobon, V.; Taidi, B.; Courot, E.; Clément, C.; Pareau, D. Resveratrol Production by Grapevine Cells in Fed-Batch Bioreactor: Experiments and Modelling. *Biochem. Eng. J.* **2018**, *131*, 9–16. [CrossRef]
111. Restrepo, M.I.; Atehortúa, L.; Rojas, L.F. Optimization of Fed-Batch Culture for Polyphenol Production from *Theobroma Cacao* Cell Culture. *Acta Hortic.* **2023**, *1359*, 281–288. [CrossRef]
112. Choi, J.W.; Cho, J.M.; Kim, Y.K.; Park, S.Y.; Kim, I.H.; Park, Y.H. Control of Glucose Concentration in a Fed—Batch Cultivation of *Scutellaria baicalensis* G. Plant Cells Using a Self—Organizing Fuzzy Logic Controller. *J. Microbiol. Biotechnol.* **2001**, *11*, 739–748.
113. Corbin, J.M.; Hashimoto, B.I.; Karuppanan, K.; Kyser, Z.R.; Wu, L.; Roberts, B.A.; Noe, A.R.; Rodriguez, R.L.; McDonald, K.A.; Nandi, S. Semicontinuous Bioreactor Production of Recombinant Butyrylcholinesterase in Transgenic Rice Cell Suspension Cultures. *Front. Plant Sci.* **2016**, *7*, 412. [CrossRef]
114. Santoyo-Garcia, J.H.; Valdivia-Cabrera, M.; Ochoa-Villarreal, M.; Casasola-Zamora, S.; Ripoll, M.; Escrich, A.; Moyano, E.; Betancor, L.; Halliday, K.J.; Loake, G.J.; et al. Increased Paclitaxel Recovery from *Taxus baccata* Vascular Stem Cells Using Novel in Situ Product Recovery Approaches. *Bioresour. Bioprocess.* **2023**, *10*, 68. [CrossRef]
115. Kreis, W.; Reinhard, E. Two-Stage Cultivation of *Digitalis lanata* Cells: Semicontinuous Production of Deacetyllanatoside C in 20-Litre Airlift Bioreactors. *J. Biotechnol.* **1990**, *16*, 123–135. [CrossRef] [PubMed]
116. Villarreal, M.L.; Arias, C.; Vega, J.; Feria-Velasco, A.; Ramírez, O.T.; Nicasio, P.; Rojas, G.; Quintero, R. Large-Scale Cultivation of *Solanum chrysotrichum* Cells: Production of the Antifungal Saponin SC-1 in 10-l Airlift Bioreactors. *Plant Cell Rep.* **1997**, *16*, 653–656. [CrossRef] [PubMed]
117. Choi, J.-W.; Kim, Y.-K.; Lee, W.H.; Pedersen, H.; Chin, C.-K. Bioreactor Operating Strategy in *Thalictrum rugosum* Plant Cell Culture for the Production of Berberine. *Biotechnol. Bioprocess Eng.* **1999**, *4*, 138–146. [CrossRef]
118. Luo, J.; Mei, X.J.; Hu, D.W. Improved Paclitaxel Production by Fed-Batch Suspension Cultures of *Taxus chinensis* in Bioreactors. *Biotechnol. Lett.* **2002**, *24*, 561–565. [CrossRef]
119. Titova, M.V.; Reshetnyak, O.V.; Osipova, E.A.; Osip'yants, A.I.; Shumilo, N.A.; Oreshnikov, A.V.; Nosov, A.M. Submerged Cultivation of *Stephania glabra* (Roxb.) Miers Cells in Different Systems: Specific Features of Growth and Accumulation of Alkaloid Stepharine. *Appl. Biochem. Microbiol.* **2012**, *48*, 645–649. [CrossRef]

120. Demidova, E.V. *Biosynthesis of Triterpene Glycosides in Suspension Cell Culture of Panax japonicus Var. Repens at Different Cultivation Conditions*; K.A. Timiryazev Institute of Plant Physiology of RAS: Moscow, Russia, 2007.
121. Titova, M.V. *Personal Communication*; Institute of Plant Physiology of RAS: Moscow, Russia, 2023.
122. Titova, M.V.; Kochkin, D.V.; Fomenkov, A.A.; Ivanov, I.M.; Kotenkova, E.A.; Kocharyan, G.L.; Dzhivishev, E.G.; Mekhtieva, N.P.; Popova, E.V.; Nosov, A.M. Obtaining and Characterization of Suspension Cell Culture of *Alhagi persarum* Boiss. et Buhse: A Producer of Isoflavonoids. *Russ. J. Plant Physiol.* **2021**, *68*, 652–660. [CrossRef]
123. Oreshnikov, A.V. *Physiology of Dioscorea Deltoidea Cell Population under Conditions of Closed Continuous Cultivation Regime*; K.A. Timiryazev Institute of Plant Physiology of Russian Academy of Sciences: Moscow, Russia, 1996.
124. Titova, M.V.; Shumilo, N.A.; Kulichenko, I.E.; Ivanov, I.M.; Sukhanova, E.S.; Nosov, A.M. Features of Respiration and Formation of Steroidal Glycosides in *Dioscorea deltoidea* Cell Suspension Culture Grown in Flasks and Bioreactors. *Russ. J. Plant Physiol.* **2015**, *62*, 557–563. [CrossRef]
125. Taticek, R.A.; Moo-Young, M.; Legge, R.L. The Scale-up of Plant Cell Culture: Engineering Considerations. *Plant Cell. Tissue Organ Cult.* **1991**, *24*, 139–158. [CrossRef]
126. Scragg, A.H. Large-Scale Plant Cell Culture: Methods, Applications and Products. *Curr. Opin. Biotechnol.* **1992**, *3*, 105–109. [CrossRef] [PubMed]
127. Marks, D.M. Equipment Design Considerations for Large Scale Cell Culture. *Cytotechnology* **2003**, *42*, 21–33. [CrossRef] [PubMed]
128. Garcia-Ochoa, F.; Gomez, E. Bioreactor Scale-up and Oxygen Transfer Rate in Microbial Processes: An Overview. *Biotechnol. Adv.* **2009**, *27*, 153–176. [CrossRef] [PubMed]
129. Ho, C.-H.; Henderson, K.A.; Rorrer, G.L. Cell Damage and Oxygen Mass Transfer during Cultivation of *Nicotiana tabacum* in a Stirred-tank Bioreactor. *Biotechnol. Prog.* **1995**, *11*, 140–145. [CrossRef]
130. Wagner, F.; Vogelmann, H. Cultivation of Plant Tissue Cultures in Bioreactors and Formation of Secondary Metabolites. In *Plant Tissue Culture and Its Biotechnological Application*; Barz, W., Reinhard, E., Zenk, M.H., Eds.; Springer: Berlin, Germany, 1977; pp. 245–255.
131. Takeda, T.; Seki, M.; Furusaki, S. Hydrodynamic Damage of Cultured Cells of *Carthamus tinctorius* in a Stirred Tank Reactor. *J. Chem. Eng. Jpn.* **1994**, *27*, 466–471. [CrossRef]
132. Jolicoeur, M.; Chavarie, C.; Carreau, P.J.; Archambault, J. Development of a Helical-ribbon Impeller Bioreactor for High-density Plant Cell Suspension Culture. *Biotechnol. Bioeng.* **1992**, *39*, 511–521. [CrossRef] [PubMed]
133. Nienow, A.W. On Impeller Circulation and Mixing Effectiveness in the Turbulent Flow Regime. *Chem. Eng. Sci.* **1997**, *52*, 2557–2565. [CrossRef]
134. Leckie, F.; Scragg, A.H.; Cliffe, K.C. The Effect of Continuous High Shear Stress on Plant Cell Suspension Cultures. In *Progress in Plant Cellular and Molecular Biology. Current Plant Science and Biotechnology in Agriculture*; Nijkamp, H.J.J., Van Der Plas, L.H.W., Van Aartrijk, J., Eds.; Springer: Dordrecht, The Netherlands, 1990; Volume 9, pp. 689–693.
135. Wilson, S.A.; Roberts, S.C. Recent Advances towards Development and Commercialization of Plant Cell Culture Processes for the Synthesis of Biomolecules. *Plant Biotechnol. J.* **2012**, *10*, 249–268. [CrossRef]
136. Pavlov, A.I.; Georgiev, M.I.; Ilieva, M.P. Production of Rosmarinic Acid by *Lavandula vera* MM Cell Suspension in Bioreactor: Effect of Dissolved Oxygen Concentration and Agitation. *World J. Microbiol. Biotechnol.* **2005**, *21*, 389–392. [CrossRef]
137. Zhong, J.; Fujiyama, K.; Seki, T.; Yoshida, T. A Quantitative Analysis of Shear Effects on Cell Suspension and Cell Culture of *Perilla frutescens* in Bioreactors. *Biotechnol. Bioeng.* **1994**, *44*, 649–654. [CrossRef] [PubMed]
138. Zhong, J.-J. Biochemical Engineering of the Production of Plant-Specific Secondary Metabolites by Cell Suspension Cultures. In *Advances in Biochemical Engineering/Biotechnology*; Zhong, J.J., Ed.; Springer: Berlin/Heidelberg, Germany, 2001; Volume 72, pp. 1–26.
139. Chattopadhyay, S.; Srivastava, A.K.; Bisaria, V.S. Production of Phytochemicals in Plant Cell Bioreactors. In *Plant Biotechnology and Molecular Markers*; Srivastava, P., Narula, A., Srivastava, S., Eds.; Springer: Dordrecht, The Netherlands, 2004; pp. 117–128.
140. Hohe, A.; Winkelmann, T.; Schwenkel, H.G. The Effect of Oxygen Partial Pressure in Bioreactors on Cell Proliferation and Subsequent Differentiation of Somatic Embryos of *Cyclamen persicum*. *Plant Cell Tissue Organ Cult.* **1999**, *59*, 39–45. [CrossRef]
141. Trung Thanh, N.; Niranjana Murthy, H.; Yu, K.-W.; Seung Jeong, C.; Hahn, E.-J.; Paek, K.-Y. Effect of Oxygen Supply on Cell Growth and Saponin Production in Bioreactor Cultures of *Panax ginseng*. *J. Plant Physiol.* **2006**, *163*, 1337–1341. [CrossRef] [PubMed]
142. Amthor, J. The McCree–de Wit–Penning de Vries–Thornley Respiration Paradigms: 30 Years Later. *Ann. Bot.* **2000**, *86*, 1–20. [CrossRef]
143. McDonald, A.E.; Sieger, S.M.; Vanlerberghe, G.C. Methods and Approaches to Study Plant Mitochondrial Alternative Oxidase. *Physiol. Plant.* **2002**, *116*, 135–143. [CrossRef]
144. Del-Saz, N.F.; Ribas-Carbo, M.; Martorell, G.; Fernie, A.R.; Florez-Sarasa, I. Measurements of Electron Partitioning between Cytochrome and Alternative Oxidase Pathways in Plant Tissues. In *Plant Respiration and Internal Oxygen. Methods in Molecular Biology*; Jagadis Gupta, K., Ed.; Humana Press: New York, NY, USA, 2017; Volume 1670, pp. 203–217.
145. Schertl, P.; Braun, H.-P. Respiratory Electron Transfer Pathways in Plant Mitochondria. *Front. Plant Sci.* **2014**, *5*, 163. [CrossRef]
146. Moore, A.L.; Albury, M.S.; Crichton, P.G.; Affourtit, C. Function of the Alternative Oxidase: Is It Still a Scavenger? *Trends Plant Sci.* **2002**, *7*, 478–481. [CrossRef] [PubMed]

147. Ferreira, A.L.; Arrabaça, J.D.; Vaz-Pinto, V.; Lima-Costa, M.E. Induction of Alternative Oxidase Chain under Salt Stress Conditions. *Biol. Plant.* **2008**, *52*, 66–71. [CrossRef]
148. Noguchi, K.; Taylor, N.L.; Millar, A.H.; Lambers, H.; Day, D.A. Response of Mitochondria to Light Intensity in the Leaves of Sun and Shade Species. *Plant. Cell Environ.* **2005**, *28*, 760–771. [CrossRef]
149. Clifton, R.; Lister, R.; Parker, K.L.; Sappl, P.G.; Elhafez, D.; Millar, A.H.; Day, D.A.; Whelan, J. Stress-Induced Co-Expression of Alternative Respiratory Chain Components in *Arabidopsis thaliana*. *Plant Mol. Biol.* **2005**, *58*, 193–212. [CrossRef] [PubMed]
150. Zhang, Q.; Soole, K.L.; Wiskich, J.T. Regulation of Respiration in Rotenone-Treated Tobacco Cell Suspension Cultures. *Planta* **2001**, *212*, 765–773. [CrossRef]
151. Duque, P.; Arrabaça, J.D. Respiratory Metabolism during Cold Storage of Apple Fruit. II. Alternative Oxidase Is Induced at the Climacteric. *Physiol. Plant.* **1999**, *107*, 24–31. [CrossRef]
152. Sieger, S.M.; Kristensen, B.K.; Robson, C.A.; Amirsadeghi, S.; Eng, E.W.Y.; Abdel-Mesih, A.; Møller, I.M.; Vanlerberghe, G.C. The Role of Alternative Oxidase in Modulating Carbon Use Efficiency and Growth during Macronutrient Stress in Tobacco Cells. *J. Exp. Bot.* **2005**, *56*, 1499–1515. [CrossRef]
153. Dong, Y.; Gao, W.-Y.; Man, S.; Zuo, B.; Wang, J.; Huang, L.; Xiao, P. Effect of Bioreactor Angle and Aeration Rate on Growth and Hydromechanics Parameters in Bioreactor Culture of Ginseng Suspension Cells. *Acta Physiol. Plant.* **2013**, *35*, 1497–1501. [CrossRef]
154. Kobayashi, Y.; Fukui, H.; Tabata, M. Effect of Oxygen Supply on Berberine Production in Cell Suspension Cultures and Immobilized Cells of *Thalictrum minus*. *Plant Cell Rep.* **1989**, *8*, 255–258. [CrossRef]
155. Gao, J.; Lee, J.M. Effect of Oxygen Supply on the Suspension Culture of Genetically Modified Tobacco Cells. *Biotechnol. Prog.* **1992**, *8*, 285–290. [CrossRef]
156. Leckie, F.; Scragg, A.H.; Cliffe, K.C. An Investigation into the Role of Initial K L a on the Growth and Alkaloid Accumulation by Cultures of *Catharanthus roseus*. *Biotechnol. Bioeng.* **1991**, *37*, 364–370. [CrossRef] [PubMed]
157. Çalık, P.; Yilgör, P.; Ayhan, P.; Demir, A.S. Oxygen Transfer Effects on Recombinant Benzaldehyde Lyase Production. *Chem. Eng. Sci.* **2004**, *59*, 5075–5083. [CrossRef]
158. Garcia-Ochoa, F.; Gomez, E. Theoretical Prediction of Gas–Liquid Mass Transfer Coefficient, Specific Area and Hold-up in Sparged Stirred Tanks. *Chem. Eng. Sci.* **2004**, *59*, 2489–2501. [CrossRef]
159. Raval, K.N.; Hellwig, S.; Prakash, G.; Ramos-Plasencia, A.; Srivastava, A.; Buchs, J. Necessity of a Two-Stage Process for the Production of Azadirachtin-Related Limonoids in Suspension Cultures of *Azadirachta indica*. *J. Biosci. Bioeng.* **2003**, *96*, 16–22. [CrossRef] [PubMed]
160. Rechmann, H.; Friedrich, A.; Forouzan, D.; Barth, S.; Schnabl, H.; Biselli, M.; Boehm, R. Characterization of Photosynthetically Active Duckweed (*Wolffia australiana*) in Vitro Culture by Respiration Activity Monitoring System (RAMOS). *Biotechnol. Lett.* **2007**, *29*, 971–977. [CrossRef]
161. Klöckner, W.; Gacem, R.; Anderlei, T.; Raven, N.; Schillberg, S.; Lattermann, C.; Büchs, J. Correlation between Mass Transfer Coefficient KLa and Relevant Operating Parameters in Cylindrical Disposable Shaken Bioreactors on a Bench-to-Pilot Scale. *J Biol. Eng.* **2013**, *7*, 28. [CrossRef]
162. García-Salas, S.; Gómez-Montes, E.O.; Ramírez-Sotelo, M.G.; Oliver-Salvador, M. del C. Shear Rate as Scale-up Criterion of the Protein Production with Enhanced Proteolytic Activity by Phosphate Addition in the *Jacaratia mexicana* Cell Culture. *Biotechnol. Biotechnol. Equip.* **2021**, *35*, 1031–1042. [CrossRef]
163. Votruba, J.; Sobotka, M. Physiological Similarity and Bioreactor Scale-Up. *Folia Microbiol.* **1992**, *37*, 331–345. [CrossRef]
164. Xia, J.; Wang, G.; Fan, M.; Chen, M.; Wang, Z.; Zhuang, Y. Understanding the Scale-up of Fermentation Processes from the Viewpoint of the Flow Field in Bioreactors and the Physiological Response of Strains. *Chin. J. Chem. Eng.* **2021**, *30*, 178–184. [CrossRef]
165. Xia, J.; Wang, G.; Lin, J.; Wang, Y.; Chu, J.; Zhuang, Y.; Zhang, S. Advances and Practices of Bioprocess Scale-Up. In *Bioreactor Engineering Research and Industrial Applications II*; Springer: Berlin/Heidelberg, Germany, 2015; pp. 137–151.
166. Lee, S.; Kim, D. Effects of Pluronic F-68 on Cell Growth of *Digitalis lanata* in Aqueous Wo-Phase Systems. *J. Microbiol. Biotechnol. Biotechnol.* **2004**, *14*, 1129–1133.
167. Zhong, J.-J.; Pan, Z.-W.; Wang, Z.-Y.; Wu, J.; Chen, F.; Takagi, M.; Yoshida, T. Effect of Mixing Time on Taxoid Production Using Suspension Cultures of *Taxus chinensis* in a Centrifugal Impeller Bioreactor. *J. Biosci. Bioeng.* **2002**, *94*, 244–250. [CrossRef]
168. Pan, Z.-W.; Zhong, J.-J.; Wu, J.-Y.; Takagi, M.; Yoshida, T. Fluid Mixing and Oxygen Transfer in Cell Suspensions of *Taxus chinensis* in a Novel Stirred Bioreactor. *Biotechnol. Bioprocess Eng.* **1999**, *4*, 269–272. [CrossRef]
169. Titova, M.V.; Shumilo, N.A.; Reshetnyak, O.V.; Glagoleva, E.S.; Nosov, A.M. Physiological Characteristics of *Panax japonicus* Suspension Cell Culture during Growth Scaling-Up. *Biotekhnologiya* **2015**, *3*, 71–80. [CrossRef]
170. Titova, M.V.; Berkovich, E.A.; Reshetnyak, O.V.; Kulichenko, I.E.; Oreshnikov, A.V.; Nosov, A.M. Respiration Activity of Suspension Cell Culture of *Polyscias filicifolia* Bailey, *Stephania glabra* (Roxb.) Miers, and *Dioscorea deltoidea* Wall. *Appl. Biochem. Microbiol.* **2011**, *47*, 87–92. [CrossRef]
171. Povydysh, M.N.; Titova, M.V.; Ivkin, D.Y.; Krasnova, M.V.; Vasilevskaya, E.R.; Fedulova, L.V.; Ivanov, I.M.; Klushin, A.G.; Popova, E.V.; Nosov, A.M. The Hypoglycemic and Hypocholesterolemic Activity of *Dioscorea deltoidea*, *Tribulus terrestris* and *Panax japonicus* Cell Culture Biomass in Rats with High-Fat Diet-Induced Obesity. *Nutrients* **2023**, *15*, 656. [CrossRef] [PubMed]

172. Povydysh, M.N.; Titova, M.V.; Ivanov, I.M.; Klushin, A.G.; Kochkin, D.V.; Galishev, B.A.; Popova, E.V.; Ivkin, D.Y.; Luzhanin, V.G.; Krasnova, M.V.; et al. Effect of Phytopreparations Based on Bioreactor-Grown Cell Biomass of *Dioscorea deltoidea*, *Tribulus terrestris* and *Panax japonicus* on Carbohydrate and Lipid Metabolism in Type 2 Diabetes Mellitus. *Nutrients* **2021**, *13*, 3811. [CrossRef]
173. Glagoleva, E.S.; Konstantinova, S.V.; Kochkin, D.V.; Ossipov, V.; Titova, M.V.; Popova, E.V.; Nosov, A.M.; Paek, K.-Y. Predominance of Oleanane-Type Ginsenoside R0 and Malonyl Esters of Protopanaxadiol-Type Ginsenosides in the 20-Year-Old Suspension Cell Culture of *Panax japonicus* C.A. Meyer. *Ind. Crop. Prod.* **2022**, *177*, 114417. [CrossRef]
174. Kotin, A.M.; Bichevaya, N.K. Anti-Teratogenic Agent. WO1996002266A1, 11 May 1996.
175. Kotin, O.A.; Kotin, A.M.; Emel'yanov, M.O. Use of a Preparation of the Plant Polyscias filicifolia for Treating Mineral Deficiencies 2022. WO2022010387A1, 13 January 2022.
176. Werner, S.; Olownia, J.; Egger, D.; Eibl, D. An Approach for Scale-Up of Geometrically Dissimilar Orbitally Shaken Single-Use Bioreactors. *Chem. Ing. Tech.* **2013**, *85*, 118–126. [CrossRef]
177. de Mello, A.F.M.; de Souza Vandenberghe, L.P.; Herrmann, L.W.; Letti, L.A.J.; Burgos, W.J.M.; Scapini, T.; Manzoki, M.C.; de Oliveira, P.Z.; Soccol, C.R. Strategies and Engineering Aspects on the Scale-up of Bioreactors for Different Bioprocesses. *Syst. Microbiol. Biomanuf.* **2023**. [CrossRef]
178. Seidel, S.; Mozaffari, F.; Maschke, R.W.; Kraume, M.; Eibl-Schindler, R.; Eibl, D. Automated Shape and Process Parameter Optimization for Scaling Up Geometrically Non-Similar Bioreactors. *Processes* **2023**, *11*, 2703. [CrossRef]

Disclaimer/Publisher's Note: The statements, opinions and data contained in all publications are solely those of the individual author(s) and contributor(s) and not of MDPI and/or the editor(s). MDPI and/or the editor(s) disclaim responsibility for any injury to people or property resulting from any ideas, methods, instructions or products referred to in the content.

Article

Production of Secondary Metabolites from Cell Cultures of *Sageretia thea* (Osbeck) M.C. Johnst. Using Balloon-Type Bubble Bioreactors

Ji-Hye Kim [1], Jong-Eun Han [1], Hosakatte Niranjana Murthy [1,2], Ja-Young Kim [3], Mi-Jin Kim [3], Taek-Kyu Jeong [3] and So-Young Park [1,*]

[1] Department of Horticultural Science, Chungbuk National University, Cheongju-si 28644, Republic of Korea
[2] Department of Botany, Karnatak University, Dharwad 580003, India
[3] Saimdang Cosmetics Co., Ltd., 143, Yangcheongsongdae-gil, Ochang-eup, Cheongwon-gu, Cheongju-si 28118, Republic of Korea
* Correspondence: soypark7@cbnu.ac.kr; Tel.: +82-43-261-2532

Abstract: *Sageretia thea* is used in the preparation of herbal medicine in China and Korea; this plant is rich in various bioactive compounds, including phenolics and flavonoids. The objective of the current study was to enhance the production of phenolic compounds in plant cell suspension cultures of *Sageretia thea*. Optimum callus was induced from cotyledon explants on MS medium containing 2,4-dichlorophenoxyacetic acid (2,4-D; 0.5 mg L^{-1}), naphthalene acetic acid (NAA, 0.5 mg L^{-1}), kinetin (KN; 0.1 mg L^{-1}) and sucrose (30 g L^{-1}). Browning of callus was successfully avoided by using 200 mg L^{-1} ascorbic acid in the callus cultures. The elicitor effect of methyl jasmonate (MeJA), salicylic acid (SA), and sodium nitroprusside (SNP) was studied in cell suspension cultures, and the addition of 200 µM MeJA was found suitable for elicitation of phenolic accumulation in the cultured cells. Phenolic and flavonoid content and antioxidant activity were determined using 2,2 Diphenyl 1 picrylhydrazyl (DPPH), 2,2′-azino-bis (3-ethybenzothiazoline-6-sulphonic acid) (ABTS), ferric reducing antioxidant power (FRAP) assays and results showed that cell cultures possessed highest phenolic and flavonoid content as well as highest DPPH, ABTS, and FRAP activities. Cell suspension cultures were established using 5 L capacity balloon-type bubble bioreactors using 2 L of MS medium 30 g L^{-1} sucrose and 0.5 mg L^{-1} 2,4-D, 0.5 mg L^{-1} NAA, and 0.1 mg L^{-1} KN. The optimum yield of 230.81 g of fresh biomass and 16.48 g of dry biomass was evident after four weeks of cultures. High-pressure liquid chromatography (HPLC) analysis showed the cell biomass produced in bioreactors possessed higher concentrations of catechin hydrate, chlorogenic acid, naringenin, and other phenolic compounds.

Keywords: antioxidant activity; bioreactors; catechin; cell suspension cultures; chlorogenic acid; elicitation; phenolics; flavonoids

Citation: Kim, J.-H.; Han, J.-E.; Murthy, H.N.; Kim, J.-Y.; Kim, M.-J.; Jeong, T.-K.; Park, S.-Y. Production of Secondary Metabolites from Cell Cultures of *Sageretia thea* (Osbeck) M.C. Johnst. Using Balloon-Type Bubble Bioreactors. *Plants* 2023, *12*, 1390. https://doi.org/10.3390/plants12061390

Academic Editor: Sofia Caretto

Received: 14 February 2023
Revised: 13 March 2023
Accepted: 15 March 2023
Published: 21 March 2023

Copyright: © 2023 by the authors. Licensee MDPI, Basel, Switzerland. This article is an open access article distributed under the terms and conditions of the Creative Commons Attribution (CC BY) license (https://creativecommons.org/licenses/by/4.0/).

1. Introduction

Sageretia thea (Osbeck) M.C. Johnst. belong to the family Rhamnaceae is an evergreen shrub that grows up to 3 m high; spines few to numerous; branchlets glabrous to pubescent. The native range of this species is Eritrea to North Somalia, Arabian Peninsula, Central and South China to Peninsula Malaysia, and Temperate East Asia (Korea and Japan). It is a scrambling shrub and grows primarily in the subtropical biome [1]. It is commonly called 'Sweet plum' or 'Chinese sweet plum' since it bears plum-like fruits that are edible and possess' higher nutritional value [2]. Fruits are rich in minerals, organic acids, and fatty acids. They are also abundant with phenolics, flavonoids, and anthocyanins [2]. *S. thea* plants are usually used in the preparation of bonsai plants that are having a very high ornamental value in China, Korea, and Japan. The leaves are used in the preparation of green tea. Traditionally *S. thea* is used in the preparation of herbal medicine for the treatment of hepatitis and fever in China and Korea [3–5]. Various bioactive compounds such as friedelin, synergic acid, betasitosterol, daucosterol, gluco-synrigic acid, and taraxerol were isolated from the aerial parts of *S. thea* [6]. Several flavanol glycosides, namely, 7-O-methylmearnsitrin, myricentrin, kaempferol 3-O-R-L-rhamonopyranoside, europetin 3-O-R-L-rhamonisde, and 7-O-methyl quercetin 3-O-R-L-rhamnopyranoside have been sequestered from leaves of *S. thea* [7]. Myrictrin and 7-O-methylmearnsitrin have demonstrated a strong in vitro antioxidant activity higher than α-tocopherol [7]. Besides, flavonoid-rich fractions obtained from leaves *S. thea* have shown a protective effect on low-density lipoprotein against oxidative modifications [3]. Pharmaceutical analysis of the fruits of *S. thea* has exhibited antioxidant, anti-diabetic, and anti-melanogenesis activities [2,8]. The leaf extracts have displayed the inhibition of oxidation of low-density lipoprotein through their antioxidant and HIV type 1 protease activities [2]. Furthermore, leaf extracts have demonstrated anticancer activities against human breast cancer (MDA-MB-231) and colorectal cancer cells (SW 480) [9,10]. In view of the above, *S. thea* raw material has been used in the pharmaceutical and cosmetic industries.

There is a tremendous demand for *S. thea* raw material by the pharmaceutical and cosmetic industries in Korea and Japan. Recently, plant cell and organ cultures have emerged as alternatives for the production of plant-based raw materials for the cosmetic and food industries [11,12]. Plant cell and organ cultures are advantageous options for the production of bioactive compounds for biomedical and cosmetic purposes because they represent standardized, contaminant-free, and bio-sustainable systems [13–16]. Cell and organ culture system has been successfully established in many plants such as *Panax ginseng* [17,18], *Echinacea* species [19], *Eleutherococcus* species [20], *Hypericum peforatum* [21], *Dendrobium candidum* [22,23] and biomass produced in bioreactor system is successfully used in nutraceutical and cosmetic industries. The hypothesis to be tested in this study is the verify the effect of growth regulators, and elicitors on cell suspension cultures of *S. thea*, on biomass and secondary metabolites production. Furthermore, to verify the possibilities of production of *S. thea* on biomass and secondary metabolites in bioreactor cultures. In the current study, we are aiming at the production of phenolic compounds from the cell culture of *S. thea*, we established a cell culture system in flasks and balloon-type bioreactor cultures. We also adopted elicitation technology for the hyperproduction of phenolics in cell cultures of *S. thea*.

2. Results

2.1. The Effect of Growth Regulators and Explant Type on the Callus Induction

The results of the effect of growth regulators and explant type on callus induction of *S. thea* are presented in Figure 1. MS medium supplemented with 2,4-D at 1 mg L^{-1} triggered 31.7%, 46.7%, 56.7%, and 20% callus induction from complete seed, seed halves, cotyledon, and leaf explants respectively. Addition of BA (0.1 mg L^{-1}) or KN (0.1 mg L^{-1}) to the 2,4-D containing medium or combination of 2,4-D (1 mg L^{-1}), NAA (0.1 mg L^{-1}), BA (0.1 mg L^{-1}) or 2,4-D (0.5 mg L^{-1}), NAA (0.5 mg L^{-1}), BA (0.1 mg L^{-1}) has not triggered a further improvement in the induction of callus (Figure 1). However, a combination of 2,4-D (1.0 mg L^{-1}), NAA (0.1 mg L^{-1}), KN (0.1 mg L^{-1}), and 2,4-D (0.5 mg L^{-1}), NAA (0.5 mg L^{-1}), BA (0.1 mg L^{-1}) has involved in the induction of higher amount of callus from various explants of *S. thea* (Figure 1). On medium supplemented 2,4-D (1.0 mg L^{-1}), NAA (0.1 mg L^{-1}), KN (0.1 mg L^{-1}) 23.3%, 80.0%, 96.7%, and 26.7% callus was induced from seed, seed halves, cotyledon, and leaf explants, respectively (Figure 1), however, along with callus adventitious root regeneration was recorded on this medium. Whereas, on MS medium supplemented with 2,4-D (0.5 mg L^{-1}), NAA (0.5 mg L^{-1}), BA (0.1 mg L^{-1}) seed, seed halves, cotyledon, and leaf explants have developed 15.0%, 43.3%, 93.3%, and 36.7% callus respectively and there was no adventitious roots development with such callus (Figures 1 and 2). Of the four explants used (seed, seed halves, cotyledons, and leaf explants), auxins and cytokinins and explants vs. culture medium showed statistical significance (F value 13.83, degree of freedom 27, and *p*-value ≤ 0.05) (Figure 1). Seed-derived callus was friable and light yellow (Figure 2), whereas the callus induced from cotyledon, both compact and friable, subsequently turned creamy in color. Callus derived from leaf explants was brown in callus; however, it did not survive in subsequent subcultures (Figure 2).

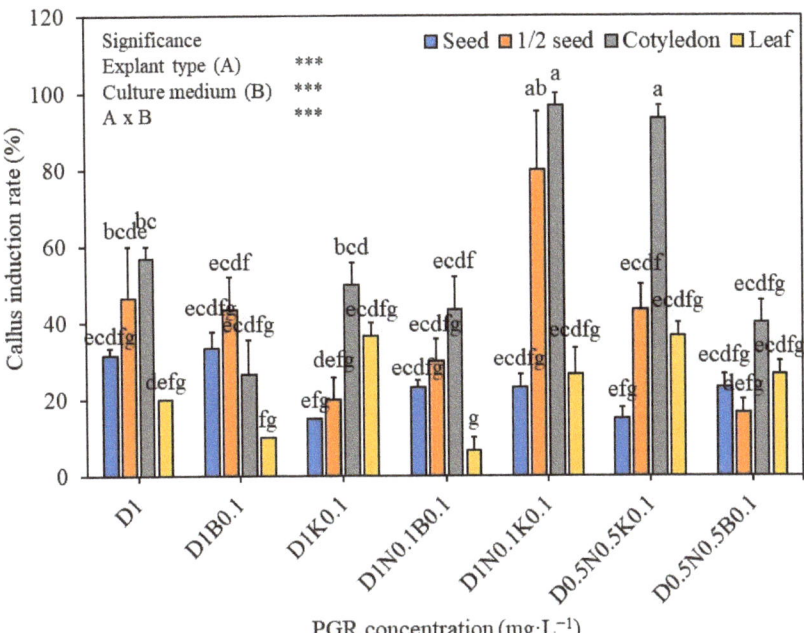

Figure 1. Effect of plant growth regulators on callus induction seed, half seed, cotyledon, and leaf explants of *S. thea* on MS medium containing 2,4-D, NAA, BA, and KN after 4 weeks of culture. Mean values with different alphabetical letters denote significant differences between the values according to Tukey's test at $p \leq 0.05$. *** Explants (A) and culture medium (B) effects are statistically significant at $p \leq 0.05$.

Figure 2. Callus induced from seed, half seed, cotyledon, and leaf explants of *S. thea* on MS medium containing 2,4-D, NAA, BA, and KN after 4 weeks of culture. Scale bars = 1 cm.

2.2. Selection of Friable Callus

Histological analysis of calli-derived from seeds, cotyledons, and leaves was of the following five different types: white, hard, soft, yellowish, and brown (Figure 3). Among these types, soft and yellowish callus possessed actively dividing cells with intercellular spaces and prominent nuclei. Whereas white, hard, and brown calli possess compactly arranged cells and had large vacuoles (Figure 3). In addition, this callus showed starch granules and thick cell walls and was not involved in cell division. Thus, the soft and yellowish callus bore rapidly dividing cells, which are composed of loosely arranged, undifferentiated cells.

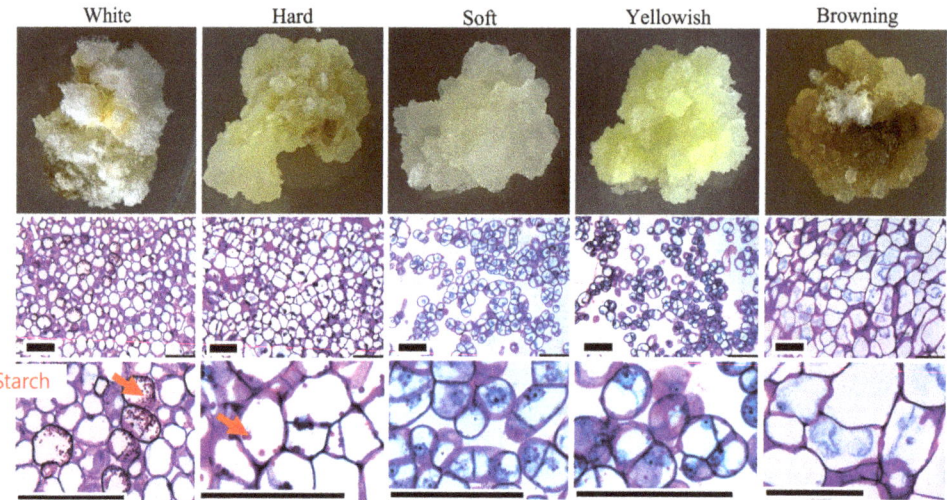

Figure 3. Morphological and histological features of callus regenerated from different explants of *S. thea*. Scale bar = 100 μm. Red arrows indicate starch grains.

2.3. Effect of Antioxidants on Callus Proliferation and Overcoming the Problems of Browning of the Medium

The callus, which was friable and actively growing, were sub-cultured on MS medium supplemented with 2,4-D (0.5 mg L^{-1}), NAA (0.5 mg L^{-1}), and KN (0.1 mg L^{-1}) for the proliferation of callus. However, the accumulation of phenolics and medium browning was a problem in subsequent cultures. Therefore, in the current study, callus of *S. thea* was sub-cultured on a medium containing 0, 100, 200 mg L^{-1} ascorbic acid (ASA) or 0, 10, 20 mg L^{-1} citric acid (CA) or 0, 10, 20 mg L^{-1} polyvinyl pyrrolidone (PVP) to overcome the problems of browning of callus and results are presented in Figure 4. The callus extract of brown callus showed an absorbance of 0.19 with a spectrophotometer at 420 nm. In contrast, the callus treated with 200 mg L^{-1} ASA did not involve browning, and such callus extract showed an optical absorbance of 0.01 at 420 nm (F value 3.83, degree of freedom 6, and p-value ≤ 0.05) (Figure 4A). Furthermore, the callus treated with 200 mg L^{-1} ASA was involved in rapid growth and attained a biomass of 1.04 g fresh weight per callus mass (F value 3.83, degree of freedom 6, and p-value ≤ 0.05) (Figure 4B). Whereas the other treatments, such as treatment with CA and PVP, were not beneficial, and they could not control the browning of callus completely (Figure 4C).

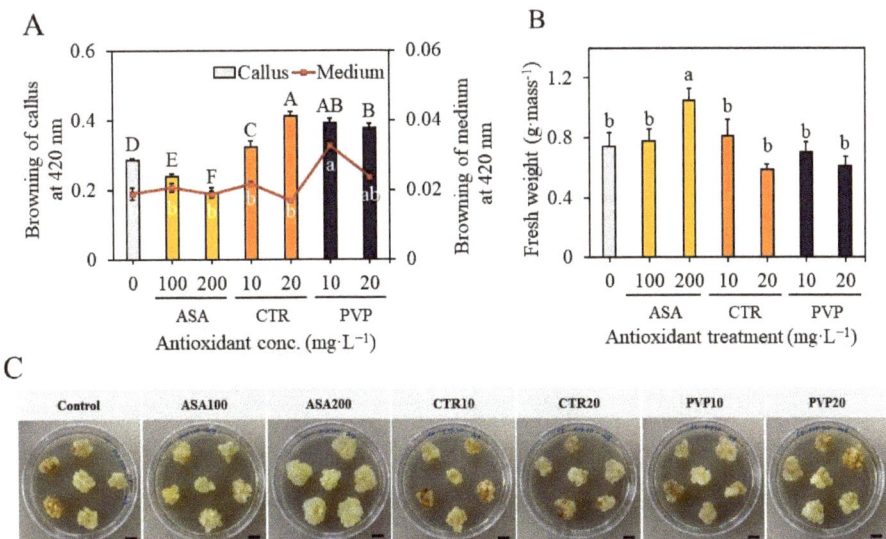

Figure 4. Morphological features of callus after treatment with 100, 200 mg L^{-1} ascorbic acid (ASA) or 10, 20 mg L^{-1} citric acid (CTR) or 10, 20 mg L^{-1} polyvinyl pyrrolidone (PVP). (**A**) Shows the absorbance of cell extract at 420 nm on a spectrophotometer, (**B**) shows the fresh weight of the callus, and (**C**) the morphology of the callus with different treatments. Different letters indicate mean values which are significantly different at $p \leq 0.05$ according to Tukey's test.

2.4. Establishment of Suspension Cultures in Erlenmeyer Flasks and Elicitation

Erlenmeyer's flasks were used for the initial cell suspension research. *S. thea* cell suspension cultures that were cultured using MS liquid medium supplemented with 2,4-D (0.5 mg L^{-1}), NAA (0.5 mg L^{-1}), and KN (0.1 mg L^{-1}) showed an initial lag phase for seven days (one week) and exponential phase from 7 to 21 days (up to three weeks) in culture, after that cell entered into stationary phase. After two weeks of culture initiation, elicitors including sodium nitroprusside (SNP; nitric oxide producer), methyl jasmonate (MeJA), and salicylic acid (SA) were added to cell suspension cultures of *S. thea* to test the concentration and effectiveness of the elicitor on the biosynthesis of phenolic compounds. The results of elicitor treatment depicted that the growth and accumulation of biomass were inhibited with the addition of elicitors SNP and MeJA (Figure 5A,B). MeJA specifically inhibited the accumulation of biomass with the increment in the concentration of MeJA (F value 3.83, degree of freedom 6, and p-value ≤ 0.05) (Figure 5A). The biomass accumulated in the control was 21.4 g after four weeks of culture, and it was reduced to 15.1 g, 13.5 g, and 12.5 g with the addition of 50, 100, and 200 µM, respectively. A similar decrease in biomass content was also noticed with SA treatments. Nevertheless, there was an increment accumulation of total phenolics and total flavonoids that was evident with MeJA treatments (Figure 6). Of the various elicitors tested, MeJA was efficient in triggering the accumulation of phenolic and flavonoid contents in the cultured cells of *S. thea* (Figure 6). The amount of phenolic content was 5.9 mg g^{-1} GAE with control cultures, whereas, in MeJA-treated cell cultures, the concentration of total phenolics was 37.5, 36.9, and 34.5 mg g^{-1} GAE (F value 3.83, degree of freedom 6, and p-value ≤ 0.05) (Figure 6). Similarly, the amount of total flavonoids was 1.6 mg g^{-1} CAT equivalents in the control, while the total flavonoid content was 18.4, 17.5, and 15.1 mg g^{-1} CAT equivalents with different concentrations of MeJA treatments (F value 3.83, degree of freedom 6, and p-value ≤ 0.05) (Figure 6). We carried out an antioxidant analysis of extracts of *S. thea* cell cultures that were treated with different elicitors, and the results are presented in Figure 7 (F value 3.83, degree of freedom 6, and

p-value ≤ 0.05). Analysis of DPPH radical scavenging activity (Figure 7A), ABTS radical scavenging activity (Figure 7B), and FRAP assay (Figure 7C) all demonstrated that extracts obtained from MeJA elicited cell cultures possessed the highest activities compared to cell extracts of SA and SNP-treated cultures and even control. The IC$_{50}$ values ABTS scavenging activity were in the range of 0.6–0.8 mg mL^{-1} with the MeJA treated extracts, whereas it was 9.3 mg mL^{-1} with the control cell extracts (Figure 7D). These results demonstrate that MeJA-treated cells are actively involved in secondary metabolism and involved in the enhanced accumulation of bioactive compounds.

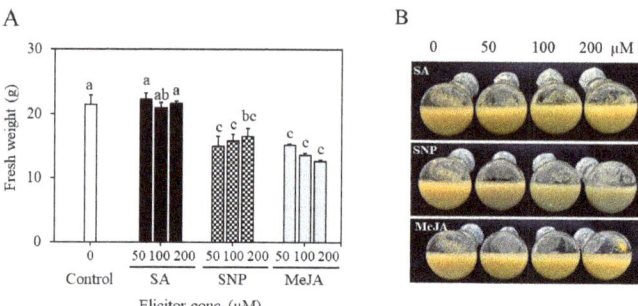

Figure 5. Effect of elicitor treatments on biomass accumulation in cell suspension cultures (**A**). Biomass growth with the increment of elicitor concentrations compared to control (**B**). SA—salicylic acid, SNP—sodium nitroprusside, MeJA—methyl jasmonate. Different letters indicate mean values which are significantly different at $p \leq 0.05$ according to Tukey's test.

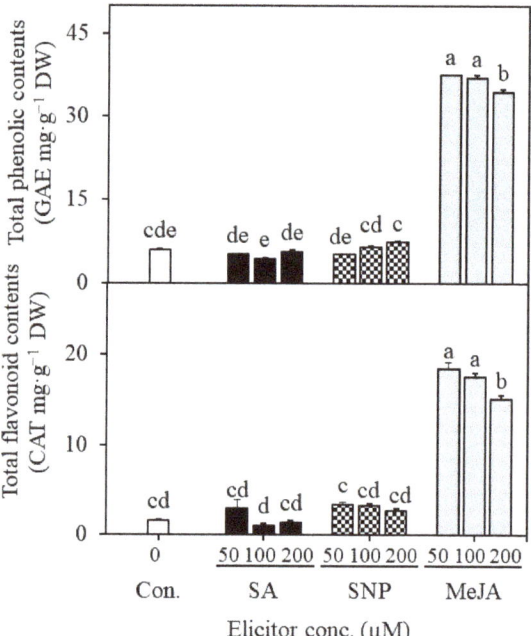

Figure 6. Effect of elicitor treatments on the accumulation of total phenolics and flavonoids in cell suspension cultures. SA—salicylic acid, SNP—sodium nitroprusside, MeJA—methyl jasmonate. Different letters indicate mean values which are significantly different at $p \leq 0.05$ according to Tukey's test.

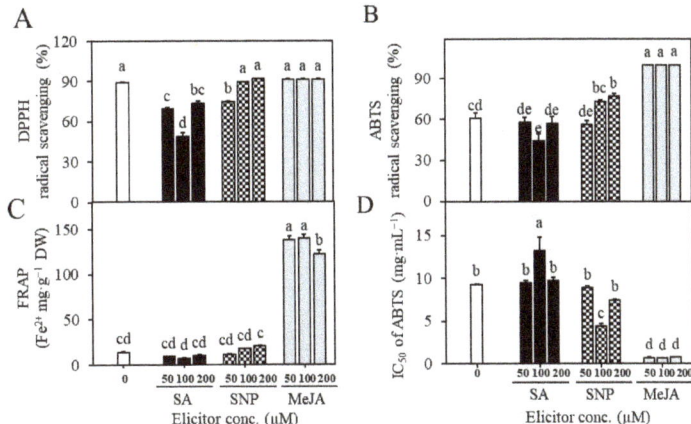

Figure 7. Effect of elicitor treatments on antioxidant activity in cell suspension cultures. SA—salicylic acid, SNP—sodium nitroprusside, MeJA—methyl jasmonate. Effect of elicitors on DPPH radical scavenging activity (**A**), ABTS radical scavenging activity (**B**), FRAP assay (**C**), and IC_{50} values of ABTS activity (**D**). Different letters indicate mean values which are significantly different at $p \leq 0.05$ according to Tukey's test.

2.5. Establishment of Bioreactor Cultures for the Production of Biomass and Phenolics

Ten g L^{-1} cells were cultured in bioreactors, cultures were aerated with 0.1 vvm, and treated with 200 µM MeJA after three weeks of culture and maintained for another week. The bioreactor cultures showed a typical lag phase for one week and an exponential phase for up to four weeks. Accumulation of 230.81 g of fresh biomass and 16.48 g of dry biomass was evident after four weeks of culture (F value 714.62, degree of freedom 4, and p-value ≤ 0.05) (Figures 8 and 9). The cell growth index was 8.2. These results depict that bioreactor cultures are efficient in the accumulation of biomass.

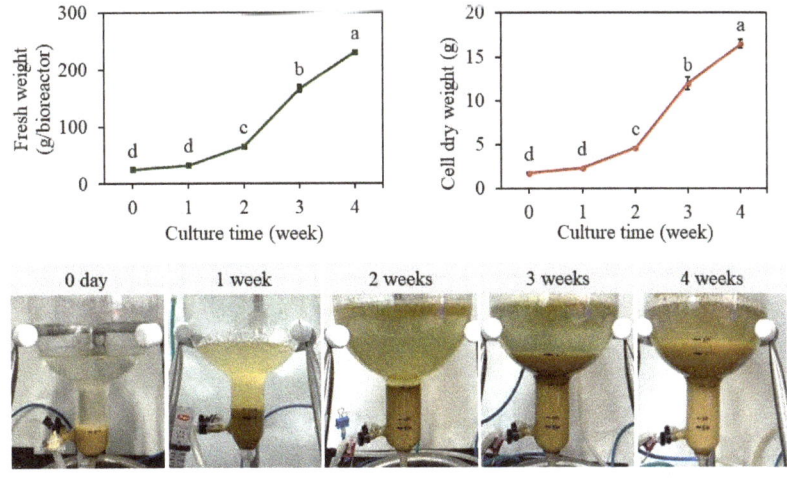

Figure 8. Cell suspension cultures in balloon-type bubble bioreactors: accumulation of fresh biomass, dry biomass over four weeks of culture Different letters indicate mean values which are significantly different at $p \leq 0.05$ according to Tukey's test.

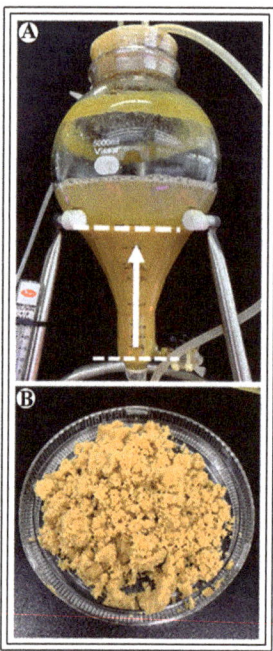

Figure 9. Cell suspension cultures in balloon-type bubble bioreactor. (**A**) Biomass accumulated after 4 weeks of culture, (**B**) Harvested cell biomass.

The findings of an HPLC examination of the phenolics present in the biomass are shown in Figures 10 and 11. Seven major phenolics were recognized in the cell extracts when compared to standards (Figure 10). The major phenolics including catechin hydrate (382.0 µg g^{-1} DW), chlorogenic acid (108.1 µg g^{-1} DW), naringin (82.9 µg g^{-1} DW), apigenin (15.5 µg g^{-1} DW), ruteolin (14.7 µg g^{-1} DW), gallic acid (13.2 µg g^{-1} DW) and ferulic acid (1.5 µg g^{-1} DW) were recorded through HPLC analysis (F value 714.62, degree of freedom 4, and p-value ≤ 0.05) (Figure 11). Comparison of cultures without elicitation and with elicitation with MeJA revealed that 28.7, 24, and 4.8-fold increments in accumulation of catechin hydrate, naringin, and chlorogenic acid are evident (Figure 11).

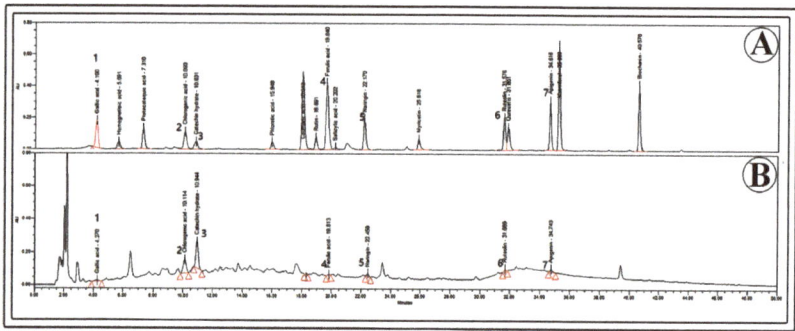

Figure 10. HPLC chromatograms of phenolic compounds. (**A**) Standards, (**B**) gallic acid (1), chlorogenic acid (2), catechin hydrate (3), ferulic acid (4), naringin (5), luteolin (6), and apigenin (7).

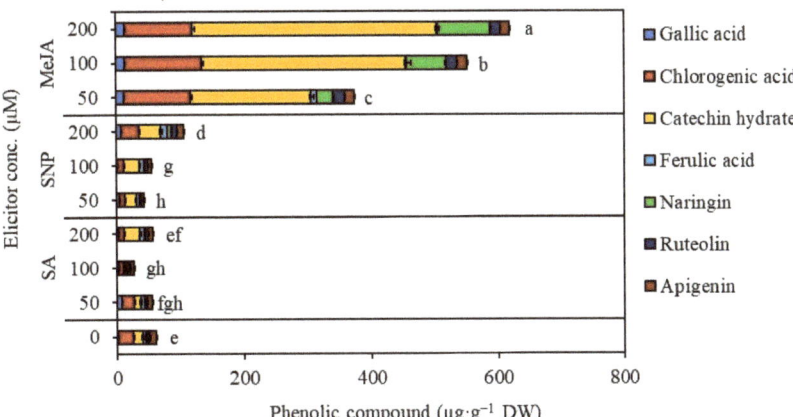

Figure 11. Quantity of phenolic compounds accumulated in cell suspension cultures with elicitor treatments. SA—salicylic acid, SNP—sodium nitroprusside, MeJA—methyl jasmonate. Different letters indicate mean values which are significantly different at $p \leq 0.05$ according to Tukey's test.

3. Discussion

Plant secondary metabolites such as phenolics and flavonoids have very high therapeutic values such as antioxidant, anti-inflammatory, anticancer, antihypertensive, antihyperglycemic, neuroprotective, and hypolipidemic activities [24]. Production of these compounds in natural plants depends upon various factors such as species, populations, and edaphic and environmental conditions where they grow. Variability in the accumulation of bioactive compounds is also evident depending on the phenological stages of development [25]. In contrast production of bioactive compounds through in vitro culture of plant cells and organs has emerged as an attractive alternative. Growth and accumulation of biomass and production of metabolites can be controlled very easily with in vitro conditions. Additionally, in vitro cultivation of plant cells allows for the manipulation of growth variables, and with additions of elicitors and precursors, hyperproduction of valuable bioactive active compounds could be achieved [14,15]. For the production of phenolic compounds; therefore, we established cell suspension cultures of *S. thea* in the current investigation.

Plant growth regulators play an important role in triggering the growth and development of explants cultured in vitro [26]. Specific explant requires a particular auxin or cytokinin or combination of auxin and cytokinin for callus induction, and it also depends on the levels of endogenous hormones present with the explants. Auxins are involved in cell division, tissue differentiation, embryogenesis, and rhizogenesis [27]. In the present study, we tested the effect of 2,4-D (0.5, 1 mg L^{-1}), NAA (0.5, 1.0 mg L^{-1}), BA (1.0 mg L^{-1}), and KN (1 mg L^{-1}) individually or in combination. A combination of 2,4-D (0.5 mg L^{-1}) with NAA (0.1 or 0.5 mg L^{-1}) and KN (0.1 mg L^{-1}) has triggered the highest amount of callus induction in seed, cotyledon, and leaf explants of *S. thea*. Similar to the present results addition of NAA along with other growth regulators has induced the highest amount of callus in *Cananga odorata* and *Scropholaria stiata* [28,29]. The selection of suitable explants is also critical in the induction of a greater amount of callus [30]. In the current study highest callus induction was recorded from seed (80%) and cotyledon (96.7%) explants of *S. thea* on the combination of medium containing 2,4-D (1.0 mg L^{-1}), NAA (0.1 mg L^{-1}) and KN (0.1 mg L^{-1}), however, the formation of roots alongside callus was noticed on both explants after four weeks of culture. Similarly, in *Hylocerus costaricensis* NAA induced rhizogenesis along with callus on NAA supplemented medium [31]. However, on the MS medium containing 2,4-D (0.5 mg L^{-1}), NAA (0.5 mg L^{-1}), and KN (0.1 mg L^{-1}), seed and cotyledon explants have produced 43.3% and 93.3% callus formation without the intervention of rhizogenesis. Variations among explant types for callus induction have also been reported in other plant species, such as *Calendula officinalis* and *Chorisia spciosa* [32,33].

The friability of callus tissue is highly desirable for establishing cell suspension culture [34]. Histological analysis is very useful to verify the process friability of the callus. The soft and yellowish callus, among the numerous forms of calli found in the current studies, had cells that were actively proliferating and had intercellular gaps and conspicuous nuclei. While dark, hard, and white calli have big vacuoles and compactly organized cells. This callus also exhibited thick cell walls, starch granules, and a lack of cell division. Similar to the present results, Betekhtin et al. [35] observed characteristic cells that are compact, thick-walled, and accumulation of insoluble starch grains in the non-friable callus. As a result, the loosely packed, undifferentiated cells that make up the rapidly dividing cells in the soft, yellowish callus were selected for further sub-culturing and callus proliferation.

Oxidative browning is a common problem in plant tissue cultures, which results in reduced growth of the cultured explants or callus [36,37]. The underlying phenomenon is the accumulation and subsequent oxidation of phenolic compounds in the tissues. Several methods have been followed to overcome the initial browning of the medium, including the use of antioxidants such as ascorbic acid and citric acid or the use of absorbents such as polyvinyl pyrrolidone and activated charcoal [37,38]. With our results, we realized callus browning during subsequent sub-cultures. To overcome the browning of the callus, we used 100, 200 mg L^{-1} ascorbic acid (ASA) or 0, 10, 20 mg L^{-1} citric acid (CA) or 0, 10, 20 mg L^{-1} polyvinyl pyrrolidone (PVP). Of the varied treatments, cultures treated with 200 mg L^{-1} ASA could be able to control the browning of the callus in subsequent subcultures. Similar to the current results, ASA has been efficiently used in overcoming the browning of callus in tissue cultures of *Themeda qundrivalis* [39], *Musa* species [40], and *Glycyrrhiza glabra* [41].

Elicitation is one of the most effective and widely employed biotechnological tools for enhanced biosynthesis and accumulation of secondary metabolite in plant tissue culture [42]. Elicitors of biotic and abiotic origin can trigger the biosynthesis of specific metabolites in cell and organ cultures [14,43]. The optimization of various parameters, such as elicitor type, concentration, duration of exposure, and treatment schedule is essential for the effectiveness of the elicitation strategies. Salicylic acid (SA), methyl jasmonate (MeJA), jasmonic acid (MJ), and other signaling molecules are frequently utilized as elicitors to stimulate secondary metabolite biosynthesis and enhancement [44–47]. In our studies, we established cell suspension cultures and tested the effect of MeJA, SA, and sodium nitroprusside (SNP; nitric oxide producer) at 50, 100, and 200 µM concentrations. MeJA was effective in causing the accumulation of phenolic and flavonoid contents in the cultured cells of *S. thea* when compared to other elicitors tested. In total, 5.9 mg g^{-1} GAE of phenols were present in control cultures, but 37.5, 36.9, and 34.5 mg g^{-1} GAE were present in cell cultures that had received MeJA treatment. Similarly, the total flavonoid content was 18.4, 17.5, and 15.1 mg g^{-1} CAT equivalents with various doses of MeJA treatments, compared to 1.6 mg g^{-1} CAT equivalents in the control. Similar to the present findings, MeJA has successfully been employed as an elicitor for the enhanced accumulation of ginsenosides in *Panax ginseng* cell and adventitious root cultures [44,45] as well as the generation of phenolic compounds in *Thevetia peruviana* cell cultures [48]. SA and SNP were not efficient in stimulating the accumulation of secondary metabolites with cell cultures of *S. thea*. However, SA and nitric oxide have been used as potent elicitors in callus/cell cultures of *Hypericum perforatum* [49] and *Catharanthus roseus* [50] for the production of hypericin and catharanthine, respectively.

The production of biomass and secondary metabolites from cell and organ cultures is excellent in bioreactor cultures, and various factors, including medium pH, aeration, and gases, can be controlled in a way that is growth-specific. Cultured cells will effectively utilize nutrients and participate in the creation and development of metabolites [14,15]. Additionally, bioreactor cultures enhance production quality, lower production costs, and allow for a process to scale up [51]. The results of the current study of bioreactor cultures showed a higher accumulation of biomass (230.81 g of fresh biomass and 16.48 g of dry biomass) as well as metabolites. The generation of biomass and secondary metabolites in a variety of plant species, such as *Panax ginseng* [17], *Echinacea* species [19], and *Hypericum perforatum* [21], has also been investigated using balloon-type bubble bioreactors. Comparing cultures with and without elicitation with MeJA, it was discovered that there was a 28.7, 24, and 4.8-fold increase in the accumulation of catechin hydrate, naringin, and chlorogenic acid. Due to its antibacterial, anti-inflammatory, antidiabetic, and anticancer activities, the bioactive molecule catechin has a very high therapeutic value. It can also effectively neutralize free radicals and has significant antioxidant properties [52]. While chlorogenic acid has a variety of physiological benefits, including hepatoprotective, gastrointestinal, renoprotective, neurological, and cardiovascular properties [53]. Naringenin, a flavonoid molecule, has been shown to have anti-inflammatory, antioxidant, antibacterial, antiadipogenic, and cardioprotective properties [54]. The bioactive compounds found in abundance in the *S. thea* cell biomass produced in bioreactors allow for the extraction of these fine chemicals as well as the use of the biomass in the production of nutraceuticals and cosmetics.

4. Materials and Methods

4.1. Plant Material and Seed Germination

Sageretia thea (Osbeck) M.C. Johnst. plants were collected from Dongbaekdongsan, which is located in Seonheul-ri, Jocheon-eub, Bukjeju-gun, Jeju-do, Republic of Korea, and maintained in the experimental garden at Chungbuk National University, Republic of Korea. Fruits were collected from the plants grown in the experimental garden and washed thoroughly in running tap water, and surface sterilized in 70% ethanol for 5 s, then in 20% sodium hypochlorite solution with one drop of Tween-20 for 30 min. Fruits were washed three times in sterile water under the laminar flow cabinet, and then the seeds were separated from the fruits. Seeds were washed in sterile distilled water three times and cultured on Murashige and Skoog (MS) [55] medium supplemented with 1.0 mg L^{-1} gibberellic acid, 0.1 mg L^{-1} kinetin, and 30 g L^{-1} sucrose for germination.

4.2. Chemicals and Reagents

Methyl jasmonate (MeJA), salicylic acid (SA), sodium nitroprusside (SNP), gallic acid (GA), catechin, Folin-Ciocalteu (FC) reagent, 2,2 Diphenyl 1 picrylhydrazyl (DPPH), 2,2'-azino-bis (3-ethybenzothiazoline-6-sulphonic acid (ABTS), 2,4,6-tripyridyl-s-triazine (TPTZ), ascorbic acid (ASA), citric acid (CA), polyvinylpyrrolidone (PVP) were procured from Sigma-Aldrich chemicals (St. Louis, MO, USA). All the tissue culture chemicals, growth regulators such as 2,4-dichlorophenoxyacetic acid (2,4-D), naphthalene acetic acid (NAA), 6-benzyladenine (BA), kinetin (KN), gibberellic acid (GA$_3$), gelrite, were obtained from Ducefa Biochemie, Haarlem, The Netherlands. High-pressure liquid chromatography (HPLC) standard phenolic compounds were procured from ChromaDex, Longmont, CO, USA.

4.3. Callus Induction

Entire seeds, seeds divided into two halves, cotyledons, and leaves obtained from young plantlets were used for callus induction. Explants were cultured on MS medium containing 30 g L^{-1} sucrose, 2,4-D (0.5, 1 mg L^{-1}), NAA (0.5, 1.0 mg L^{-1}), BA (1.0 mg L^{-1}), KN (1 mg L^{-1}) individually or in combination. Medium pH was adjusted to 5.8 and then added with 2.4 g L^{-1} gelrite and autoclaved at 121 °C and 121 Kilopascals for 15 min. The cultures were maintained in dark at 25 ± 1 °C for 4 weeks in culture rooms.

4.4. Histological Analysis of Callus

For the histological analysis of callus, each type of callus (hard, soft, callus of different colors such as white, yellowish, brown, irrespective of their origin from seeds, cotyledons, and leaves) 0.5 mm^3 callus was fixed in formalin, glacial acetic acid, and 95% ethyl alcohol and water (10: 5: 50: 35) solution for two days. Then callus was degassed and dehydrated with an alcohol series. After infiltration with Technovit 7100 (Kulzer Technik, Wertheim, Germany), a mold was made by polymerizing with an embedding solution. Microtome (Leica, Nussloch, Germany) sections (5 µm) were taken and attached to glass slides and stained with periodic acid Schiff reagent and toluidine blue O. The preparations were observed using an optical microscope (Leica, Germany).

4.5. Raising of Callus Cultures with Supplementation of Antioxidants

The callus developed from the different explants invariably use to accumulate phenolic accumulation that hindered the callus growth. To overcome initial phenolic accumulation and to facilitate the growth of the callus, callus cultures sub-cultured MS medium with growth regulators 0.5 mg L^{-1} 2,4-D, 0.5 mg L^{-1} NAA, and 0.1 mg L^{-1} KN and various antioxidants such as 100, 200 mg L^{-1} ASA or 10, 20 mg L^{-1} citric acid or 10, 20 mg L^{-1} PVP. After 4 weeks of culture, callus growth and browning of cultures with phenolics were investigated.

4.6. Establishment of Cell Suspension Culture in Erlenmeyer's Flasks and Elicitation of Cultures

S. thea cell suspensions were established in 100 mL Erlenmeyer's flasks with 50 mL of MS liquid medium, 30 g L^{-1} of sucrose, 0.5 mg L^{-1} of 2,4-D, 0.5 mg L^{-1} of NAA, and 0.1 mg L^{-1} of KN. In total, 3 g of friable callus was added to 50 mL of liquid medium, and the cultures were maintained on an orbital shaker at 120 rpm for four weeks. The cultures were added with elicitors such as salicylic acid (SA), methyl jasmonate (MeJA), or sodium nitroprusside (SNP) at a concentration of 50, 100, and 200 µM after two weeks of culture and maintained for another two weeks.

4.7. Establishment of Cell Suspension Cultures in Balloon-Type Bubble Bioreactors

For the production of S. thea biomass, cell suspension cultures were established in 5 L balloon-type bubble bioreactors (Samsung Biotech, Seoul, Republic of Korea) containing 2 L of MS liquid medium containing 30 g L^{-1} sucrose and 0.5 mg L^{-1} 2,4-D, 0.5 mg L^{-1} NAA and 0.1 mg L^{-1} KN. In total, 20 g of cells were inoculated to the medium. After three weeks of culture initiation, 100 µM MeJA was added to elicit the cells to involve in secondary metabolism and maintained for another week. The bioreactor cultures were maintained in a culture room, and the dark were aerated with sterile air at 0.1 vvm (air volume/medium volume/min). After four weeks of culture, cells were harvested by bypassing the medium through a stainless-steel sieve, and cells were washed thoroughly with sterile distilled water. The fresh weight (FW) was determined after air drying cells, dry weight of the cell was determined by drying the cell biomass in a freeze-dryer at −80 °C for three days. The growth index (Gi) was calculated on the basis of weight of dry tissue according to the following formula: Gi = (DW1 − DW0)/DW0: where DW0 was the weight of inoculum and DW1 was the final weight of tissue after a culture growth period. The total phenolic content (TPC), total flavonoid content (TFC), and antioxidant analysis were all calculated using the cell biomass.

4.8. Preparation of Cell Extracts

The freeze-dried cells (0.1 g) were taken in 10 mL of ethanol and subjected to ultrasonic waves at 36 °C for 1 h. The ethanol extract was filtered through Whatman (grade 2) filter paper, and the filtrate was used for the analysis of phenolics, flavonoids, and antioxidant activities.

4.9. Estimation of Total Phenolic Content

Total phenolic content (TPC) was estimated by using the Folin Ciocalteu reagent method, as described by Murthy et al. [56], with some modifications. Briefly, a known amount of sample was taken and made up to 3 mL with distilled water, and 0.1 mL of 2 N Folin Ciocalteu reagent was added, followed by incubation for 6 min, and then 0.5 mL of 20% Na_2CO_3 was added to each tube. Tubes were kept in warm water for 30 min, and the absorbance was read at 760 nm using a UV-Visible spectrophotometer. Gallic acid was used as the standard compound.

4.10. Estimation of Total Flavonoid Content

The flavonoid content of extracts was analyzed as described by Pekal and Pyrzynskaet [57]. To brief, 0.1 mL of extract was taken and made up the volume to 3 mL by using distilled water, followed by the addition of 0.15 mL of 10% $AlCl_3$ and 2 mL of 1 M NaOH after 5 min of incubation at room temperature. Solutions were vortexed, and absorbance was measured at 510 nm. Catechin was used as standard.

4.11. Analysis of Antioxidant Activities

4.11.1. 2,2 Diphenyl 1 picrylhydrazyl (DPPH) Radical Scavenging Assay

Extract (0.1 mL) was added with 1.9 mL of 0.1 mM DPPH solution prepared in ethanol. The tubes were vortexed and incubated in the dark for 15 min. The discoloration of the DPPH solution was measured at 517 nm against ethanol as blank using a UV-visible spectrophotometer. Gallic acid was used as standard, and the activity of the extracts was expressed as mg gallic acid equivalent (GAE)/g extract [58].

4.11.2. 2,2′-Azino-bis (3-ethybenzothiazoline-6-sulphonic Acid (ABTS) Assay

The ABTS assay was carried out as per the method of Re [59]. The ABTS solution was prepared by mixing 7 mM of ABTS and 2.45 mM potassium persulfate in a ratio of 1:1 and stored in the dark for 24 h. At the time of analysis, the ABTS solution was diluted with phosphate buffer (pH 7.3) to obtain the value of 0.70 at 732 nm. Fifty microliters of the extract were added to 950 microliters of diluted ABTS solution, and the mixture was allowed to stay in the dark for 10 min then absorbance was measured at 732 nm using UV-visible spectrophotometry. Antioxidant activity was expressed in percentage, i.e., ABTS radical scavenging activity = absorbance of control solution-absorbance of sample solution/absorbance of control solution × 100.

4.11.3. Ferric Reducing Antioxidant Power (FRAP) Assay

FRAP assay was carried out according to the method described by Benzie and Strain [60]. FRAP reagent was prepared by mixing 300 mM acetate buffer of pH 3.6, 10 mM 2,4,6-tripyridyl-s-triazine (TPTZ) in 40 mM HCl and 20 mM $FeCl_2.6H_2O$ in the ratio 10:1:1. In total, 0.2 mL of extract was added with 3 mL of FRAP reagent, tubes were vortexed and incubated for 6 min at room temperature, and absorbance was measured at 593 nm using a UV-visible spectrophotometer. Ascorbic acid was used as standard, and activity is expressed as mg ascorbic acid equivalent (AAE)/g extract.

4.12. Quantification of Phenolic Compounds Using High-Performance Liquid Chromatography (HPLC)

The cell biomass obtained from the cultures was ground in a sterilized mortar. The powdered sample (0.1 g) was mixed with 10 mL of 80% ethanol, and the extract was obtained by ultrasonication, as explained above. The extract was concentrated using nitrogen gas and dissolved in 0.5 mL 80% ethanol and used for analysis. The extract was filtered through a membrane filter (0.45 µm) and used for analysis. HPLC equipment (2690 Separation Module, Waters Chromatography, Milford, CT, USA) included a photodiode array detector (PDA), and compound separation was performed using a Fortis C18 column (5 µm, 150 × 4.6 mm). The mobile phase consisted of acetic acid and water (1:99 v/v) (solvent A) and acetic acid and acetonitrile (1:99 v/v) (solvent B) and was filtered using Whatman Glass microfiber filters before use. The flow rate was 1.0 mL.min-1, and the column temperature was 25 °C. The peaks were detected at 280 nm, and compounds were identified and quantified based on the retention time of standards and peak areas.

4.13. Statistical Analysis

The results are presented as mean values and standard errors. One-way analysis of variance (ANOVA) was used to determine whether the groups differed significantly. Statistical assessments of the difference between mean values were then assessed using Tukey's test. A value of $p = 0.05$ was considered to indicate statistically significant differences. All the data were analyzed using a SAS program (Software Version 9.4; SAS Institute, Cary, NC, USA).

5. Conclusions

Plant cell culture techniques can be used as alternatives for the production of biomass that are abundant in bioactive materials. In the present study, we established stepwise protocols for the induction of callus and establishment of suspension cultures for the production of phenolic compounds in *S. thea*. Methyl jasmonate elicited cell cultures of *S. thea*, has the highest content of total phenolics and flavonoids, and also demonstrated increased antioxidant activity. The cell biomass was also rich in specific bioactive compounds, including catechin hydrate, chlorogenic acid, naringenin, and others.

Author Contributions: S.-Y.P. planned, executed, and supervised the research work. J.-H.K. conducted the experiments. J.-E.H., J.-Y.K., M.-J.K. and T.-K.J. helped with the phytochemical analysis. H.N.M. has interpreted the data and written the manuscript. All authors have read and agreed to the published version of the manuscript.

Funding: This work was supported by a National Research Foundation of Korea (NRF) grant funded by the Korean government (MSIT) (No. NRF-2020R1A2C2102401) and Technology Innovation Program (Grant number P0018148) funded by the Ministry of Trade, Industry & Energy (MOTIE, Korea).

Data Availability Statement: The datasets used and/or analyzed during this study are available from the corresponding author upon reasonable request.

Acknowledgments: Hosakatte Niranjana Murthy is thankful for the "Brain Pool" (BP) program, Grant No. 415 2022H1D3A2A02056665.

Conflicts of Interest: The authors declare no competing interests.

References

1. POWO (Plants of the World Online). Facilitated by the Royal Botanic Gardens, Kew. 2023. Available online: http://www.plantsoftheworldonline.org/ (accessed on 10 January 2023).
2. Hyun, T.K.; Sang, S.C.; Sang, C.K.; Kim, J.S. Nutritional and nutraceutical characteristics of *Sageretia theezans* fruit. *J. Food Drug Anal.* **2015**, *23*, 742–749. [CrossRef]
3. Chung, S.K.; Chen, C.Y.O.; Blumberg, J.B. Flavonoid-rich fraction from *Sageretia theezans* leaves scavenges reactive oxygen radical species and increases the resistance of low-density lipoprotein in oxidation. *J. Med. Food* **2009**, *12*, 1310–1315. [CrossRef] [PubMed]
4. Song, S.C.; Song, C.K.; Kim, J.S. Vegetation and habitat environment of *Sageretia thea* in Jeju island. *Korean J. Med. Crop Sci.* **2014**, *22*, 301–305. [CrossRef]
5. Song, S.C.; Song, C.K.; Kim, J.S. Characteristics of seed germination and fruit for *Segeretia thea* in Jeju region. *Korean J. Med. Crop Sci.* **2015**, *23*, 8–12. [CrossRef]
6. Xu, L.; Yang, X.; Li, B. Chemical constituents of *Sageretia theezans* Brongn. *Zhongguo Zhong Yao Za Zhi* **1994**, *19*, 675–676.
7. Chung, S.K.; Kim, Y.C.; Takay, Y.; Terahima, K.; Niwa, M. Novel flavanol glycoside, 7,O-methyl mearnsitrin, from *Sageretia theezans* and its antioxidant effect. *J. Agric. Food Chem.* **2004**, *52*, 4664–4668. [CrossRef]
8. Ko, G.A.; Shrestha, S.; Cho, S.K. *Sageretia thea* fruit extracts rich in methyl linoleate and methyl linolenate downregulate melanogenesis via the Akt/GSK3 β signalling pathway. *Nutr. Res. Pract.* **2018**, *12*, 3–12. [CrossRef] [PubMed]
9. Ko, G.A.; Son, M.; Cho, S.K. Comparative evaluation of free radical scavenging activities and cytotoxicity of various solvent fractions of Sandoong *Sageretia thea* (Osbeck) M.C. Johnst. Branches. *Food Sci. Bioethanol.* **2016**, *25*, 1683–1691. [CrossRef] [PubMed]
10. Kim, H.N.; Park, H.G.; Park, S.B.; Kim, J.D.; Eo, H.J.; Song, J.H.; Jeong, J.B. Extracts from *Sageretia thea* reduce cell viability inducing cyclin D1 proteasomal degradation and HO-1 expression in human colorectal cancer cells. *BMC Complemen. Altern. Med.* **2019**, *19*, 43. [CrossRef]
11. Eibl, R.; Meier, P.; Stutz, I.; Schildberger, D.; Huhn, T.; Eibl, D. Plant cell culture technology in the cosmetics and food industries: Current state and future trends. *Appl. Microbiol. Biotechnol.* **2018**, *102*, 8661–8675. [CrossRef]
12. Krasteva, G.; Georgiev, V.; Pavlov, A. Recent applications of plant cell culture technology in cosmetics and foods. *Eng. Life Sci.* **2021**, *21*, 68–76. [CrossRef]
13. Barbulova, A.; Apone, F.; Colucci, G. Plant cell cultures as source of cosmetic active ingredients. *Cosmetics* **2014**, *1*, 94–104. [CrossRef]
14. Murthy, H.N.; Lee, E.J.; Paek, K.Y. Production of secondary metabolites from cell and organ cultures: Strategies and approaches for biomass improvement and metabolite accumulation. *Plant Cell Tissue Organ Cult.* **2014**, *118*, 1–16. [CrossRef]

15. Murthy, H.N.; Dalawai, D.; Bhat, M.A.; Dandin, V.S.; Paek, K.Y.; Park, S.Y. Biotechnological Production of Useful Phytochemicals from Adventitious Root Cultures. In *Plant Cell and Tissue Differentiation and Secondary Metabolites: Fundamentals and Applications*; Ramawat, K.G., Ekiert, H.M., Goyal, S., Eds.; Springer Nature: Geneva, Switzerland, 2021; pp. 469–486.
16. Murthy, H.N.; Joseph, K.S.; Paek, K.Y.; Park, S.Y. Anthraquinone production from cell and organ cultures of *Rubia* species: An overview. *Metabolites* **2023**, *13*, 39. [CrossRef]
17. Thanh, N.T.; Murthy, H.N.; Paek, K.Y. Optimization of ginseng cell culture in airlift bioreactors and developing the large-scale production system. *Ind. Crops Prod.* **2014**, *60*, 343–348. [CrossRef]
18. Murthy, H.N.; Georgiev, M.I.; Kim, Y.S.; Jeong, C.C.; Kim, S.J.; Park, S.Y.; Paek, K.Y. Ginsenosides: Perspective for sustainable biotechnological production. *Appl. Microbiol. Biotechnol.* **2014**, *98*, 6243–6295. [CrossRef]
19. Murthy, H.N.; Kim, Y.S.; Park, S.Y.; Paek, K.Y. Biotechnological production of caffeic acid derivatives from cell and organ cultures of *Echinacea* species. *Appl. Microbiol. Biotechnol.* **2014**, *98*, 7707–7717. [CrossRef] [PubMed]
20. Murthy, H.N.; Kim, Y.S.; Georgiev, M.I.; Paek, K.Y. Biotechnological production of eleutherosides: Current state and perspectives. *Appl. Microbiol. Biotechnol.* **2014**, *98*, 7319–7329. [CrossRef] [PubMed]
21. Murthy, H.N.; Kim, Y.S.; Park, S.Y.; Paek, K.Y. Hypericins: Biotechnological production from cell and organ cultures. *Appl. Microbiol. Biotechnol.* **2014**, *98*, 9187–9198. [CrossRef]
22. Cui, H.Y.; Murthy, H.N.; Moh, S.H.; Cui, Y.; Lee, E.J.; Paek, K.Y. Protocrom culture of *Dendrobium candidum* in balloon type bubble bioreactors. *Biochem. Eng. J.* **2014**, *88*, 26–29. [CrossRef]
23. Cui, H.Y.; Murthy, H.N.; Moh, S.H.; Cui, Y.Y.; Lee, E.J.; Paek, K.Y. Production of biomass and bioactive compounds in protocorm cultures of *Dendrobium candidum* Wall ex Lindl. using balloon type bubble bioreactors. *Ind. Crops Prod.* **2014**, *53*, 28–33. [CrossRef]
24. Thungmunnithum, D.; Thongboonyou, A.; Pholboon, P.; Yangsabi, A. Flavonoids and other phenolic compounds from medicinal plants for pharmaceutical and medical aspects: An overview. *Medicines* **2018**, *5*, 93. [CrossRef] [PubMed]
25. Ncube, B.; Finnie, J.F.; Van Staden, J. Quality from the field: The impact of environmental factors as quality determinants in medicinal plants. *S. Afr. J. Bot.* **2012**, *82*, 11–20. [CrossRef]
26. Sehgal, H.; Joshi, M. The journey and new breakthroughs of plant growth regulators in tissue cultures. In *Advances in Plant Tissue Culture, Current Development and Future Trends*; Academic Press: Cambridge, MA, USA, 2022; pp. 85–108. [CrossRef]
27. Machakova, I.; Zazimolova, E.; George, E.F. Plant growth regulators: Introduction; auxins, their analogues and inhibitors. In *Plant Propagation by Tissue Culture*, 3rd ed.; George, E.F., Hall, M.A., De Klerk, G.J., Eds.; Springer: Dordrecht, The Netherlands, 2008; pp. 175–204. [CrossRef]
28. Nurazah, Z.; Radzali, M.; Syahida, A.; Maziah, M. Effects of plant growth regulators on callus induction from *Cananga odorata* flower petal explant. *Afr. J. Biotechnol.* **2009**, *8*, 2740–2743.
29. Khanpour-Ardesteni, N.; Sharifi, M.; Behmanesh, M. Establishment of callus and cell suspension culture of *Scrophularia striata* Boiss.: An in vitro approach for acetoside production. *Cytobiotechnology* **2015**, *67*, 475–485. [CrossRef] [PubMed]
30. Cheng, H.; Yu, L.J.; Hu, Q.Y.; Chen, S.C.; Sun, Y.P. Establishment of callus and cell suspension cultures of *Corydalis saxicola* Bunting, a rare medicinal plant. *Z. Naturforsch.* **2006**, *61*, 251–256. [CrossRef]
31. Winson, K.W.S.; Chew, B.L.; Sathasivam, K.; Subramaniam, S. The establishment of callus and cell suspension cultures of *Hylocerus costaricensis* for the production of betalain pigments with antioxidant potential. *Ind. Crops Prod.* **2020**, *155*, 112750. [CrossRef]
32. Legha, M.R.; Prasad, K.V.; Singh, S.K.; Kaur, C.; Arora, A.; Kumar, S. Induction of carotenoid pigments in callus cultures of *Calendula officinals* L. in response to nitrogen and sucrose levels. *In Vitro Cell. Dev. Biol. Plant* **2012**, *28*, 39–45. [CrossRef]
33. Fahim, J.R.; Hegazi, G.A.E.M.; El-Fadi, R.E.S.A.; Alhady, M.R.A.A.A.; Desoukey, S.Y.; Rmadan, M.A.; Kamel, M.S. Production of rhoifolin and tiliroside from callus cultures of *Chorisia chodatii* and *Chorisia spciosa*. *Phytochem. Lett.* **2015**, *13*, 218–227. [CrossRef]
34. Akaneme, F.I.; Eneobong, E.E. Tissue culture of *Pinus carbaea* Mor. var. *hondurensis* barr. And golf. II. Effects of two auxins and two cytokinins on callus growth habits and subsequent organogenesis. *Afr. J. Biotechnol.* **2008**, *7*, 757–765.
35. Betekhtin, A.; Rojek, M.; Jaskowiak, J.; Milewaska-Handel, A.; Kwasniewaska, J.; Kostyukova, Y.; Hasterok, R. Nuclear genome stability in long-term cultivated callus lines of *Fogopyrum tataricum* (L.) Gaetn. *PLoS ONE* **2017**, *12*, e0173537. [CrossRef] [PubMed]
36. Laukkanen, H.; Rautiainen, L.; Taulavuori, E.; Hohtola, A. Changes in cellular structures and enzymatic acuities during browning of Scots pine callus derived from mature buds. *Tree Physiol.* **2000**, *20*, 467. [CrossRef] [PubMed]
37. Uchendu, E.E.; Paliyath, G.; Brown, D.C.; Saxena, P.K. In vitro propagation of North American ginseng (*Panax quinquefolius* L.). *In Vitro Cellular Dev. Biol. Plant* **2011**, *47*, 710–718. [CrossRef]
38. Tang, W.; Newton, R.J.; Outhavong, V. Exogenously added polyamines recover browning tissue into normal callus cultures and improve plant regeneration in pine. *Physiol. Plant.* **2004**, *124*, 386–395. [CrossRef]
39. Habibi, N.; Sutar, R.K.; Purohit, S.D. Role of PGRs and inhibitors in induction and control of somatic embryogenesis in *Themeda quadrivalvis*. *Indian J. Exp. Biol.* **2009**, *47*, 198–203.
40. Nagomuo, M.; Mnene, E.; Ndakidemi, P. Control of lethal browning by using ascorbic acid on shoot tip cultures of a local *Musa* spp. (banana) cv. Mzuzu in Tanzania. *Afr. J. Biotechnol.* **2014**, *13*, 1721–1725. [CrossRef]
41. Vijayalakshmi, U.; Shourie, A. Remedial effect of ascorbic acid and citric acid on oxidative browning of *Glycyrrhiza glabra* callus cultures. *BioTechnolgia* **2016**, *97*, 179–186. [CrossRef]
42. Ho, T.T.; Murthy, H.N.; Park, S.Y. Methyl jasmonate induced odixative stress and accumulation of secondary metabolites in plant cell and organ cultures. *Int. J. Mol. Sci.* **2020**, *21*, 716. [CrossRef]

43. Halder, M.; Sarkar, S.; Jha, S. Elicitation: A biotechnological tool for enhancement of secondary metabolites in hairy root cultures. *Eng. Life Sci.* **2019**, *19*, 880–895. [CrossRef]
44. Thanh, N.T.; Murthy, H.N.; Yu, K.Y.; Hahn, E.J.; Paek, K.Y. Methyl jasmonate elicitation enhanced synthesis of ginsenoside by cell suspension cultures of *Panax ginseng* in 5-l balloon type bubble bioreactors. *Appl. Microbiol. Biotechnol.* **2005**, *67*, 197–201. [CrossRef]
45. Kim, Y.S.; Hahn, E.J.; Murthy, H.N.; Paek, K.Y. Adventitious root growth and ginsenoside accumulation of *Panax ginseng* cultures as affected by methyl jasmonate. *Biotechnol. Lett.* **2004**, *26*, 1619–1622. [CrossRef]
46. Wang, J.W.; Wu, J.Y. Nitric oxide is involved in methyl jasmonate-induced defense responses and secondary metabolism activities in *Taxus* cells. *Plant Cell Physiol.* **2005**, *46*, 923–930. [CrossRef] [PubMed]
47. Zhang, B.; Zheng, P.; Wang, J.W. Nitric oxide elicitation for secondary metabolite production in cultured plant cell. *Appl. Microbiol. Biotechnol.* **2012**, *93*, 455–466. [CrossRef] [PubMed]
48. Mendoza, D.; Cuaspud, O.; Arias, P.; Ruiz, O.; Arias, M. Effect of salicylic acid and methyl jasmonate in the production of phenolic compounds in plant cell suspension cultures of *Thevetia peruviana*. *Biotechnol. Rep.* **2018**, *19*, e00273. [CrossRef]
49. Gadzovaska, S.; Maury, S.; Delaunay, A.; Spasenoski, M.; Hagege, D.; Courtois, D.; Joseph, C. The influence of salicylic acid elicitation on shoots, callus, and cell suspension cultures on production of naphtodianthrones and phenylpropanoids in *Hypericum perforatum* L. *Plant Cell Tissue Organ Cult.* **2013**, *113*, 25–39. [CrossRef]
50. Xu, M.J.; Dong, J.F.; Zhu, M.Y. Effect of nitric oxide on catharanthine production and growth of *Catharanthus roseus* suspension cells. *Biotechnol. Bioeng.* **2005**, *89*, 367–372. [CrossRef]
51. Marchev, A.S.; Yordanov, Z.P.; Georgiev, M.I. Green (cell) factories for advanced production of plant secondary metabolites. *Crit. Rev. Biotechnol.* **2020**, *40*, 443–458. [CrossRef]
52. Chen, G.; Yi, Z.; Chen, X.; Su, W.; Li, X. Polyphenol nonoparticals from commonly consumed tea for scavenging free radicals, stabilizing, pickering, emulsions, and inhibiting cancer cells. *ACS Appl. Nano Mater.* **2021**, *4*, 652–665. [CrossRef]
53. Lu, H.; Tian, Z.; Cui, Y.; Liu, Z.; Ma, X. Chlorogenic acid: A comprehensive review of the dietary sources, processing effects, bioavailability, beneficial properties, mechanism of action, and future directions. *Compr. Rev. Food Sci. Food Saf.* **2020**, *19*, 3130–3158. [CrossRef]
54. Salehi, B.; Fokou, P.V.T.; Sharifi-Rad, M.; Zucca, P.; Pezzani, R.; Martins, N.; Sharifi-Rad, J. The therapeutic potential of naringenin: A review of clinical trials. *Pharmaceuticals* **2019**, *12*, 11. [CrossRef]
55. Murasige, T.; Skoog, F. A revised medium for growth and bioassays with tobacco tissue cultures. *Physiol. Plant.* **1962**, *15*, 473–497. [CrossRef]
56. Murthy, H.N.; Dewir, Y.H.; Dalawai, D.; Al-Suhaibani, N. Comparative physicochemical analysis of seed oils of wild cucumber (*Cucumis sativus* var. *hardwickii* (Royle) Alef.), cucumber (*Cucumis sativus* L. var. sativus), and gherkin (*Cucumis aunguira* L.). *S. Afr. J. Bot.* **2022**, *145*, 186–191. [CrossRef]
57. Pekal, A.; Pyrzynska, K. Evaluation of aluminum complexation reaction for flavonoid content assay. *Food Anal. Methods* **2014**, *7*, 1776–1782. [CrossRef]
58. Yadav, G.G.; Murthy, H.N.; Dewir, Y.H. Nutritional composition and in vitro antioxidant activities of seed kernel and seed oil of *Balanites roxburghii*: An underutilized species. *Horticulturae* **2022**, *8*, 798. [CrossRef]
59. Re, R.; Pellegrini, N.; Proteggente, A.; Pannala, A.; Yang, M.; Rice-Evans, C. Antidominant activity applying an improved ABTS radical cation decolourization assay. *Free Radic. Biol. Med.* **1999**, *26*, 1231–1237. [CrossRef] [PubMed]
60. Benzie, I.F.F.; Strain, J.J. Ferric reducing antioxidant power assay: Direct measure of total antioxidant activity of biological fluids and modified version for simultaneous measurement of total antioxidant power and ascorbic acid concentration. *Methods Enzymol.* **1999**, *299*, 15–27. [CrossRef]

Disclaimer/Publisher's Note: The statements, opinions and data contained in all publications are solely those of the individual author(s) and contributor(s) and not of MDPI and/or the editor(s). MDPI and/or the editor(s) disclaim responsibility for any injury to people or property resulting from any ideas, methods, instructions or products referred to in the content.

Article

Metabolic Discrimination between Adventitious Roots and Standard Medicinal Part of *Atractylodes macrocephala* Koidz. Using FT-IR Spectroscopy

So Yeon Choi [1,2,†], Seong Sub Ku [1,3,†], Myung Suk Ahn [4], Eun Jin So [1], HyeRan Kim [1], Sang Un Park [2], Moon-Soon Lee [3], Young Min Kang [5], Sung Ran Min [1,*] and Suk Weon Kim [6,*]

[1] Plant Systems Engineering Research Center, Korea Research Institute of Bioscience and Biotechnology (KRIBB), 125 Gwahak-ro, Yuseong–gu, Daejeon 34141, Republic of Korea
[2] Department of Crop Science, Chungnam National University, 99 Daehak-ro, Yuseong-gu, Daejeon 34134, Republic of Korea
[3] Department of Industrial Plant Science and Technology, Chungbuk National University, 1 Cheongdae-ro, Seowon-gu, Cheongju 28644, Republic of Korea
[4] Floriculture Research Division, National Institute of Horticultural and Herbal Science, RDA, Wanju 55365, Republic of Korea
[5] Herbal Medicine Resources Research Center, Korea Institute of Oriental Medicine (KIOM), 111 Geonjae-ro, Naju 58245, Republic of Korea
[6] Biological Resources Center, Korea Research Institute of Bioscience and Biotechnology, 181 Ipsingil, Jeongeup 56212, Republic of Korea
* Correspondence: srmin@kribb.re.kr (S.R.M.); kimsw@kribb.re.kr (S.W.K.)
† These authors contributed equally to this work.

Abstract: This study aims to examine the metabolic discrimination between in vitro grown adventitious roots and the standard medicinal parts of *Atractylodes macrocephala*. To achieve this goal, firstly, in vitro culture conditions of adventitious roots such as indole-3-butyric acid (IBA) concentrations, types of media, inorganic salt strength of culture medium, and elicitor types and concentrations were optimized. The optimal culture conditions for proliferation of adventitious roots was found to consist of Murashige and Skoog (MS) medium containing 5 mg L^{-1} IBA. Whole cell extracts from adventitious roots and the standard medicinal parts of *A. macrocephala* were subjected to Fourier transform infrared spectroscopy (FT-IR). Principal component analysis (PCA) and partial least square discriminant analysis (PLS-DA) from FT-IR spectral data showed that adventitious roots and standard medicinal parts were clearly distinguished in the PCA and PLS-DA score plot. Furthermore, the overall metabolite pattern from adventitious roots was changed depending on the dose-dependent manner of chemicals. These results suggest that FT-IR spectroscopy can be applied as an alternative tool for the screening of higher metabolic root lines and for discriminating metabolic similarity between in vitro grown adventitious roots and the standard medicinal parts. In addition, the adventitious roots proliferation system established in this study can be directly applied as an alternative means for the commercial production of *A. macrocephala*.

Keywords: indole-3-butyric acid (IBA); liquid culture; medicinal plants; partial least square discriminant analysis (PLS-DA); principal component analysis (PCA)

1. Introduction

Baekchul is a rhizome of *Atractylodes macrocephala* Koidz. or *Atractylodes japonica* Koidz., a perennial herbaceous plant belonging to the Asteraceae family. It is a medicinal plant widely used in East Asia including Korea, China, and Japan [1]. Over 200 chemical compounds have been isolated from *Atractylodis* Rhizoma (AR) including a series of sesquiterpenoids, alkynes, triterpenoids, aromatic glycosides, oligosaccharides, and polysaccharides [2–4]. In particular, *A. macrocephala* contains volatile components in its essential oil, and the main volatile components are sesquiterpenoids such as atractylenolide (I, II, and III), actractylon, 3β-acetoxyatractylone, and 3β-hydroxyatractylone [5]. Although, recent research has revealed that the extracts from AR could help to improve gastrointestinal function, anti-tumor activity, immunomodulatory activity, anti-inflammatory activity, and anti-bacterial activity [4]. However, the toxicity and safety of the chemicals have not been fully defined [6].

Despite the valuable pharmacological properties of AR, *A. japonica* presents several horticultural challenges that limit potential for the mass production of the medicinal rhizome. Most significantly, it has a low seed set rate and a slow growth rate for its rhizomes [7]. Baekchul (*A. macrocephala*) is also difficult to cultivate due to low resistance to root rot disease [8]. Therefore, the development of novel approaches for the mass production of medicinal AR is an indispensable topic. Plant cell and tissue culture systems could be applied as a novel approach for the mass supply of medicinal plant resources under aseptic culture conditions. In particular, adventitious roots proliferated in vitro offer an attractive source for useful phytochemicals due to their genetic stability and biosynthetic capacity [9]. Adventitious roots can be easily induced from several differentiated plant organs, such as the leaf, stem, and root.

To date, tissue culture studies of *A. macrocephala* have mainly focused on the plant regeneration system, including clonal multiplication of *Atractylodes lancea* by shoot tip culture [10] and the effect of plant growth regulators on plant regeneration from the auxiliary bud of the *Atractylodes* species [1,11]. However, there are only a few studies published that treat the establishment of a hairy roots culture system using leaf explants of *A. japonica* [12]. There has not been a published study of the induction and mass propagation system of adventitious roots from *A. macrocephala*.

To employ the in vitro proliferated adventitious roots as a herbal medicine resource, it is necessary to investigate differences in quantitative and qualitative patterns. In order to compare metabolic similarity between standard medicinal parts of *A. macrocephala* and their adventitious roots, the metabolite fingerprinting approach using 1HNMR (proton nuclear magnetic resonance spectroscopy), mass spectrometry, and FT-IR (Fourier transform infrared spectroscopy) is very efficient [13,14]. In particular, FT-IR spectroscopy, combined with multivariate statistical analysis from whole plant cell extracts, can be applied with high reproducibility and sensitivity. FT-IR spectral analysis is widely used to identify closely related microbial species [15,16], for the classification of plant species [17], and for cultivar identification [18,19]. Moreover, FT-IR spectral analysis can be applied in the comparison of the metabolic similarity between standard medicinal parts and their in vitro proliferated adventitious roots [20,21].

Therefore, we attempted to establish the optimal culture conditions for the mass production of adventitious roots of *A. macrocephala* in this study. First, we investigated the effects of IBA concentrations, the impact of types of media, inorganic salt concentrations, and elicitor types and concentrations on the growth of adventitious roots. Secondly, a quantitative analysis system using HPLC analysis to investigate the content of active compounds between adventitious roots after elicitor treatment and standard roots. Furthermore, a rapid metabolic discrimination system was developed using FT-IR spectroscopy for the investigation of the metabolic equivalence between in vitro grown adventitious roots and standard medicinal parts.

2. Results

2.1. Effect of IBA Concentrations on Growth of Adventitious Roots

In order to investigate the effect of the optimal IBA concentration on the growth of adventitious roots, adventitious roots were transferred to MS [22] medium containing 30 g L^{-1} sucrose and several concentrations of IBA (Figure 1A). After 10 days of culture, longitudinal growth and root branching from inoculated adventitious roots began to be seen in most IBA treatments except for those subjected to treatment of lower than 0.1 mg L^{-1} IBA. In the 5–10 mg L^{-1} IBA treatments, adventitious roots rapidly proliferated from the

20th to 40th days of culture (Figure 1B,C). After 40 days of culture, the changes in fresh and dry weight from adventitious roots were examined. The fresh weight from dventitious roots was the highest at 7136.7 ± 395.5 mg (Figure 1B) in the 10 mg L^{-1} IBA treatment. Additionally, the dry weight from adventitious roots was the highest at 556.7 ± 5.8 mg in 5 mg L^{-1} IBA treatment (Figure 1C). Overall, there was no significant difference in the growth rate of adventitious roots between those subjected to the 5 mg L^{-1} IBA and the 10 mg L^{-1} IBA treatments. However, overall, the increase in fresh and dry weight from adventitious roots appeared to be dependent on the IBA concentration. In light of these results, the optimum concentration of IBA required for adventitious root proliferation of *A. macrocephala* was determined to be 5 mg L^{-1} IBA. In subsequent experiments, the IBA was set at a concentration of 5 mg L^{-1}.

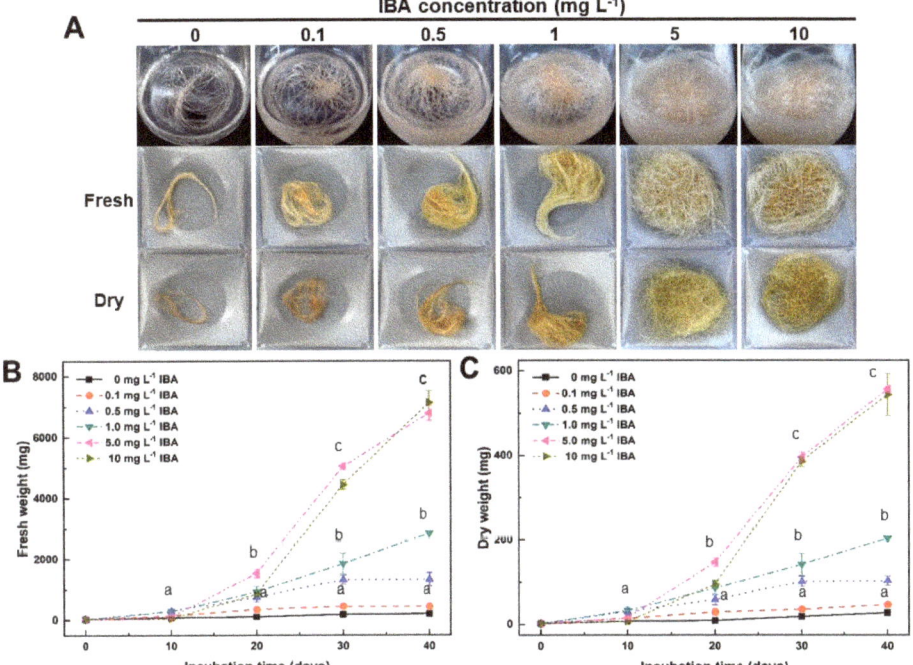

Figure 1. Effect of IBA concentrations on adventitious root growth of *A. macrocephala* after 40 days of culture in Murashige and Skoog medium containing various concentrations of IBA. (**A**) Root morphology. (**B**) Change in fresh weight of adventitious roots. (**C**) Change in dry weight of adventitious roots. Results are expressed as mean ± SD (n = 3). Different superscript letters indicate a significant difference ($p < 0.05$).

The analysis of the FT-IR spectral data, combined with multivariate analysis of the adventitious roots treated with IBA concentrations and the standard medicinal parts of *A. macrocephala*, revealed that there had been a quantitative and qualitative change between samples in the 1700–1500, 1450–1200, and 1100–900 cm^{-1} regions of FT-IR spectra (see Supplementary Materials, Figure S1A,C). The PCA and PLS-DA score plot also showed that adventitious roots treated with IBA concentrations and the standard medicinal parts of *A. macrocephala* could be clearly discriminated (see Supplementary Materials, Figure S1B,D). These results clearly confirm that there is a significant difference between the adventitious roots and the standard medicinal parts of *A. macrocephala* at the whole metabolite level. Furthermore, these results also reveal that the whole metabolite pattern of adventitious roots changes in accord with the increase in the concentration of IBA.

2.2. Effect of Culture Media and Inorganic Salt Concentrations in MS Medium on Growth of Adventitious Roots

The effect of culture media and inorganic salt concentrations in MS medium on the growth of adventitious roots was examined (Figures 2 and 3). As part of the investigation of the effect of culture media on growth of adventitious roots, roots were transferred to MS, Schenk and Hildebrandt (SH) [23], and Gamborg et al. (B5) [24] media containing 5 mg L^{-1} IBA (Figure 2). After 10 days of culture, the adventitious roots cultured in SH and B5 medium began to brown slightly more than those in MS medium. However, adventitious roots continued to grow vigorously until 40 days of culture in all culture media (Figure 2A). After 40 days of culture, the change in the dry weight of the adventitious roots was examined. The dry weight from adventitious roots was the highest at 630 ± 54.4 mg in MS medium (Figure 2B). However, there was no significant difference in the growth rate of adventitious roots from that observed in culture media types. Considering these results, the optimal culture media required for adventitious root proliferation of *A. macrocephala* was determined to be the MS medium. In the subsequent experiments, the MS medium was used to encourage adventitious root proliferation.

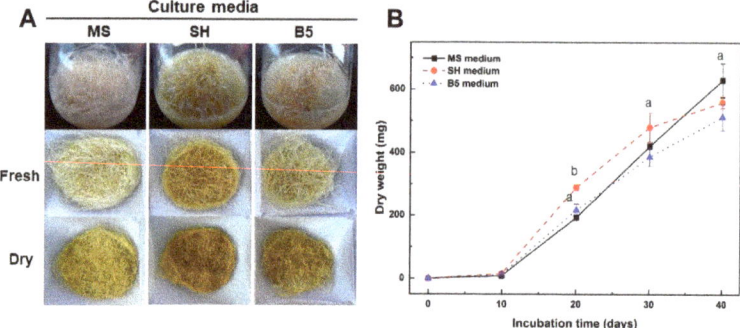

Figure 2. Effect of culture media on adventitious root growth of *A. macrocephala* after 40 days of culture in various media containing 5 mg L^{-1} IBA. (**A**) Root morphology. (**B**) Change in dry weight of adventitious roots. Results are expressed as mean ± SD (n = 3). Different superscript letters indicate a significant difference ($p < 0.05$).

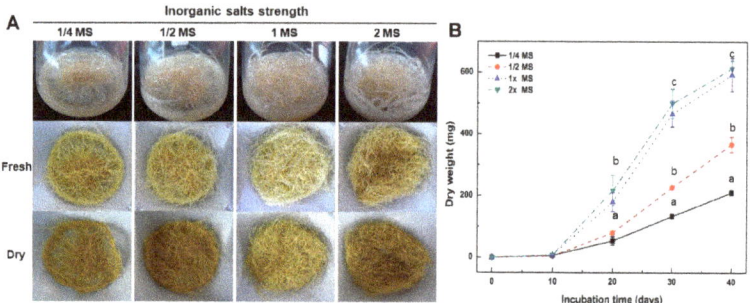

Figure 3. Effect of inorganic salt concentrations on adventitious root growth of A. macrocephala after 40 days of culture in 1/4, 1/2, 1, and 2× MS inorganic salt medium containing 5 mg L^{-1} IBA. (**A**) Root morphology. (**B**) Change in dry weight of adventitious roots. Results are expressed as mean ± SD (n = 3). Different superscript letters indicate a significant difference ($p < 0.05$).

Whole cell extracts were subjected to FT-IR spectroscopy (see Supplementary Materials, Figure S2) so as to compare the overall metabolic changes from the adventitious roots cultured on three different culture media and the standard medicinal parts of *A. macrocephala*.

Major spectral changes were observed in the 1700–1500, 1450–1200, and 1100–900 cm^{-1} regions of FT-IR spectra (see Supplementary Materials, Figure S2A,C). The PCA and PLS-DA score plot also revealed that adventitious roots and the standard medicinal parts were clearly discriminated (see Supplementary Materials, Figure S2B,D). These results suggest that there is a significant difference between adventitious roots and standard medicinal parts at the whole metabolite level. Furthermore, these results also show that the whole metabolite pattern of adventitious roots from the SH medium is more similar to the B5 medium than to the MS medium.

To investigate the effect of the concentration of inorganic salts in the MS medium on the growth of adventitious roots, roots were transferred to a medium of 1/4, 1/2, 1, and 2 times strength of MS inorganic salts containing 5 mg L^{-1} IBA and 30 g L^{-1} sucrose (Figure 3). After 10 days in the culture, the adventitious roots cultured in low concentrations (1/4 and 1/2) of MS inorganic salts did not branch lateral roots. In particular, adventitious roots did not proliferate in the 1/4 strength MS inorganic salts medium. After 20 days in the culture, lateral roots were induced from adventitious roots into all treatments. Lateral root development from the adventitious roots increased in accord with increasing MS inorganic salt concentrations. However, the proliferation of adventitious roots in the 1- and 2-times strength MS medium was much higher than that in the low concentration (1/4 and 1/2) MS medium for the first 40 days of culture (Figure 3A). After 40 days of culture, the change in dry weight from adventitious roots was examined. The dry weight from adventitious roots was the highest at 613.3 ± 25.8 mg in 2× MS medium (Figure 3B). However, there was no significant difference in the increase in dry weight in the 1× MS medium. These results suggest that the optimal inorganic salt concentration for adventitious root proliferation of A. macrocephala should be 1× MS medium. In the subsequent experiments, the 1× MS medium was used to promote adventitious root proliferation.

Similar to FT-IR analysis of IBA concentrations, FT-IR based metabolic changes of adventitious roots in MS inorganic salt concentration treatments revealed the same concentration dependency manner (see Supplementary Materials, Figure S3). The PCA and PLS-DA score plot showed that adventitious roots treated with MS inorganic salt concentrations and the standard medicinal parts of *A. macrocephala* were clearly discriminated (see Supplementary Materials, Figure S3B,D). These results clearly reveal a significant difference between adventitious roots and standard medicinal parts of *A. macrocephala* at the whole metabolite level. Furthermore, these results also show that the whole metabolite pattern for adventitious roots changes as the concentration of MS inorganic salts increases.

2.3. Effect of Elicitor Types and Concentrations on Growth and Metabolic Change in Adventitious Roots

The effect of elicitor types and concentrations on the growth of adventitious roots was examined (Figure 4). To investigate the effect of elicitor types and concentrations on the growth of adventitious roots, the roots were transferred to MS medium containing several concentrations of salicylic acid (SA) and methyl jasmonate (MeJA) during the last week of the whole growth process. After one additional week of incubation, the change in the growth of the adventitious roots was examined. In SA treatments, there was no significant difference in growth rates of adventitious roots regardless of concentrations (Figure 4A, see Supplementary Materials, Figure S4A,C). Similar to SA treatments, there was no significant difference in growth rates of adventitious roots from MeJA treatments (Figure 4B, see Supplementary Materials, Figure S4B,D). However, the scent of adventitious roots became stronger, and their color changed to dark brown in the 44.8 mg L^{-1} MeJA treatment. To quantify the atractylenolide I and III contents from adventitious roots, HPLC analysis was conducted (Figure 5, see Supplementary Materials, Figure S5A,B). Elicitor (MeJA, SA)-treated adventitious root samples showed a peak at the retention time of 12.375 (atractylenolide III) and 32.111 (atractylenolide I) on the HPLC chromatogram (Figure 5A). In the 44.8 mg L^{-1} MeJA treatment, the atractylenolide I content was the highest at 0.35 mg g^{-1}, and it was 1.7 times higher than that of the control (0.21 mg g^{-1}).

The contents of Atractylenolide I and III were the highest at 0.87 mg g^{-1} in the 44.8 mg L^{-1} MeJA treatment, which was 5.8 times higher than that of the control treatment (0.15 mg g^{-1}). In the case of the MeJA treatment, the atractylenolide content increased as the MeJA concentration increased. However, the SA treatment did not significantly affect the change in the contents of atractylenolide I and III in adventitious roots of *A. macrocephala*, even though the concentration increased (Figure 5B). To check whether the adventitious root could be used as an alternative means for standard medicinal parts of *A. macrocephala*, the content of atractylenolides I and III from two standard samples (Std1 and Std2) was also investigated (see Supplementary Materials, Table S1). The content of atractylenolide I from Std1 and Std2 was 1.16 ± 0.08 and 2.1 ± 0.23, and atractylenolide III was 2.99 ± 0.03 and 6.19 ± 0.05, respectively. These results clearly show that the content of atractylenolides I and III from adventitious roots was much lower than those of two standard samples, even though elicitation treatments has been conducted. However, the contents of atractylenolide I and III was different by 1.8 to 2 times between standard samples Std1 and Std2.

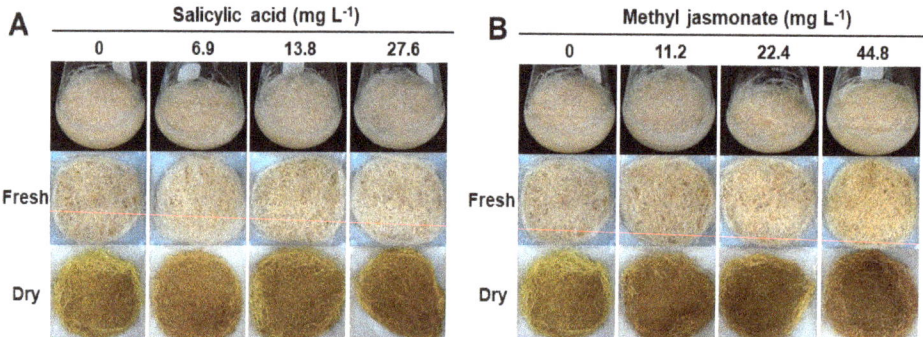

Figure 4. Effect of salicylic acid (**A**) and methyl jasmonate (**B**) concentrations on adventitious root growth of *A. macrocephala* after 40 days of culture on MS medium containing 5 mg L^{-1} IBA.

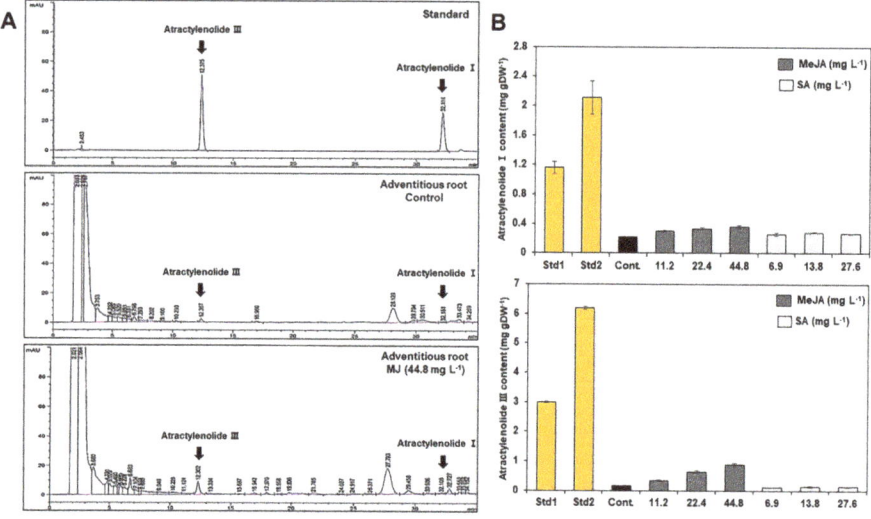

Figure 5. Typical HPLC chromatograms (**A**) and quantification of atractylenolide I and III (**B**) from adventitious roots of *A. macrocephala* after elicitor treatments. Std1, Std2: standard medicinal parts. Cont.: non-treated elicitors of adventitious roots.

To investigate the effect of elicitor types and concentrations on the metabolic change in adventitious roots, whole cell extracts from elicitor-treated adventitious roots were subjected to FT-IR spectroscopy (Figure 6). Similar to the FT-IR analysis of IBA concentrations, major spectral changes from elicitor-treated adventitious roots were observed in the 1700–1500, 1450–1200, and 1100–900 cm^{-1} regions of FT-IR spectra (Figure 6A). The PC loading plot indicated the major spectral variations that have an important role in the discrete clustering pattern on the PCA score plot were the 1700–1500, 1250–1100, and 1100–900 cm^{-1} regions of FT-IR spectra (Figure 6C).

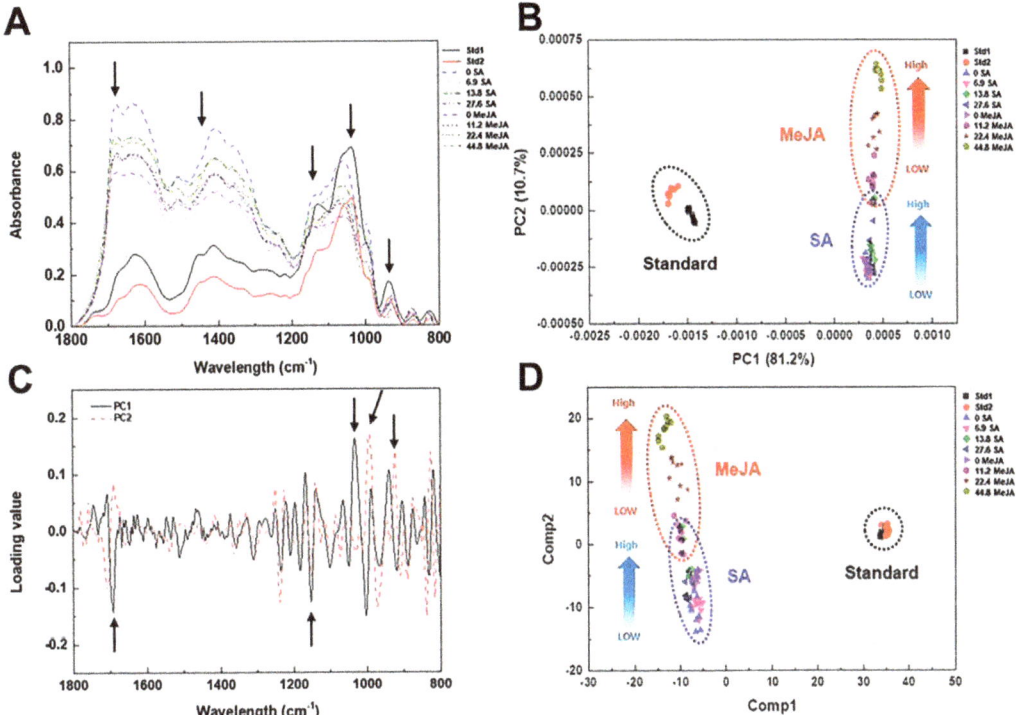

Figure 6. Multivariate analysis of FT-IR spectral data from cell extracts of standard rhizome parts and elicitor-treated adventitious roots of *A. macrocephala*. (**A**) Comparison of FT-IR spectral data from adventitious roots and standard rhizome parts. (**B**) PCA score plot of the FT-IR spectral data from adventitious roots and standard rhizome parts. (**C**) PCA loading plot based on PCA data from adventitious roots and standard rhizome parts. (**D**) PLS-DA score plot of FT-IR spectral data from adventitious roots and standard rhizome parts. Arrows indicate the FT-IR spectral regions showing significant variations between spectral data (**A**,**C**). Dotted circles represent each cluster for each elicitor treatment and standard rhizome parts (**B**,**D**).

The PCA and PLS-DA score plot also showed that adventitious roots and the standard medicinal parts were clearly discriminated (Figure 6B,D). The PCA score plot showed that PC1 and PC2 accounted for 81.2% and 10.7% of the total variation, respectively (Figure 6B). The replicated samples from each treatment were grouped in discrete clusters on the PCA score plot. The PC1 axis of the PCA score plot showed a separation pattern between adventitious roots and the standard medicinal parts, but the PC2 axis showed a discrete separation pattern into two groups corresponding to SA and MeJA treatments (Figure 6B). Similar to the PCA score plot, adventitious roots and the standard medicinal parts were clearly discriminated on the PLS-DA score plot (Figure 6D). Furthermore, the whole metabolite pattern from adventitious roots within SA and MeJA treatments changed as the concentration of elicitor increased. These results clearly suggest that there is a significant difference between adventitious roots and standard medicinal parts at the whole metabolite level.

3. Discussion

Adventitious roots proliferated in in vitro culture conditions that contained suitable phytohormones and produced important secondary metabolites [25]. Thus, the research established the optimal culture conditions for mass proliferation of adventitious roots of *A. macrocephala*. To achieve this goal, the effect of several environmental conditions in the culture, including IBA concentrations, types of media, inorganic salt concentrations of MS medium, and elicitor types and concentrations on adventitious root growth was examined.

The exogenous supply of plant growth regulators are required not only for the growth of adventitious roots, but also for the induction of in vitro morphogenesis. Auxin and ethylene act as the primary activators, while cytokinin and ethylene play roles primarily as inhibitors [26]. Many studies have also reported that the efficacy of diverse auxins for the establishment and propagation of in vitro grown adventitious roots varies according to the family and the species of the plant [27]. IBA is the most suitable effective auxin for the induction and development of adventitious root culture of medicinal plants [25]. Many studies report that IBA had a stimulatory role in root growth. Moreover, the optimal IBA concentration was different from species to species. Low concentrations of IBA (less than 2 mg L^{-1}) are more effective for adventitious root growth of *Couroupita guianensis* [28] and *Hypericum perforatum* [29]. High concentration IBA treatment (more than 3 mg L^{-1}) was effective for adventitious root growth in some plants, including *Cynanchum wilfordii* [21], *Orthosiphon stamineus* [30], *Panax quinquefolius* [31], and *Panax ginseng* [32]. In this study, the adventitious roots of *A. macrocephala* proliferated more rapidly at a high IBA concentration (5–10 mg L^{-1}) than at 1 mg L^{-1} IBA or less (Figure 1).

In general, media properties such as media type and inorganic salt concentrations have an impact on the induction and proliferation of adventitious root culture of medicinal plants. The choice of culture media for establishment and growth of adventitious roots in medicinal plants is determined by the species of the plant itself [25]. It has been reported that MS medium was highly suitable for the proliferation of adventitious roots of *Rumex crispus* [33], *Echinacea purpurea* [34], and *Andrographis paniculata* [35]. In the present study, the adventitious roots of *A. macrocephala* proliferated more rapidly in MS medium than in SH and B5 media (Figure 2). However, Kim et al. [36] insisted that B5 medium was much better than any other culture media for adventitious root cultures of *Scopolia parviflora*. In addition, the inorganic salt concentration in the MS medium also affects adventitious root growth. It has been reported that the full strength MS medium was suitable for the proliferation of adventitious roots in *Rumex crispus* [33] and *Oplopanax elatus* [37], while the half-strength MS medium was optimal for *Echinacea purpurea* [34] and *Echinacea angustifolia* [38]. This study found that the adventitious roots of *A. macrocephala* proliferated more rapidly in higher concentrations (1 and 2×) of MS inorganic salt medium rather than lower concentration treatments (Figure 3).

Among the various types of abiotic elicitors, MeJA was the most representative chemical elicitor for adventitious root cultures of medicinal plants such as *Hypericum perforatum*, *Panax ginseng*, *Polygonum multiflorum*, and *Scopolia parviflora* [39–42]. MeJA can increase the synthesis of secondary metabolites in adventitious root cultures through signal transduction which accelerates the enzyme catalysis process, thus forming bioactive compounds such as alkaloids, flavonoids, terpenoids, and polyphenols [43]. Similar to MeJA, SA also has the potential to increase secondary metabolite accumulation in the adventitious root culture of medicinal plants. Furthermore, SA is a plant hormone that can stimulate the expression of the biosynthetic pathway genes of secondary metabolites. Moreover, SA and MeJA can activate the defense mechanism of plants. Thus, SA and MeJA can have an adverse effect on the growth of adventitious roots in increased concentrations [41,42]. There was no significant difference in growth rates in adventitious roots in SA treatments regardless of the concentrations (Figure 4A, see Supplementary Materials, Figure S4A,C). Similar to SA treatments, there was no significant difference in growth rates of adventitious roots in MeJA treatments (Figure 4B, see Supplementary Materials, Figure S4B,D). The SA and MeJA treatment did not significantly affect the growth of adventitious roots of *A. macrocephala* because the SA and MeJA treatment was performed after the proliferation of adventitious roots of *A. macrocephala*. In addition, the SA and MeJA treatments were performed for only one week in this study. Thus, it is assumed that the short treatment period did not affect the proliferation of adventitious growth.

To investigate the effect of elicitor types and the concentrations on metabolic change in adventitious roots, whole cell extracts from elicitor-treated adventitious roots were subjected to FT-IR spectroscopy (Figure 6). There have been many previous studies of FT-IR peak assignment and chemical composition. Spectroscopic techniques yield spectra that contain key bands that are characteristic of individual components. The data provide information about the chemical composition of the sample, including both primary and secondary metabolites [44,45]. Carbohydrates, including cellulosic fibers, monosaccharides, and polysaccharides, give a complex fingerprint in the characteristic bands visible in the 900–1200 cm^{-1} region of the infrared spectrum. Cellular proteins and amino acids give characteristic peaks of 1500–1700 cm^{-1} and 1200–1500 cm^{-1}, that are designated as amide I, amide II, and amide III, respectively [46]. In this study, FT-IR combined with multivariate analysis was capable of discerning metabolite differences between adventitious roots and the standard medicinal parts of *A. macrocephala*. The greatest variation observed was in the carbohydrate, amide, and phospholipid/DNA/RNA regions (900–1200, 1500–1700, and 1300–1500 cm^{-1}, respectively) of FT-IR (Figure 6). These results indicate that qualitative and quantitative metabolic changes corresponding to the polysaccharide, protein amide I and II, and polyphenol regions were important for discrimination between adventitious roots and the standard medicinal parts of *A. macrocephala*.

In this study, the content of atractylenolide I from the adventitious root was 5.2–9.5 times lower than that of Std1 and Std2. In addition, the content of atractylenolide III was also 19–39 times lower than that of Std1 and Std2 (see Supplementary Materials, Table S1). However, the contents of atractylenolide I and III from two standard samples changed depending on different collection areas. The content of metabolites from plant samples can be easily changed by modifications in physicochemical conditions such as cultivation period and areas, soil conditions, or environmental conditions. Unfortunately, the content of atractylenolide I and III from adventitious roots was insufficiently low compared to the standard medicinal parts. In this study, we firstly examined the tissue cultural factors related to the mass proliferation of adventitious roots except for chemical elicitation. Thus, it is inferred that those tissue cultural factors may not significantly affect the change in the productivity of secondary metabolites from adventitious roots of *A. macrocephala*. Moreover, the whole incubation period of the adventitious roots was only 40 days. Therefore, if the culture period of adventitious roots is fully extended or the physical conditions such as culture temperature and light conditions are changed, it is expected that the productivity of atractylenolide from the adventitious roots can be increased. Furthermore, FT-IR analysis showed that overall metabolite content, including polysaccharide, protein, and polyphenol from adventitious roots, was significantly different from that of the standard medicinal parts (Figure 6). In the future, we are going to analyze various physiological activities, including antioxidant, anti-bacterial, and cosmetic activity, using adventitious roots of *A. macrocephala*. These studies will clarify whether the adventitious roots could be used as an alternative means for standard medicinal parts of *A. macrocephala*.

The present study evaluated the impact of several cultural environment factors, including IBA concentration, media types, MS inorganic salt concentrations, and elicitor types and concentrations, on the establishment of the in vitro proliferation system of adventitious roots. Furthermore, the study also revealed the metabolic discrimination system between adventitious roots and the standard medicinal parts by FT-IR, combined with multivariate analysis after the application of several cultural environment factors and elicitors. This research is critical because current agriculture demands a new approach that can overcome the harmful factors affecting the cultivation of crops such as biological or abiotic stress, pesticides, heavy metals, and poor soil conditions. The impact of the present study suggests an alternative form of cellular agriculture for the production of biomass and bioactive compounds employing *A. macrocephala*.

4. Materials and Methods

4.1. Plant Materials and Induction of Adventitious Roots from Leaf Explants

The in vitro shoots of *A. macrocephala* Koidz. were provided by Dr. Jae Whune Kim of the Plant Biotechnology Park (9-4, Hari-gil, Bokheung-myeon, Sunchang-gun, Jeollabuk-do 595-833, Republic of Korea). The shoots were subcultured onto MS basal medium at four-week intervals. The basal medium consisted of MS salts and the pH was adjusted to 5.8 using 1N NaOH, and then it was sterilized at 121 °C for 15 min. To induce adventitious roots, root explants were cut into small segments (approximately 1 cm in length). The root explants were then cultured onto a MS medium supplemented with 1 mg L^{-1} indole-3-butyric acid (IBA), 30 g/L sucrose, and 0.6% Phyto agar at 25 °C in the dark. After six weeks of incubation, the adventitious roots derived from root explants were cut and subcultured onto the same fresh medium every 4 weeks for a total of 2 months. Unless stated otherwise, all culture media contained 30 g L^{-1} sucrose and 0.6% Phyto agar. For liquid media, Phyto agar was omitted. All cultures were maintained in the dark at 25 °C.

To establish the suspension culture system of adventitious roots, roots were cut into small segments (approximately 2 cm in length) and then transferred to a 250 mL Erlenmeyer flask containing 50 mL of liquid culture medium of the same composition. Suspension culture of adventitious roots was maintained in a shaking incubator (JSSI-300CL, JSR, Republic of Korea) at 100 rpm agitation at 25 °C in the dark.

4.2. Effect of IBA Concentrations, Types of Culture Media, and Inorganic Salt Concentrations in Culture Medium on Growth of Adventitious Roots

To accelerate the proliferation of adventitious roots in the suspension culture system, the effect of IBA concentrations, types of culture media, and inorganic salt concentrations in the culture media on the amount of fresh and dry weight in adventitious roots was examined. Approximately 30 mg (fresh weight) of suspension cultured adventitious roots were transferred to an Erlenmeyer flask (100 mL) containing 25 mL of liquid culture medium, supplemented with 30 g L^{-1} sucrose and 0, 0.1, 0.5, 1, 5, and 10 mg L^{-1} IBA, respectively. The adventitious roots were collected and washed with distilled water every 10 days in the culture. After removing sufficient moisture from the surface, the fresh weight was measured and then the roots were dried for 48 h in a dry oven at 60 °C so as to measure the dry weight of adventitious roots. The changes in fresh weight and dry weight of adventitious roots were investigated for 40 days of the culture. All treatments were repeated 3 times. Adventitious root samples from each treatment were also freeze-dried and pulverized into fine powder, using a pestle and mortar, for FT-IR analysis.

To examine the effect of culture media on the proliferation of adventitious roots, suspension-cultured roots were transferred to liquid MS, SH, and B5 media that were supplemented with 5 mg L^{-1} IBA and 30 g L^{-1} sucrose. All culture conditions, inoculation methods, and the measurements of fresh and dry weight were performed in accord with the above procedures.

The effect of inorganic salts in the culture medium on growth of adventitious roots was also examined in the same manner. The suspension cultured roots were transferred to solutions with 1/4, 1/2, 1, and 2 times the strength of MS liquid medium, supplemented with 5 mg L^{-1} IBA and 30 g L^{-1} sucrose. All culture conditions, inoculation methods, and measurements of fresh and dry weight were performed in accord with the above procedures.

4.3. Effect of Elicitor Types and Concentrations on Growth and Metabolic Change in Adventitious Roots

To investigate the effect of elicitor types and concentrations on metabolic change in adventitious roots, each elicitor was treated during the last week of the whole growth process. The concentrations of salicylic acid (SA) and methyl jasmonate (MeJA) treatments were adjusted to 0, 6.9, 13.8, and 27.6 mg L^{-1}, and 0, 11.2, 22.4, and 44.8 mg L^{-1}, respectively. After one additional week in the culture, adventitious roots were collected for examination of their fresh weight and dry weight. All culture conditions, inoculation methods, and measurements of fresh and dry weight were performed in accord with the above procedures. All treatments were repeated 3 times. Adventitious root samples collected from each treatment were also freeze-dried and pulverized into fine powder using a pestle and mortar for HPLC and FT-IR analysis.

4.4. Quantification of Atractylenolide from Adventitious Roots by HPLC Analysis

Before analysis, the freeze-dried standard medicinal parts and adventitious roots treated with elicitors of *A. macrocephala* were pulverized, powdered, and stored at −70 °C. The medicinal parts, two standard samples (Std1 and Std2), were collected from A and B, which are the representative production areas of *A. macrocephala* in the Republic of Korea. These medicinal parts were kindly provided by the Korea Institute of Oriental Medicine, and each elicitor-treated lyophilized adventitial root and standard medicinal parts were quantified using HPLC for quantitative analysis of atractylenolide I and III. Each one gram of root powder was extracted into 95% ethanol (1 g × 30 mL) for 4 h using an ultrasonic extractor (Fisher Scientific Sonic Homogenizer Mod.60, Waltham, MA, USA) to ensure the complete extraction of bioactive compounds. All samples were centrifuged at 3000 rpm for 15 min and filtered through a 0.22 μm PTFE syringe filter. To obtain the extract, the ethanol solution was evaporated in an oven at 40 °C, dissolved in sterile distilled water, and then freeze-dried. The standard solutions of atractylenolide I and III were prepared using 100% methanol (*v/v*). The extracts from adventitious roots were prepared with the same solvent used in the standard solution and analyzed using HPLC analysis. To construct a standard calibration curve, atractylenolide I and III solutions were prepared in methanol. The concentration to plot standard curves were 6.25, 12.5, 25, 50, and 100 μg mL^{-1} and high linearity of $R^2 > 0.999$ was obtained from each calibration curve equation.

The HPLC analysis for quantification of atractylenolide I and III from elicitor-treated lyophilized adventitious roots and standard medicinal parts of *A. macrocephala* was performed using an Agilent 1200 series HPLC system (Agilent Technologies, Palo Alto, CA, USA), equipped with a diode array detector, binary gradient pump, autosampler, and vacuum degasser. The indicated compounds were eluted using a ZORBAX Eclipse Plus C18 column (C18, 4.6 mm I.D. × 250 mm, 5 μm particle size) at 25 °C with a flow rate of 1.0 mL/min. The injection volume was 10 μL with needle wash. The mobile phase comprised a mixture of acetonitrile and water. Atractylenolide I and III as standard compounds were eluted under gradient conditions (0–15 min, 50:50; 15–25 min, 60:40; and 25–35 min, 70:30). Data were collected and integrated using Agilent Chemstation B.04.01 software.

4.5. Whole Cell Extract Preparation for FT-IR Spectroscopy

To prepare crude whole cell extracts from root samples, 20 mg of powder derived from each root sample was mixed with 200 μL of 20% (*v/v*) methanol in a 1.5 mL Eppendorf tube, as described below. The mixtures were incubated in a 50 °C water bath for 20 min, followed by centrifugation at 13,000 rpm for 10 min. The resulting supernatants were then transferred to a fresh Eppendorf tube. These crude whole cell extracts obtained from adventitious roots and standard medicinal parts of *A. macrocephala* were stored at −20 °C until analysis by FT-IR spectroscopy.

4.6. FT-IR Spectroscopy and Multivariate Statistical Analysis

FT-IR spectroscopy analysis was performed according to the method of Kim et al. [17]. For the infrared measurements, a Tensor 27 FT-IR spectrometer (Bruker Optics GmbH, Ettlingen, Germany), equipped with a deuterated triglycine sulfate (DTGS) detector was used. After preheating the 384-well silicon plate to 37 °C, a supernatant sample (5 µL) was dropped on it and allowed to dry for 20 min. After drying the plate, it was placed in a microplate reader device (HTS-XT; Bruker Optics GbH, Ettlingen, Germany) and infrared spectra were obtained using the Tensor 27 spectrometer. Additionally, all FT-IR spectra were acquired with OPUS software (ver. 6.5, Bruker Optics Inc., Billerica, MA, USA). Infrared spectra were recorded from 4000 to 400 cm^{-1} using a spectral resolution of 4 cm^{-1} interval. A signal-to-noise ratio of 128 interferograms was co-added and averaged with the analytical results. The infrared spectra were obtained by subtracting the background spectra of the plate used for sample deposition. The original digital FT-IR spectra were collected from a spectral range of 1800 to 800 cm^{-1} for multivariate analysis. The pre-processing steps included baseline correction, spectral intensity normalization, and smoothing (ver. 6.5, Bruker Optics Inc.), as well as differentiation using Python software (version 2.7.15; Python Software Foundation). The processed spectral data were subjected to multivariate statistical analysis.

The FT-IR spectral data were imported into the R statistical analysis program (version 2.15.0; R Development Core Team) for the multivariate statistical analysis. The spectral data were limited to the 1800–800 cm^{-1} regions and then subjected to principal component analysis (PCA) and partial least square discriminant analysis (PLS-DA). PCA was performed using the non-linear iterative partial least squares (NIPALS) algorithm. The PCA loadings were then examined to identify the variables that were most important for discriminating between the root samples. To obtain a more distinct clustering of metabolic variation between the root samples, PLS-DA was applied.

4.7. Statistical Analysis

Statistical analysis between different groups was evaluated with a *t*-test. At least three biological replicates were performed for each analysis. Quantitative data are expressed as mean ± standard deviation (SD). A Student's *t*-test was conducted in Excel.

Supplementary Materials: The following supporting information can be downloaded at: https://www.mdpi.com/article/10.3390/plants12091821/s1. Table S1: Quantification of atractylenolide I and III from standard medicinal parts and adventitious roots with elicitor-treated of *A. macrocephala*. Figure S1: Multivariate analysis of FT-IR spectral data from cell extracts of adventitious roots and standard rhizome parts of *A. macrocephala* after incubation on different IBA concentrations; Figure S2: Multivariate analysis of FT-IR spectral data from cell extracts of adventitious roots and standard rhizome parts of *A. macrocephala* after incubation on different culture media; Figure S3: Multivariate analysis of FT-IR spectral data from cell extracts of adventitious roots and standard rhizome parts of *A. macrocephala* after incubation on different MS inorganic salt concentrations; Figure S4: Effect of salicylic acid concentrations on change of fresh weight (A) and dry weight (C) and methyl jasmonate concentrations on change of fresh weight (B) and dry weight (D) from adventitious roots of *A. macroscephala* after 40 days of culture on MS medium containing 5 mg L^{-1} IBA; Figure S5: Calibration curve used for atractylenolide I (A) and III (B) quantification by HPLC chromatography.

Author Contributions: S.Y.C. and S.S.K. performed and analyzed the experiments and wrote the draft of the manuscript; M.S.A. performed the experiment involving FT-IR analysis; E.J.S., H.K., S.U.P., M.-S.L. and Y.M.K. analyzed the data and revised the manuscript; S.R.M. and S.W.K. conceptualized and supervised the study. All authors have read and agreed to the published version of the manuscript.

Funding: This research was supported by the Korea Research Institute of Bioscience and Biotechnology (KRIBB) Research Initiative Program (KGM5372322) and the New Breeding Technologies Development Program (Project No. PJ01653001) of the Rural Development Administration, Republic of Korea.

Data Availability Statement: The data presented are contained within the article or Supplementary Materials.

Acknowledgments: The authors thank Jae Whune Kim for providing in vitro shoots of *A. macrocephala*.

Conflicts of Interest: The authors declare that they have no conflict of interest.

References

1. Koo, W.-L.; Cho, J.-H.; Park, C.-G.; Ahn, Y.-S.; Park, C.-B. Effect of plant growth regulators on in vitro cultured Atractylodes hybrid 'Dachul' (*A. macrocephala* × *A. japonica*). *Korean J. Plant Resour.* **2011**, *24*, 591–598. [CrossRef]
2. Sakurai, T.; Yamada, H.; Saito, K.; Kano, Y. Enzyme inhibitory activities of acetylene and sesquiterpene compounds in Atractylodes rhizome. *Biol. Pharm. Bull.* **1993**, *16*, 142–145. [CrossRef] [PubMed]
3. Jun, X.; Fu, P.; Lei, Y.; Cheng, P. Pharmacological effects of medicinal components of *Atractylodes lancea* (Thunb.) DC. *Chin. Med.* **2018**, *13*, 59. [CrossRef] [PubMed]
4. Zhang, W.; Zhao, Z.; Chang, L.; Cao, Y.; Wang, S.; Kang, C.; Wang, H.; Zhou, L.; Huang, L.; Guo, L. Atractylodis Rhizoma: A review of its traditional uses, phytochemistry, pharmacology, toxicology and quality control. *J. Ethnopharmacol.* **2021**, *266*, 113415. [CrossRef]
5. Chung, H.G.; Bang, K.H.; Bang, J.K.; Lee, S.E.; Seong, N.S.; Cho, J.H.; Han, B.S.; Kim, S.M. Comparison of volatile components in essential oil from different origin of *Atractylodes* spp. *Korean J. Med. Crop Sci.* **2004**, *12*, 149–153.
6. Zhu, B.; Zhang, Q.L.; Hua, J.W.; Cheng, W.L.; Qin, L.-P. The traditional uses, phytochemistry, and pharmacology of *Atractylodes macrocephala* Koidz.: A review. *J. Ethnopharmacol.* **2018**, *226*, 143–167. [CrossRef] [PubMed]
7. Hwang, M.H.; Seo, J.W.; Park, B.J.; Han, K.J.; Lee, J.G.; Kim, N.Y.; Kim, M.J.; Seong, E.S. Evaluation of growth characters and biological activities of 'Dachul', a hybrid medicinal plant of *Atractylodes macrocephala* × *Atractylodes japonica*, under different artificial light sources. *Plants* **2022**, *11*, 2035. [CrossRef] [PubMed]
8. Cho, J.H.; Kim, Y.W.; Bang, K.H.; Park, C.G.; Seong, N.S. Isolation of the Phytophthora root rot pathogen of *Atractylodes macrocephala*, *Phytophthora drechsleri*, and bioassay of the isolates with seedlings. *Korean J. Med. Crop Sci.* **2002**, *10*, 155–161.
9. Murthy, H.N.; Dandin, V.S.; Paek, K.-Y. Tools for biotechnological production of useful phytochemicals from adventitious root cultures. *Phytochem. Rev.* **2016**, *15*, 129–145. [CrossRef]
10. Hiraoka, N.; Yamada, N.; Kodama, T.; Tomita, Y. In vitro propagation of *Atractylodes lancea*. *Plant Cell Rep.* **1984**, *3*, 85–87. [CrossRef]
11. Mao, B.; He, B.; Chen, Z.; Wang, B.; Pan, H.; Li, D. Effects of plant growth regulators on the rapid proliferation of shoots and root induction in the Chinese traditional medicinal plant *Atractylodes macrocephala*. *Front. Biol. China* **2009**, *4*, 217–221. [CrossRef]
12. Lu, Z.; Zhang, S.L.; Lu, F.; Yang, S.H. Establishment of culture system of *Atractylodes* Koidz. ez Kitam. hairy roots and determination of polysaccharide. *J. Chin. Pharm. Sci.* **2014**, *49*, 1386–1392.
13. Goodacre, R.; Timmins, É.M.; Burton, R.; Kaderbhai, N.; Woodward, A.M.; Kell, D.B.; Rooney, P.J. Rapid identification of urinary tract infection bacteria using hyperspectral whole-organism fingerprinting and artificial neural networks. *Microbiology* **1998**, *144*, 1157–1170. [CrossRef]
14. Ward, J.L.; Harris, C.; Lewis, J.; Beale, M.H. Assessment of ^{1}H NMR spectroscopy and multivariate analysis as a technique for metabolite fingerprinting of *Arabidopsis thaliana*. *Phytochemistry* **2003**, *62*, 949–957. [CrossRef]
15. Timmins, É.M.; Howell, S.A.; Alsberg, B.K.; Noble, W.C.; Goodacre, R. Rapid differentiation of closely related *Candida* species and strains by pyrolysis-mass spectrometry and fourier transform-infrared spectroscopy. *J. Clin. Microbiol.* **1998**, *36*, 367–374. [CrossRef]
16. Wenning, M.; Seiler, H.; Scherer, S. Fourier-transform infrared microspectroscopy, a novel and rapid tool for identification of yeasts. *Appl. Environ. Microbiol.* **2002**, *68*, 4717–4721. [CrossRef]
17. Kim, S.W.; Ban, S.H.; Chung, H.; Cho, S.H.; Chung, H.J.; Choi, P.S.; Yoo, O.J.; Liu, J.R. Taxonomic discrimination of higher plants by multivariate analysis of Fourier transform infrared spectroscopy data. *Plant Cell Rep.* **2004**, *23*, 246–250. [CrossRef]
18. Kim, S.W.; Min, S.R.; Kim, J.; Park, S.K.; Kim, T.I.; Liu, J.R. Rapid discrimination of commercial strawberry cultivars using Fourier transform infrared spectroscopy data combined by multivariate analysis. *Plant Biotechnol. Rep.* **2009**, *3*, 87–93. [CrossRef]
19. Kwon, Y.-K.; Ahn, M.S.; Park, J.S.; Liu, J.R.; In, D.S.; Min, B.W.; Kim, S.W. Discrimination of cultivation ages and cultivars of ginseng leaves using Fourier transform infrared spectroscopy combined with multivariate analysis. *J. Ginseng Res.* **2014**, *38*, 52–58. [CrossRef]
20. Ahn, M.S.; Min, S.R.; Jie, E.Y.; So, E.J.; Choi, S.Y.; Moon, B.C.; Kang, Y.M.; Park, S.-Y.; Kim, S.W. Rapid comparison of metabolic equivalence of standard medicinal parts from medicinal plants and their in vitro-generated adventitious roots using FT-IR spectroscopy. *J. Plant Biotechnol.* **2015**, *42*, 257–264. [CrossRef]
21. Ahn, M.S.; So, E.J.; Jie, E.Y.; Choi, S.Y.; Park, S.U.; Moon, B.C.; Kang, Y.M.; Min, S.R.; Kim, S.W. Metabolic comparison between standard medicinal parts and their adventitious roots of *Cynanchum wilfordii* (Maxim.) Hemsl. using FT-IR spectroscopy after IBA and elicitor treatment. *J. Plant Biotechnol.* **2018**, *45*, 250–256. [CrossRef]
22. Murashige, T.; Skoog, F. A revised medium for rapid growth and bioassay with tobacco tissue cultures. *Physiol. Plant.* **1962**, *15*, 473–497. [CrossRef]

23. Schenk, R.U.; Hildebrandt, A.C. Medium and techniques for induction and growth of monocotyledonous and dicotyledonous plant cell cultures. *Can. J. Bot.* **1971**, *50*, 199–204. [CrossRef]
24. Gamborg, O.L.; Miller, R.A.; Ojima, K. Nutrient requirement of suspension cultures of soybean root cells. *Exp. Cell Res.* **1968**, *50*, 151–158. [CrossRef]
25. Rahmat, E.; Kang, Y. Adventitious root culture for secondary metabolite production in medicinal plants: A review. *J. Plant Biotechnol.* **2019**, *46*, 143–157. [CrossRef]
26. Pop, T.I.; Pamfil, D.; Bellini, C. Auxin control in the formation of adventitious roots. *Not. Bot. Horti Agrobot. Cluj-Napoca* **2011**, *39*, 307–316. [CrossRef]
27. Wei, K.; Ruan, L.; Wang, L.; Cheng, H. Auxin-induced adventitious roots formation in nodal cuttings of *Camellia sinensis*. *Int. J. Mol. Sci.* **2019**, *20*, 4817. [CrossRef]
28. Manokari, M.; Shekhawat, M. Implications of auxins in induction of adventitious roots from leaf explants of cannon ball tree (*Couroupita guianensis* Aubl.). *World Sci. News.* **2016**, *33*, 109–121.
29. Cui, X.-H.; Murthy, H.N.; Wu, C.-H.; Paek, K.-Y. Sucrose-induced osmotic stress affects biomass, metabolite, and antioxidant levels in root suspension cultures of *Hypericum perforatum* L. *Plant Cell Tiss. Organ Cult.* **2010**, *103*, 7–14. [CrossRef]
30. Ling, A.P.K.; Kok, K.M.; Hussein, S.; Ong, S.L. Effects of plant growth regulators on adventitious roots induction from different explants of *Orthosiphon stamineus*. *Am.-Eurasian J. Sustain. Agric.* **2009**, *3*, 493–501.
31. Liu, H.; Wang, J.; Gao, W.; Wang, Q.; Zhang, L.; Man, S. Optimization and quality assessment of adventitious roots culture in *Panax quinquefolius* L. *Acta Physiol. Plant.* **2014**, *36*, 713–719. [CrossRef]
32. Wu, C.-H.; Murthy, H.N.; Hahn, E.-J.; Paek, K.-Y. Establishment of adventitious root co-culture of Ginseng and Echinacea for the production of secondary metabolites. *Acta Physiol. Plant.* **2008**, *30*, 891–896. [CrossRef]
33. Chang, S.W.; Kim, I.H.; Han, T.J. Anthraquinone productivity by the cultures of adventitious roots and hairy roots from curled dock (*Rumex crispus*). *Korean J. Plant Tiss. Cult.* **1999**, *26*, 7–14.
34. Wu, C.-H.; Murthy, H.N.; Hahn, E.-J.; Paek, K.-Y. Improved production of caffeic acid derivatives in suspension cultures of *Echinacea purpurea* by medium replenishment strategy. *Arch. Pharm. Res.* **2007**, *30*, 945–949. [CrossRef]
35. Sharma, S.N.; Jha, Z.; Sinha, R.K. Establishment of in vitro adventitious root cultures and analysis of andrographolide in *Andrographis paniculata*. *Nat. Prod. Commun.* **2013**, *8*, 1045–1047. [CrossRef] [PubMed]
36. Kim, W.J.; Jung, H.Y.; Min, J.Y.; Chung, Y.G.; Lee, C.H.; Choi, M.S. Production of tropane alkaloids by two-stage culture of *Scopolia parviflora* Nakai adventitious root. *Korean J. Med. Crop Sci.* **2004**, *12*, 372–377.
37. Jiang, X.L.; Jin, M.Y.; Piao, X.C.; Yin, C.R.; Lian, M.L. Fed-batch culture of *Oplopanax elatus* adventitious roots: Feeding medium selection through comprehensive evaluation using an analytic hierarchy process. *Biochem. Eng. J.* **2021**, *167*, 107927. [CrossRef]
38. Wu, C.-H.; Dewir, Y.H.; Hahn, E.-J.; Paek, K.-Y. Optimizing of culturing conditions for the production of biomass and phenolics from adventitious roots of *Echinacea angustifolia*. *J. Plant Biol.* **2006**, *49*, 193–199. [CrossRef]
39. Wu, S.-Q.; Yu, X.-K.; Lian, M.-L.; Park, S.-Y.; Piao, X.-C. Several factors affecting hypericin production of *Hypericum perforatum* during adventitious root culture in airlift bioreactors. *Acta Physiol. Plant.* **2014**, *36*, 975–981. [CrossRef]
40. Sivakumar, G.; Paek, K.Y. Methyl jasmonate induce enhanced production of soluble biophenols in *Panax ginseng* adventitious roots from commercial scale bioreactors. *Chem. Nat. Compd.* **2005**, *41*, 669–673. [CrossRef]
41. Ho, T.-T.; Lee, J.-D.; Jee, C.-S.; Paek, K.-Y.; Park, S.-Y. Improvement of biosynthesis and accumulation of bioactive compounds by elicitation in adventitious root cultures of *Polygonum multiforum*. *Appl. Microbiol. Biotechnol.* **2018**, *102*, 199–209. [CrossRef] [PubMed]
42. Kang, S.-M.; Jung, H.-Y.; Kang, Y.-M.; Yun, D.-J.; Bahk, J.-D.; Yang, J.-K.; Choi, M.-S. Effects of methyl jasmonate and salicylic acid on the production of tropane alkaloids and the expression of PMT and H6H in adventitious root cultures of *Scopolia parviflora*. *Plant Sci.* **2004**, *166*, 745–751. [CrossRef]
43. Zhao, J.; Davis, L.C.; Verpoorte, R. Elicitors signal transduction leading to production of plant secondary metabolites. *Biotechnol. Adv.* **2005**, *23*, 283–333.
44. Baranski, R.; Baranska, M.; Schulz, H.; Simon, P.W.; Nothnagel, T. Single seed Raman measurements allow taxonomical discrimination of Apiaceae accessions collected in gene banks. *Biopolymers* **2006**, *81*, 497–505. [CrossRef] [PubMed]
45. Schulz, H.; Baranska, M. Identification and quantification of valuable plant substances by IR and Raman spectroscopy. *Vib. Spectrosc.* **2007**, *43*, 13–25. [CrossRef]
46. Schwinté, P.; Foerstendorf, H.; Hussain, Z.; Gärtner, W.; Mronginski, M.-A.; Hildebrandt, P.; Siebert, F. FTIR study of the photoinduced processes of plant phytochrome phyA using isotope-labeled bilins and density functional theory calculations. *Biophys. J.* **2008**, *95*, 1256–1267. [CrossRef] [PubMed]

Disclaimer/Publisher's Note: The statements, opinions and data contained in all publications are solely those of the individual author(s) and contributor(s) and not of MDPI and/or the editor(s). MDPI and/or the editor(s) disclaim responsibility for any injury to people or property resulting from any ideas, methods, instructions or products referred to in the content.

Article

Improving Flavonoid Accumulation of Bioreactor-Cultured Adventitious Roots in *Oplopanax elatus* Using Yeast Extract

Mei-Yu Jin [1], Miao Wang [2], Xiao-Han Wu [2], Ming-Zhi Fan [2], Han-Xi Li [2], Yu-Qing Guo [2], Jun Jiang [2], Cheng-Ri Yin [1,*] and Mei-Lan Lian [2,*]

1 Department of Chemistry, Yanbian University, Park Road 977, Yanji 133002, China; jinmeiyu@ybu.edu.cn
2 Agricultural College, Yanbian University, Park Road 977, Yanji 133002, China; 17704338616@163.com (M.W.); 18651676603@163.com (X.-H.W.); fan6252019@163.com (M.-Z.F.); 13614351819@163.com (H.-X.L.); 15204849165@163.com (Y.-Q.G.); jiangjun@ybu.edu.cn (J.J.)
* Correspondence: cryin@ybu.edu.cn (C.-R.Y.); lianmeilan2001@163.com (M.-L.L.);
Tel.: +86-43-273-2042 (C.-R.Y.); +86-43-243-5625 (M.-L.L.)

Abstract: *Oplopanax elatus* is an endangered medicinal plant, and adventitious root (AR) culture is an effective way to obtain its raw materials. Yeast extract (YE) is a lower-price elicitor and can efficiently promote metabolite synthesis. In this study, the bioreactor-cultured *O. elatus* ARs were treated with YE in a suspension culture system to investigate the elicitation effect of YE on flavonoid accumulation, serving for further industrial production. Among YE concentrations (25–250 mg/L), 100 mg/L YE was the most suitable for increasing the flavonoid accumulation. The ARs with various ages (35-, 40-, and 45-day-old) responded differently to YE stimulation, where the highest flavonoid accumulation was found when 35-day-old ARs were treated with 100 mg/L YE. After YE treatment, the flavonoid content increased, peaked at 4 days, and then decreased. By comparison, the flavonoid content and antioxidant activities in the YE group were obviously higher than those in the control. Subsequently, the flavonoids of ARs were extracted by flash extraction, where the optimized extraction process was: 63% ethanol, 69 s of extraction time, and a 57 mL/g liquid–material ratio. The findings provide a reference for the further industrial production of flavonoid-enriched *O. elatus* ARs, and the cultured ARs have potential application for the future production of products.

Keywords: *Oplopanax elatus*; yeast extract; elicitor concentration; elicitation time; elicitation duration; flash extraction

1. Introduction

Oplopanax elatus Nakai is a perennial shrubby plant of the Araliaceae family, mainly distributed in Changbai Mountain area, with some distribution in Russia and Korea [1]. *O. elatus* has high pharmacological effects, such as antioxidant, anti-inflammatory, anti-cancer, and antidepression [1]. However, the exploitation and utilization of *O. elatus* are delayed due to the shortage of raw materials caused by an uncontrollable collection of wild resources and the immature technology of artificial cultivation [2]. Adventitious roots (ARs) of plants can be mass and rapidly produced using plant tissue culture technology [2], and AR culture is an efficient way to obtain raw materials of rare medicinal plants [3]. Recently, *O. elatus* ARs have been successfully cultured in batch [3–5] or fed-batch bioreactor culture systems [2,6], and the fed-batch culture shows a high efficiency and has promising application value [6]. In addition, *O. elatus* AR cultures contain various useful metabolites, and flavonoids, as one of the main bioactive compounds, have important biological effects, especially antioxidant activity [5]. Therefore, the production of *O. elatus* flavonoids using AR culture could contribute to the development of antioxidant products in the future.

Bioreactor culture technology has been extensively applied in plant AR culture [2,6–10], and the ARs of various plant species have been produced in large-scale bioreactor systems [11–13]. In AR bioreactor culture of medicinal plants, metabolite synthesis can be controlled by various strategies, such as improving the culture medium and air volume or inoculation density, and implementing elicitation, among which the elicitation strategy is the most efficient for enhancing metabolite synthesis [14]. For example, Ho et al. [15] indicated that 50 µM methyl jasmonate (MeJA) greatly improves the phenolic and flavonoid accumulation of *Polygonum multiflorum* ARs. Le et al. [16] demonstrated that the saponin content obviously increases when *Panax ginseng* ARs are treated with a biotic elicitor (*Mesorhizobium amorphae*, GS3037). For *O. elatus* AR culture, we previously treated *O. elatus* ARs with MeJA (200 µM) and found that flavonoid and phenolic contents were considerably enhanced in the batch culture system [3]. However, the increased cost of AR culture caused by the high price of elicitors, such as MeJA, must be considered in large-scale industrial production. Consequently, excavating effective and economic elicitors is crucial for the production of useful metabolites in a culture system. An elicitor, as a trigger factor, can promote metabolite synthesis by regulating several enzyme activities; the elicitation effects and their underlying mechanisms vary among the different elicitors [14]. Yeast extract (YE) is a natural product obtained from yeast by separation and purification, and contains a variety of components, such as polypeptide, amino acid, flavor nucleotide, trace elements, and B vitamins [17]. The price of YE is much cheaper compared with MeJA, but its enhancement effect on bioactive compound accumulation in plant cell/organ culture has been repeatedly mentioned [18,19]. For instance, Goncharuk et al. [18] elucidated that YE treatment could increase phenolic compounds accumulation during *Linum grandiflorum* cell culture. Maqsood and Abdul [19] indicated that YE is beneficial for alkaloid accumulation in the hairy root culture of *Catharanthus roseus*. However, the elicitation effect of YE has not been investigated in *O. elatus* AR culture. Therefore, in this study, *O. elatus* ARs produced in a fed-batch bioreactor culture system were treated with YE in a suspension flask culture system, and the suitable YE concentration, AR age contacting YE, and YE treatment duration were selected for the enhancement of flavonoid accumulation. Then, the elicitation effect of YE was verified in the fed-batch bioreactor culture system by comparing flavonoid contents and antioxidant activities of YE-treated and -untreated ARs. This study aimed to confirm a suitable YE treatment method for obtaining large amounts of flavonoids through *O. elatus* AR culture and to provide a reference for the production of antioxidant products.

The optimization of the extraction process is an important step before the production of products. Flash extraction was firstly proposed in 1993 by Liu et al. [20], who indicated that flash extraction has many advantages, such as sufficient extraction and saving time, solvent, and energy. In this study, the flavonoids of YE-treated fed-batch bioreactor-cultured ARs were extracted by flash extraction, and the extraction process was optimized using the response surface method (RSM) to provide a reference for the further utilization of *O. elatus* ARs.

2. Results

2.1. Effect of YE Concentration on Flavonoid Accumulation of Fed-Batch Bioreactor-Cultured ARs

After 45 days of fed-batch bioreactor culture, ARs were treated with different concentrations of YE for 8 days, and then AR dry weight (DW) and flavonoid content were determined to select a suitable YE concentration. Figure 1a shows that the AR biomass did not change after YE treatment with different concentrations (25–150 mg/L). However, the total flavonoid content increased significantly ($p < 0.05$) at 75 mg/L YE and 100 mg/L YE, and the highest total flavonoid content was determined at 100 mg/L YE (Figure 1b). YE lower than 75 mg/L or higher than 100 mg/L did not exhibit an enhancement effect on flavonoid accumulation (Figure 1b). At the optimal YE concentration (100 mg/L), the total flavonoid productivity exceeded 3 g/L, which was significantly ($p < 0.05$) higher than that of the control and other YE concentration groups (Figure 1c).

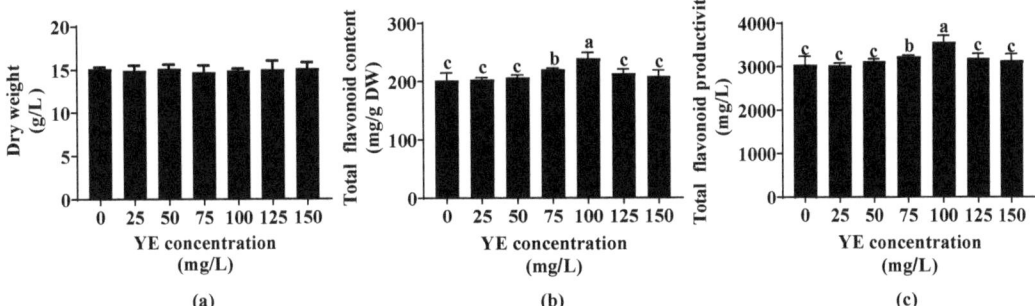

Figure 1. Effect of yeast extract (YE) concentration on biomass (**a**) and flavonoid content (**b**) and productivity (**c**) of *Oplopanax elatus* adventitious roots (ARs). The 45-day-old fed-batch bioreactor-cultured ARs were transferred to flasks and treated with YE for 8 days. Data represent the mean value ± standard deviation (*n* = 3). The different letters within the same column indicate significant difference by Duncan's multiple range test at $p < 0.05$.

2.2. Effect of YE on Flavonoid Accumulation of Fed-Batch Bioreactor-Cultured ARs at Different Ages

The fed-batch bioreactor-cultured ARs at different ages (35-, 40-, and 45-day-old) were treated with 100 mg/L YE (selected in the above YE concentration experiment) for 8 days, and the elicitation effect of YE on flavonoid accumulation was investigated. AR DW did not show a significant difference among the different groups (Figure 2a). Compared with the YE-untreated control group, the total flavonoid contents in all YE groups increased significantly ($p < 0.05$) (Figure 2b). The highest total flavonoid content was found when the 30-day-old ARs were treated with YE (100 mg/L), which was approximately 70 mg/g DW more than that in the other YE treatment groups (Figure 2b). In the optimal YE treatment time, i.e., treating 35-day-old ARs with 100 mg/L YE for 8 days, the flavonoid productivity reached 4.7 g/L (Figure 2c), which was significantly higher than other YE groups ($p < 0.05$) and the control ($p < 0.01$).

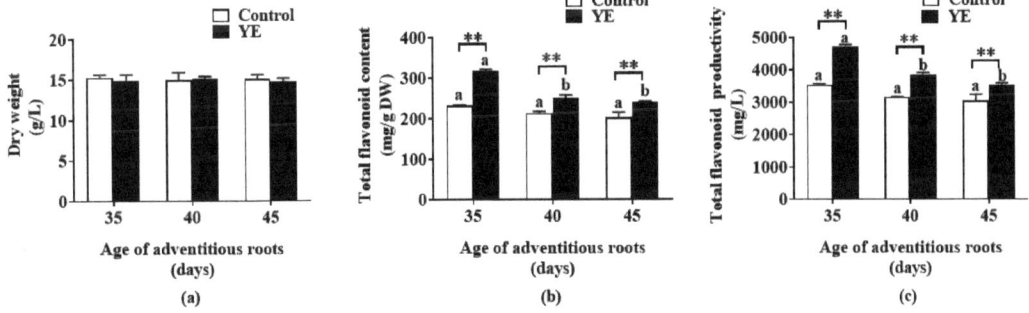

Figure 2. Effect of yeast extract (YE) on biomass (**a**) and flavonoid content (**b**) and productivity (**c**) of *Oplopanax elatus* adventitious roots (AR) at different ages. The 35-, 40, and 45-day-old fed-batch cultured *O. elatus* ARs were transferred to flasks and treated with 100 mg/L YE (the control group was added with equal amount of the medium) for 8 days in the YE group. Data represent the mean value ± standard deviation (*n* = 3). The different letters within the same color column indicate significant difference by Duncan's multiple range test at $p < 0.05$. ** indicates significant difference between the groups of YE and control by Student's *t*-test at $p < 0.01$.

2.3. Effect of YE Treatment Duration on Flavonoid Accumulation of Fed-Batch Bioreactor-Cultured ARs

The 35-day-old fed-batch bioreactor-cultured ARs were treated with YE (100 mg/L) on different days to investigate the elicitation effect of YE treatment duration. Figure 3a shows that AR DW did not change from 1 day to 10 days in the control and YE treatment groups. The flavonoid accumulation exerted an increase trend from 1 day to 4 days after YE treatment, peaked at 4 days (354 mg/g DW), and then decreased in the following days (Figure 3b). Compared with the control group, the total flavonoid content in the YE group was significantly ($p < 0.01$) increased on all treatment days. After 4 days of YE treatment, the flavonoid productivity reached 5.4 g/L (Figure 3c), which was 1.5 g/L higher than that of the control.

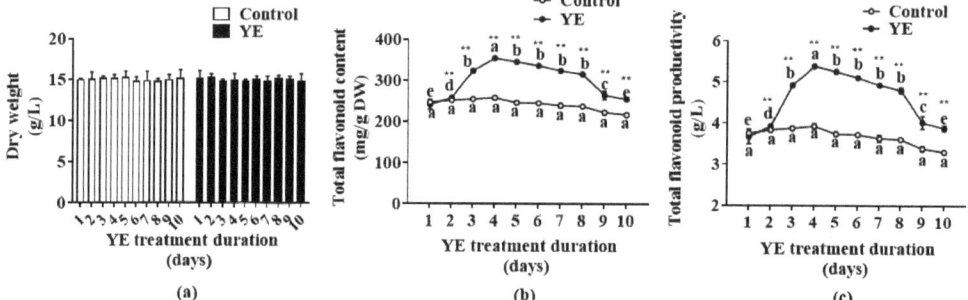

Figure 3. Effect of yeast extract (YE) treatment duration on biomass (**a**) and flavonoid content (**b**) and productivity (**c**) of *Oplopanax elatus* adventitious roots (ARs). The 35-day-old fed-batch bioreactor-cultured ARs were transferred to flasks and treated with 100 mg/L YE (the control group was added with equal amount of the medium) in the YE group. Data represent the mean value ± standard deviation ($n = 3$). The different letters within the same line indicate significant difference by Duncan's multiple range test at $p < 0.05$. ** indicates significant difference between the groups of YE and control at each time point by Student's *t*-test at $p < 0.01$.

2.4. Comparison of Flavonoid Contents and Antioxidant Activities of YE-Treated and YE-Untreated ARs in Fed-Batch Bioreactor Culture System

The ARs were cultured in the fed-batch bioreactor (5 L) system and treated with 100 mg/L YE for 4 days on day 35 according to the result of the above YE elicitation experiments, and the contents of total flavonoids and rutin, quercetin, and kaempferide were compared with those in the control. Table 1 shows that, in the YE group, the contents of total flavonoids (582.2 mg/g DW), rutin (2.9 mg/g DW), and kaempferide (161.9 µg/g DW) were significantly ($p < 0.05$) higher than those in the control; however, the quercetin content was not significantly ($p < 0.05$) affected by the YE treatment. The AR DW in the YE (67.17 ± 5.6 g/bioreactor) and control (65.14 ± 7.1 g/bioreactor) groups did not show a significant ($p < 0.05$) difference. The total flavonoid productivity in the YE treatment group reached 7.8 g/L, whereas that in the control group was only 4.6 g/L.

Table 1. Comparison of flavonoid contents between yeast extract (YE)-treated and YE-untreated adventitious roots (ARs) of *Oplopanax elatus* in fed-batch bioreactor system.

Treatment	Total Flavonoids (mg/g DW)	Rutin (mg/g DW)	Quercetin (µg/g DW)	Kaempferide (µg/g DW)
control	357.7 ± 38.9	2.4 ± 0.1	234.5 ± 10.3	117.4 ± 7.5
YE	582.2 ± 11.7 *	2.9 ± 0.1 *	266.6 ± 11.7	161.9 ± 9.7 *

The ARs in the YE group were treated with 100 mg/L YE for 4 days on day 35 of the fed-batch bioreactor culture. The ARs in the control group were cultured for 49 days in the fed-batch bioreactor culture system. Data represent the mean value ± standard deviation ($n = 3$). * indicates significant difference between the groups of YE and control by Student's *t*-test at $p < 0.05$.

Subsequently, antioxidant activities of extracts from YE-treated (YE-ARE) and YE-untreated ARs (C-ARE) were compared by evaluating abilities of scavenging 2,2-diphenyl-1-picrylhydrazyl (DPPH) and 2'2'-azino-bis(3-ethylbenzothiazoline-6-sulfonic acid) (ABTS)$^+$ radicals and chelating Fe^{2+}. As shown in Figure 4, when YE-ARE concentrations increased from 15.6 µg/mL to 31.3 µg/mL, the DPPH scavenging rate dramatically increased, and then gently rose until 500 µg/mL YE-ARE; the DPPH scavenging rate in the control group increased when increasing the C-ARE concentrations (Figure 4a). At all extract concentrations, DPPH scavenging rates in the YE-ARE group were higher than those in the C-ARE group. The concentration for 50% of the maximal effect (EC_{50}) in the YE-ARE group was only 19.8 µg/mL, whereas that in the C-ARE was 27.6 µg/mL. The $ABTS^+$ scavenging rate (Figure 4b) and Fe^{2+} chelating rate (Figure 4c) in the YE-ARE and C-ARE groups exerted a similar pattern, which increased when the AR extract concentration increased. The EC_{50} values in the YE-ARE group regarding $ABTS^+$ scavenging (0.85 mg/mL) and Fe^{2+} chelating rates (4.61 mg/mL) were lower than those in the C-ARE group. The results of DPPH and $ABTS^+$ scavenging and Fe^{2+} chelating rates indicated the high antioxidant activity of YE-treated ARs, which may be related to the high contents of flavonoids contained in the YE-treated ARs (Table 1). The findings elucidate that YE-treated *O. elatus* ARs have potential value for use in the development of antioxidant products.

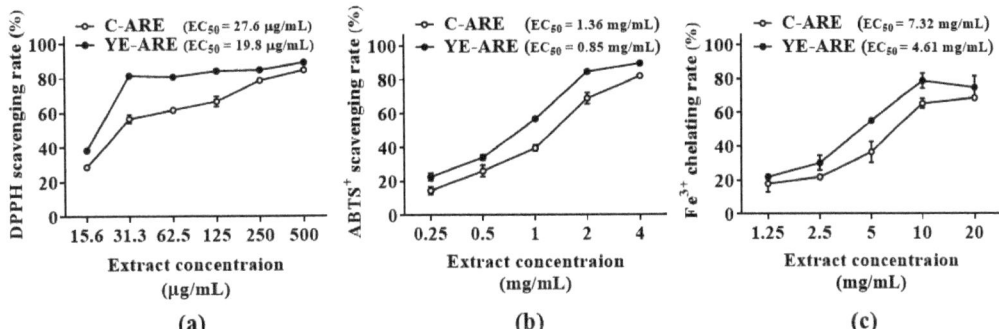

Figure 4. Comparison of antioxidant ability of extracts from YE-untreated (C-ARE) and -treated (YE-ARE) adventitious roots (ARs) of *Oplopanax elatus* from the fed-batch bioreactor culture. (a) DPPH scavenging rate. (b) $ABTS^+$ scavenging rate. (c) Fe^{3+} chelating rate. The ARs in YE group were treated with 100 mg/L YE for 4 days on day 35 of fed-batch bioreactor culture. The ARs in the control group were cultured for 49 days in fed-batch bioreactor culture system. DPPH = 2,2-diphenyl-1-picrylhydrazyl; ABTS = 2'2'-azino-bis (3-ethylbenzothiazoline-6-sulfonic acid). EC_{50} = concentration for 50% of maximal effect. Data represent the mean value ± standard deviation (n = 3).

2.5. Flash Extraction Process Optimization of Flavonoids from YE-Treated Fed-Batch Bioreactor-Cultured ARs

To future utilize the *O. elatus* AR cultures, this study used the flash extraction method to extract the flavonoids of YE-treated ARs from the fed-batch bioreactor culture, and the extraction process was optimized by RSM. To provide a basis for the RSM experiment, single-factor experiments were implemented. Figure 5 shows that flavonoid yields were significantly different among the three solvent groups (water, ethanol, and methanol), and the highest value was found when ethanol was used as the solvent (Figure 5a). Then, 60% ethanol, 70 s of extraction time, and a 60 mL/g liquid–material ratio were selected through conducting the experiments of ethanol concentration (Figure 5b), extraction time (Figure 5c), and liquid–material ratio (Figure 5d). The single experiment results were used in the following RSM experiment.

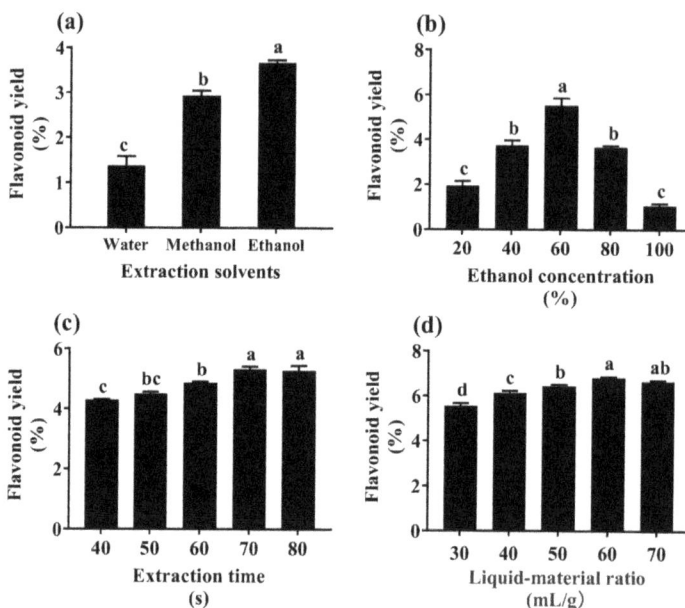

Figure 5. Effect of extraction solvent (**a**), solvent (ethanol) concentration (**b**), extraction time (**c**), and liquid–material ratio (**d**) on flavonoid yield of *Oplopanax elatus* adventitious roots from fed-batch bioreactor culture. Data represent the mean value ± standard deviation (n = 3). The different letters within the same column indicate significant difference by Duncan's multiple range test at $p < 0.05$.

The RSM experiment was designed based on a Box–Behnken design with three factors (ethanol concentration [X_1], extract time [X_2], and liquid–material ratio [X_3]) at three levels. Each independent variable was coded at three levels between −1, 0, and +1 corresponding to the low, mid, and high level (Table 2), which included a total of 17 combination groups (Table 3). ARs were extracted depending on the extraction conditions of each combination group, and the extraction efficiency was evaluated with the flavonoid yield as the index. Table 3 shows that the flavonoid yield considerably differed among the groups, where high flavonoid yields (>7%) occurred at the central points (group 2, 5, 16, and 17).

Table 2. Coded and experimental levels of independent variables used in response surface method experiment.

Independent Variables	Coded Level		
	−1	0	+1
ethanol concentration (X_1)	50%	60%	70%
extraction time (X_2)	60 s	70 s	80 s
liquid–material ratio (X_3)	50 mL/g	60 mL/g	70 mL/g

Further, the experimental data were fitted with various models (linear model, interactive model, quadratic model, and cubic model), and the regression equation was obtained, which was Y (flavonoid yield) = $7.06 + 0.3X_1 + 0.096X_2 - 0.17X_3 - 0.23X_1X_2 - 0.37X_1X_3 + 0.36X_2X_3 - 0.78X_1^2 - 0.61X_2^2 - 0.52X_3^2$. The predicted flavonoid yield in each group was calculated according to the regression equation, showing that it did not deviate much from the actual data of the experiment (Table 3).

Table 3. Box–Behnken experimental design matrix and experimental responses.

Groups	Codes (Levels)			Flavonoid Yield (%)	
	Ethanol Concentration (X_1)	Extraction Time (X_2)	Liquid–Material Ratio (X_3)	Actual Value	Predicted Value
1	0 (60%)	−1 (60 s)	1 (70 mL/g)	5.56 ± 0.12	5.66
2 [a]	0 (60%)	0 (70 s)	0 (60 mL/g)	7.00 ± 0.03	7.06
3	−1 (50%)	−1 (60 s)	0 (60 mL/g)	4.99 ± 0.08	5.04
4	−1 (50%)	0 (70 s)	1 (70 mL/g)	5.45 ± 0.11	5.30
5 [a]	0 (60%)	0 (70 s)	0 (60 mL/g)	6.92 ± 0.07	7.06
6	1 (70%)	0 (70 s)	1 (70 mL/g)	6.39 ± 0.16	6.22
7	0 (60%)	1 (80 s)	−1 (50 mL/g)	6.69 ± 0.01	6.60
8	−1 (50%)	1 (80 s)	0 (60 mL/g)	6.19 ± 0.05	6.10
9	−1 (50%)	0 (70 s)	−1 (50 mL/g)	6.18 ± 0.15	6.36
10	1 (70%)	1 (80 s)	0 (60 mL/g)	5.90 ± 0.06	5.84
11	0 (60%)	0 (70 s)	0 (60 mL/g)	6.82 ± 0.04	7.06
12	0 (60%)	−1 (60 s)	−1 (50 mL/g)	5.50 ± 0.11	5.26
13	0 (60%)	1 (80 s)	1 (70 mL/g)	5.29 ± 0.12	5.52
14	1 (70%)	−1 (60 s)	0 (60 mL/g)	5.60 ± 0.12	5.70
15	1 (70%)	0 (70 s)	−1 (50 mL/g)	5.69 ± 0.07	5.84
16 [a]	0 (60%)	0 (70 s)	0 (60 mL/g)	7.34 ± 0.10	7.06
17 [a]	0 (60%)	0 (70 s)	0 (60 mL/g)	7.21 ± 0.01	7.06

Data represent the mean value ± standard deviation ($n = 3$). [a] Central points (used to determine the experimental error).

The result of the model adequacy test shows that the lack of fit (0.2712) was not significant and that the correlation coefficient (R^2) was high (0.9487) (Table 4), indicating that the model was valid [21]. Ethanol concentrations significantly ($p < 0.01$) affected the flavonoid yield, whereas the extraction time or liquid–material ratio did not show a significant effect. The order of F values was: ethanol concentration (11.64) > liquid–material ratio (3.73) > extraction time (1.18), indicating that the solvent concentration had the greatest influence on the extraction efficiency, followed by the liquid–material ratio and extraction time. In addition, Table 4 also shows that the interactions between the ethanol concentration and liquid–material ratio ($p = 0.0226$) and the extraction time and liquid–material ratio ($p = 0.0247$) were significant ($p < 0.05$), whereas that between the ethanol concentration and extraction time was not significant ($p = 0.1159$).

Table 4. Model adequacy test and ANOVA analysis.

Variables	Sun of Squares	Degree of Freedom	Mean Square	F	p
model	8.14	9	0.90	14.37	0.0010 **
X_1	0.73	1	0.73	11.64	0.0113 *
X_2	0.074	1	0.074	1.18	0.3137
X_3	0.23	1	0.23	3.73	0.0948
$X_1 X_2$	0.20	1	0.20	3.22	0.1159
$X_1 X_3$	0.53	1	0.53	8.47	0.0226 *
$X_2 X_3$	0.51	1	0.51	8.13	0.0247 *
X_1^2	2.55	1	2.55	40.49	0.0004 **
X_2^2	1.57	1	1.57	24.93	0.0016 **
X_3^2	1.14	1	1.14	18.12	0.0038 **
residual	0.44	7	0.063	-	-
lack of fit	0.26	3	0.086	1.90	0.2712
pure error	0.18	4	0.045	-	-
cor total	8.58	16	-	-	-
R^2	0.9487				

X_1 = ethanol concentration. X_2 = extraction time. X_3 = liquid–material ratio. * and ** indicate significant differences at $p < 0.05$ and $p < 0.01$, respectively. R^2 = correlation coefficient.

The response surface plots also exhibited an interaction between both independent variables (Figure 6). A nearly circular contour line is found in Figure 6a, indicating a minor interaction between the ethanol concentration and extraction time; by contrast, oval contour lines are found in Figure 6b,c, indicating high interactions between the ethanol concentration and liquid−material ratio and the extraction time and liquid−material ratio [22], which is consistent with Table 4.

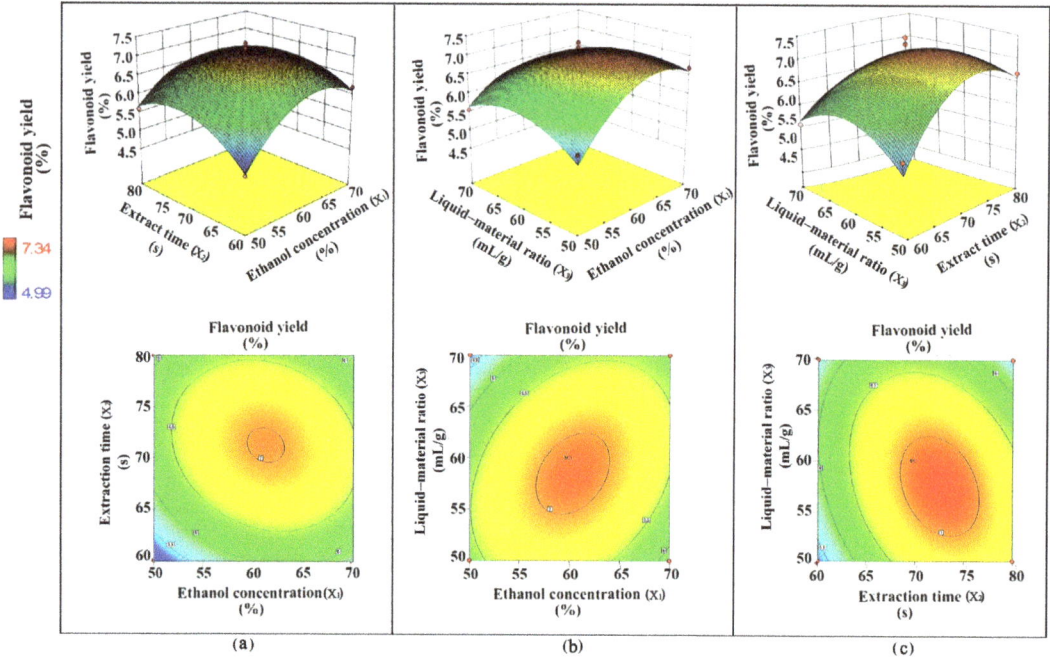

Figure 6. Interaction effect of two factors on flavonoid yield of extract from *Oplopanax elatus* adventitious roots (ARs). (**a**) interaction between ethanol concentration and extraction time. (**b**) interaction between ethanol concentration and liquid−material ratio. (**c**) interaction between extraction time and liquid−material ratio. The ARs were harvested from bioreactors after 4 days of yeast extract (100 mg/L) treatment on day 35 of fed-batch bioreactor culture.

The optimal ethanol concentration (63%), extraction time (69 s), and liquid−material ratio (57 mL/g) were obtained by applying the methodology of the desired function, and the predicted flavonoid yield was 7.12%. Then, the optimized extraction conditions were verified through three repeated tests. Table 5 shows that the flavonoid yields in the triplet repeated tests were 7.04, 7.14, and 7.06%, with an average value of 7.08% and small relative standard deviation (1.58), indicating that the result of the triplet repeated tests was stable and close to the predicted value (7.12%) and proving that the optimized extraction process was feasible.

Table 5. Verification result.

Experiment Repetition	Flavonoid Yield (%)	Average Flavonoid Yield (%)	RSD (%)
1st	7.04		
2nd	7.14	7.08	1.58
3th	7.06		

RSD = relative standard deviation.

3. Discussion

Bioreactor culture of *O. elatus* ARs has been repeatedly studied in recent years using batch and fed-batch culture methods [2,4–6,10]. Compared with batch culture, biomass and bioactive compounds increase dramatically in fed-batch bioreactor culture because the medium inhibition in the early culture stage can be avoided and the nutrient depletion in mid or late culture stage can be replenished. However, efforts to increase bioactive compounds need to be continued during the culture. MeJA as an elicitor has a good elicitation role in many culture systems [4,23–26], including *O. elatus* AR culture. Jiang et al. [4] indicated that MeJA (200 µM) increases bioactive compound production in the batch culture of *O. elatus* ARs. However, the use of MeJA could greatly increase the production cost. Thus, the selection of economical and efficient elicitors is vital for commercial production. In this study, YE was selected as an elicitor to investigate its improvement effect on flavonoid accumulation, aiming to apply the elicitation method on the further large-scale fed-batch bioreactor culture for the mass production of flavonoid-enriched *O. elatus* ARs.

3.1. Optimization of YE Elicitation

Studies have indicated that plant metabolite synthesis can be promoted by various elicitors, including YE [27–29]. However, no elicitation method is suitable for all culture systems. The elicitation efficiency is affected by several factors, such as elicitor types, elicitor concentrations, elicitation treatment time, and duration [30]. Therefore, the selection of an appropriate elicitation strategy is essential for improving the production of useful metabolites.

The concentration effect of elicitors can be divided into two types, i.e., the reaction saturation type and optimum concentration type [31]. For the reaction saturation type, secondary metabolite synthesis increases with an increase in elicitor concentrations, and remains stable after reaching the maximum value. For the optimum concentration type, secondary metabolite synthesis reaches the highest value at a certain level of elicitor concentration, but decreases when the elicitor concentration continues to increase, where most plant culture is of this type [31]. The effect of YE concentration has been investigated by several researchers. For instance, Vijayalakshmi and Shourie [29] indicated that, among the tested YE concentrations (25–175 mg/L), treating with 75 mg/L YE for 10 days is the most suitable for promoting the flavonoid accumulation of *Glycyrrhiza glabra* calluses, where the flavonoid yield is increased more than two-fold. Kochan et al. [32] obtained the maximum yield of ginsenosides (Rb1, Rb2, Rc, Rd, Re, and Rg1) in hairy roots of *Panax quinquefolium* at 50 mg/mL YE, which is obviously higher than that of the other tested YE concentrations (50–2000 mg/L). In this study, we found that the total flavonoid content of *O. elatus* ARs increased when increasing the YE concentrations, peaked at 100 mg/L YE, and then decreased when the YE concentration was higher than 100 mg/L, elucidating that the concentration effect of YE on the flavonoid synthesis of *O. elatus* ARs belongs to the optimum concentration type. These findings prove that the suitable YE concentration is not equal in different culture systems and indicates the importance of selecting elicitor concentrations.

At different stages of growth and development, the sensitivity of plant cells to elicitors varies, and cells that accumulate a certain biomass can effectively receive the elicitor signals and show a high activity [8]. The elicitation efficiency depends on the ages of cultures, and generally shows a good effect when an elicitor stimulates plant cell/organ cultures at the exponential or stationary growth stage. For example, Loc and Giang [28] treated cell cultures of *Centella asiatica* with YE (4 g/L) on day 5, day 10, and day 15 of culture, and suggested that YE treating 10-day-old cells is the most suitable for the production of asiaticosides. In this study, compared with YE treating 40- and 45-day-old ARs, flavonoid accumulation significantly ($p < 0.01$) increased when 35-day-old *O. elatus* ARs were treated with YE (100 mg/L).

The effect of elicitation duration is also critically important. Because of the different defense response rates of plant cells, the suitable elicitation duration differs in each culture system [8]. In plant cell/organ culture, metabolite accumulation tends to increase after a certain elicitation duration and then decrease with an extension of the duration [33]. Farjaminezhad and Garoosi [27] demonstrated that a high azadirachtin content (16.1 mg/g DW) occurs when *Azadirachta indica* cells are treated with YE (25 mg/L) for 2 days. Chen and Chen [34] indicated that an increase in cryptotanshinone is achieved after 7 days of YE treatment in *Ti*-transformed *Salvia miltiorrhiza* cell culture. In this study, we found that treating 35-day-old fed-batch cultured *O. elatus* ARs with YE (100 mg/L) for 4 days was most favorable for enhancing flavonoid synthesis.

The elicitation effect of YE in fed-batch *O. elatus* AR culture was clarified in this study. Meanwhile, our recent study found that salicylic acid could also greatly increase the bioactive compound accumulation of fed-batch cultured *O. elatus* ARs [35]. Therefore, to maximize the production of the useful bioactive compounds, the combined use of elicitors should be investigated in further studies.

3.2. Flavonoid Extraction Using Flash Extraction

Flash extraction is a new extraction technology with many advantages, such as being fast and safe, saving energy, environmental protection, and having a high efficiency. Recently, flash extraction has been used to extract various primary (such as polysaccharides, oils, proteins) and secondary metabolites (such as phenols, alkaloids, and terpenoids) of plants [36].

In flash extraction, many factors, such as solvent types and concentrations, the extraction time, the extraction temperature, and the liquid–material ratio, affect the extraction efficiency [37]. To optimize the flash extraction process, various experimental design methods have been applied, among which RSM is commonly used [38]. For example, Zhang et al. [39] optimized the flash extraction process using RSM for the extraction of phenolics from *Eriobotrya japonica* leaves and indicated that the optimal extraction conditions are 62% ethanol as the solvent, a 32 mL/g liquid–material ratio, and 127 s of extraction time, extracting twice. Li et al. [40] extracted flavonoids from *Crotalaria ferruginea* and applied RSM to optimize the flash extraction process, and the optimal extraction process was 60% ethanol as the solvent, 92 s of extraction time, and a 35 mL/g liquid–material ratio. In this study, we extracted flavonoids of *O. elatus* ARs using flash extraction and used RSM to optimize the extraction process. The result indicates that the optimal extraction process was: 67% ethanol as the solvent, 67 s of extraction time, and a 57 mL/g liquid–material ratio. Under this optimal extraction conditions, the flavonoid yield was higher than 7%. This study proved that the flash extraction method was suitable for the extraction of *O. elatus* ARs, which provides a theoretical basis for rapid and convenient extraction in further large-scale production.

4. Materials and Methods

4.1. Plant Materials and YE Solution Preparation

O. elatus ARs were induced from in vitro cultured seedling roots. The induced ARs were cut into approximately 1 cm length and batch cultured in a 5 L balloon-type airlift bioreactor according to the method of Jiang et al. [5]. After 30 days of culture, the ARs were used for fed-batch bioreactor culture.

YE powder (Shanghai Yuanye Bio-Technology Co., Ltd., Shanghai, China) was dissolved and diluted with deionized water. The YE solution was sterilized through a membrane filter (0.22 µm) and then used in the following YE treatment experiments.

4.2. Fed-Batch AR Culture

The 20 g (fresh weight, FW) ARs were inoculated in a 5 L bioreactor with 3 L of half-strength Murashige and Skoog (MS) medium [41] supplemented with 3 mg/L indolebutyric acid (IBA) (Shanghai Yuanye Bio-Technology Co., Ltd., Shanghai, China) and 40 g/L sucrose. After 15 days of culture, 2 L feeding medium (MS + 3 mg/L IBA + 60 g/L sucrose) was added to the bioreactor. The pH of culture medium was adjusted to 5.8 prior to autoclaving.

The bioreactor was aerated at 0.1 vvm (air volume/medium volume/min) and maintained at 25 °C in the dark.

4.3. YE Elicitation Experiment Design

The experiment process is shown in Figure 7, namely ARs and culture medium were collected from the bioreactor of fed-batch culture and transferred to flasks to conduct YE treatment experiments; the confirmed YE treatment method was verified in the fed-batch bioreactor system.

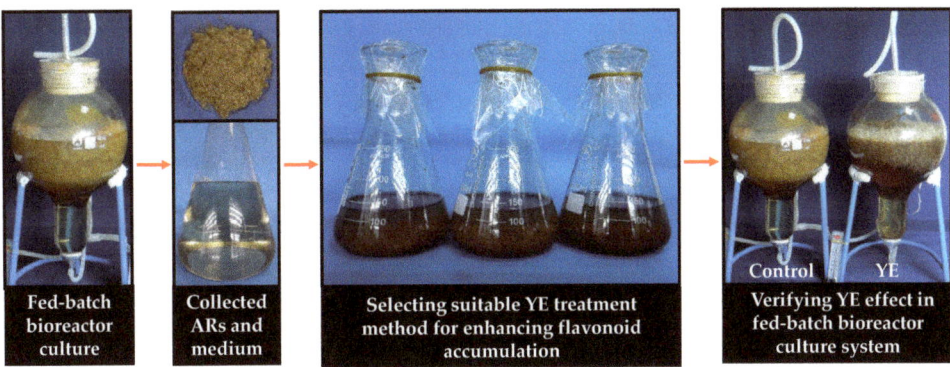

Figure 7. Experiment process.

Elicitation experiments were performed using 250 mL Erlenmeyer flasks. In the first experiment, the effect of YE concentration was investigated. ARs (20 g, FW) were inoculated in the 5 L bioreactor and feeding medium was added after 15 days of initial culture. After 45 days (total culture days) of culture, approximately 600 g fresh ARs and 3 L culture medium were collected. A total of 100 mL medium and 20 g fresh ARs was added to the flask, as well as different concentrations of YE (25, 50, 75, 100, 125, and 150 mg/L); an equal amount of the medium was added to the control group (YE = 0 mg/L). After 8 days of YE treatment, ARs were harvested from the flasks, and the AR biomass (DW) and total flavonoid content were determined. In the second experiment, the effect of YE on flavonoid accumulation of ARs at different ages was investigated. ARs (20 g, FW) were inoculated in 5 L bioreactors, and feeding medium was added after 15 days of initial culture. ARs and culture medium were separately collected from bioreactors after 35, 40, and 45 days (total culture days) of culture. A total of 20 g fresh ARs and 100 mL medium were added to the flask, and then 100 mg/L YE was added according to the result of YE concentration experiment; equal amount of medium was added to the control group. After 8 days of YE treatment, ARs were harvested from the flasks, and AR DW and total flavonoid content were determined. In the third experiment, the effect of YE treatment duration was investigated. ARs (20 g, FW) were inoculated in 5 L bioreactors, and feeding medium was added after 15 days of initial culture. ARs and culture medium were collected from bioreactors after 35 days (total culture days) of culture according to the result of above experiment. A total of 20 g fresh ARs and 100 mL medium were added to the flasks, and then 100 mg/L YE was added according to the result of YE concentration experiment; an equal amount of medium was added to the control group. The ARs were sampled at 1-day intervals, and the AR DW and total flavonoid content were determined. All flasks were kept at 100 rpm on a shaker at 25 °C in the dark.

4.4. Comparison of Flavonoid Contents and Antioxidant Activities between YE-Treated and YE-Untreated ARs in Fed-Batch Bioreactor Culture

The elicitation effect of YE was verified in the fed-batch bioreactor culture system, and the flavonoid contents and antioxidant activities between YE-treated and YE-untreated ARs were compared.

In the YE group, 100 mg/L YE was added to the bioreactor on day 35 of fed-batch bioreactor culture, and the ARs were harvested after 4 days of YE treatment (39 days of total culture period). In the control group, the ARs were harvested after 39 days of fed-batch bioreactor culture. At the end of the bioreactor culture, AR dry weight, the contents of total flavonoids, rutin, quercetin, and kaempferide, and rates of scavenging DPPH, $ABTS^+$ radicals, and chelating Fe^{2+} were determined.

4.5. Optimization of Extraction Process

To extract flavonoids, the YE-treated fed-batch bioreactor-cultured *O. elatus* ARs were extracted using a flash extractor (Shanghai Precision Equipment Co., Ltd., Shanghai, China) with a voltage of 220 V and a revolution of 10,000 rpm.

A total of 3 g dry ARs and solvent were added into an extraction vessel (500 mL) and then extracted for the schedule time. The mixture solution was passed through a sieve (38 μm) and the filtrate was collected. The filtrate was concentrated with a rotary evaporator and then lyophilized to obtain the dry extract. The dry extract weight and the flavonoid content were determined and then flavonoid yield was calculated according to the following formula and used as an evaluation index.

Flavonoid yield (%) = (extract DW (g)/(AR DW [g]) × flavonoid content (mg/g DW) × 0.1

To optimize extraction process, single-factor experiments (Table 6) were firstly conducted to select suitable solvent type and concentration, extract time, and liquid−material ratio. On the basis of the result of the single-factor experiments, the RSM was used to optimize the extraction process by adjusting ethanol (solvent) concentration, extraction time, and liquid−material ratio (Table 2). A total of 17 combination groups are shown in Table 3, and the ARs were extracted according to the relevant extraction condition of each group.

Table 6. Single factor experiment design.

Experiments	Solvent Type	Solvent Concentration (%)	Extraction Time (s)	Liquid−Material Ratio (mL/g)
1	Water, ethanol, methanol	80	60	40
2	Ethanol	40, 50, 60, 70, 80	60	40
3	Ethanol	80	40, 50, 60, 70, 80	40
4	Ethanol	80	70	30, 40, 50, 60, 70

4.6. Determination of AR Dry Weight

The harvested ARs were washed with distilled water trice and dried at 45 °C until a constant weight was achieved after the AR surface water was removed, and then AR DW was recorded.

4.7. Determination of Flavonoid Content

Total flavonoid content was determined according to the method of Jin et al. [6]. In brief, the dry AR sample (0.1 g) was soaked in 10 mL of 70% ethanol, heated at 60 °C for 3 h, and filtered with a filter paper. The filtrate was used to determine the total flavonoid content via the aluminum nitrate colorimetric method at 510 nm using a spectrophotometer; rutin was used as the standard. The result was expressed as the equivalent of rutin per gram of the DW sample.

The method of Zhang et al. [42] was used to determine the contents of rutin, quercetin, and kaempferide using high-performance liquid chromatography with a C_{15} column (4.6 × 250 mm, 5 μm, Thermo Scientific, Waltham, MA, USA) and ultraviolet–visible detector (SPD-15C, Shimadzu Co., Kyoto, Japan). The mobile phases were methanol (A) and 0.1% phosphorus solution (B). Gradient elution was performed as follows: 55% B for 0–15 min and 20% B for 15–30 min. The flow rate was 0.8 mL/min. Rutin, quercetin, and kaempferide were detected at 366 nm (Figure 8). Standards of rutin (purity > 98%), quercetin (purity > 98%), and kaempferide (purity > 98%) were purchased from Shanghai Yuanye Bio-Technology Co., Ltd. (Shanghai, China).

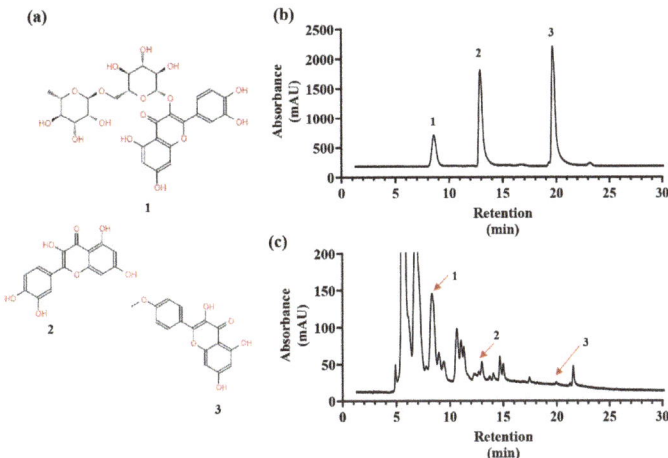

Figure 8. Chemical structure (**a**) and high-performance liquid chromatography profiles of rutin, quercetin, and kaempferide standards (**b**) and adventitious root sample of *Oplopanax elatus* (**c**). (1) Rutin. (2) Quercetin. (3) Kaempferide.

4.8. Determination of Antioxidant Activities

The dry samples of YE-treated ARs were soaked in 70% ethanol and heated at 60 °C for 3 h. After filtration through a filter paper, the filtrate was lyophilized after concentrating under a rotary evaporation. The dry extract was dissolved in deionized water and the different concentrations of the AR extract were prepared using double dilution method. The antioxidant activities of AR extracts of different concentrations were evaluated by determining rates of scavenging DPPH and ATBS$^+$ radicals and chelating Fe^{2+}.

The method of Jiang et al. [5] was used to determine the DPPH radical scavenging rate. In brief, a 96-well plate was added with100 μL of 10 mM DPPH (Notales Biotechnology Co., Ltd., Beijing, China) solution and 100 μL of AR extract (15.6–500 μg/mL). After 30 min of reaction in the dark, the absorbance of the mixture was determined at 517 nm using a microplate reader (iMark, Bio-Rad Laboratories, Inc., Hercules, CA, USA). To determine ABTS$^+$ scavenging rate, the ABTS$^+$ solution was prepared according to Fu et al. [43]. A total of 3.9 mL ABTS$^+$ solution and 0.1 mL of AR extract (0.25–4 mg/mL) were mixed and reacted for 6 min in the dark. The absorbance of the mixture was determined at 734 nm (UV-T6, Beijing Purkinje General Instrument Co., Ltd., Beijing, China). The Fe^{2+} chelating rate was determined using the method of Fu et al. [43]. In brief, a 96-well plate was added with 50 μL of AR extract (1.25–20 mg/mL), 2.5 μL $FeCl_2$, 10 L ferrozine, and 137.5 L deionized water, and was incubated for 10 min in the dark. The absorbance of the mixture was determined at 562 mm (iMark, Bio-Rad Laboratories, Inc., Hercules, CA, USA).

4.9. Statistical Analysis

All data are presented as the mean ± standard deviation of three independent replicates. Data were analyzed using Duncan's multiple range test or Student's *t*-test by using SPSS statistics 22.0 software (IBM Institute, Almonk, NY, USA). Values of $p < 0.05$ were considered statistically significant.

5. Conclusions

The YE concentration, AR age of YE treatment, and YE treatment duration critically affected the flavonoid accumulation of fed-batch bioreactor-cultured *O. elatus* ARs. The concentration effect of YE belonged to the optimum concentration type, where the optimal YE concentration was 100 mg/L; flavonoid accumulation was the most favorable when 35-day-old fed-batch cultured ARs were treated with 100 mg/L YE for 4 days, in which the total flavonoid content was 224.5 mg/g DW higher than the control; the contents of rutin and kaempferide were also greatly increased after YE treatment, whereas YE did not significantly affect the quercetin content. Flavonoid accumulation was enhanced by YE treatment, and the antioxidant activity was correspondingly increased; the rates of scavenging DPPH and ABTS and chelating Fe^{2+} in YE-ARE were higher than those in C-ARE. Flash extraction efficiently extracted flavonoids from *O. elatus* ARs, where the optimized extraction process was: 63% ethanol, 69 s of extraction time, and a 57 mL/g liquid−material ratio. The findings of this study provide a useful method for increasing the flavonoid production of *O. elatus* ARs and form a theoretical basis for the utilization of ARs in the future development of *O. elatus* ARs.

Author Contributions: Conceptualization, M.-Y.J.; methodology, M.W. and X.-H.W.; validation, M.-Z.F.; investigation, H.-X.L.; data curation, Y.-Q.G. and J.J.; supervision and funding acquisition, C.-R.Y.; writing—review and editing, and funding acquisition, M.-L.L. All authors have read and agreed to the published version of the manuscript.

Funding: This research was funded by National Natural Science Foundation of China, grant number 81960685 and the Key Projects of Science and Technology Development Plan of Jilin Province, grant number 20210204171YY.

Data Availability Statement: The datasets used and/or analyzed during this study are available from the corresponding author upon reasonable request.

Conflicts of Interest: The authors declare no conflict of interest.

References

1. Shikov, A.N.; Pozharitskaya, O.N.; Makarov, V.G.; Yang, W.Z.; Guo, D.A. *Oplopanax elatus* (Nakai) Nakai: Chemistry, traditional use and pharmacology. *Chin. J. Nat. Med.* **2014**, *12*, 721–729. [CrossRef] [PubMed]
2. Jiang, X.L.; Jin, M.Y.; Piao, X.C.; Yin, C.R.; Lian, M.L. Fed-batch culture of *Oplopanax elatus* adventitious roots: Feeding medium selection through comprehensive evaluation using an analytic hierarchy process. *Biochem. Eng. J.* **2021**, *167*, 107927. [CrossRef]
3. Han, L.; Piao, X.C.; Jiang, Y.J.; Jiang, X.L.; Yin, C.R.; Lian, M.L. A high production of flavonoids and anthraquinones via adventitious root culture of *Oplopanax elatus* and evaluating antioxidant activity. *Plant Cell Tiss. Organ Cult.* **2019**, *137*, 173–179. [CrossRef]
4. Jiang, X.L.; Piao, X.C.; Gao, R.; Jin, M.Y.; Jin, X.H.; Lian, M.L. Improvement of bioactive compound accumulation in adventitious root cultures of an endangered plant species, *Oplopanax elatus*. *Acta Physiol. Plant.* **2017**, *39*, 226. [CrossRef]
5. Jiang, Y.J.; Piao, X.C.; Liu, J.S.; Lian, Z.X.; Kim, M.J.; Lian, M.L. Bioactive compound production by adventitious root culture of *Oplopanax elatus* in balloon-type airlift bioreactor systems and bioactivity property. *Plant Cell Tiss. Organ Cult.* **2015**, *123*, 413–425. [CrossRef]
6. Jin, M.Y.; Piao, X.C.; Wu, X.H.; Fan, M.Z.; Li, X.F.; Yin, C.R.; Lian, M.L. *Oplopanax elatus* adventitious root production through fed-batch culture and their anti-bacterial effects. *Plant Cell Tiss. Organ Cult.* **2020**, *140*, 447–457. [CrossRef]
7. Lee, K.J.; Park, Y.; Kim, J.Y.; Jeong, T.K.; Yun, K.S.; Paek, K.; Park, S.Y. Production of biomass and bioactive compounds from adventitious root cultures of *Polygonum multiflorum* using air-lift bioreactors. *J. Plant Biotechnol.* **2015**, *42*, 34–42. [CrossRef]
8. Linh, N.T.N.; Cuong, L.K.; Tam, H.T.; Tung, H.T.; Luan, V.Q.; Hien, V.T.; Loc, N.H.; Nhut, D.T. Improvement of bioactive saponin accumulation in adventitious root cultures of *Panax vietnamensis* via culture periods and elicitation. *Plant Cell Tiss. Organ Cult.* **2019**, *137*, 101–113. [CrossRef]

9. Wu, C.H.; Tang, J.; Jin, Z.X.; Wang, M.; Liu, Z.Q.; Huang, T.; Lian, M.L. Optimizing co-culture conditions of adventitious roots of *Echinacea pallida* and *Echinacea purpurea* in air-lift bioreactor systems. *Biochem. Eng. J.* **2018**, *132*, 206–216. [CrossRef]
10. Jin, M.Y.; Hao, Y.J.; Zhang, K.X.; Yin, C.R.; Jiang, J.; Piao, X.C.; Lian, M.L. Fed-batch culture of *Oplopanax elatus* adventitious roots: Establishment of a complete culture system. *Biochem. Eng. J.* **2023**, *194*, 108898. [CrossRef]
11. Choi, S.; Son, S.; Yun, S.; Kwon, O.; Seon, J.; Paek, K. Pilot-scale culture of adventitious roots of ginseng in a bioreactor system. *Plant Cell Tiss. Organ Cult.* **2000**, *62*, 187–193. [CrossRef]
12. Cui, H.Y.; Baque, M.A.; Lee, E.J.; Paek, K. Scale-up of adventitious root cultures of *Echinacea angustifolia* in a pilot-scale bioreactor for the production of biomass and caffeic acid derivatives. *Plant Biotechnol. Rep.* **2013**, *7*, 297–308. [CrossRef]
13. Ho, T.T.; Lee, K.J.; Lee, J.D.; Bhushan, S.; Paek, K.; Park, S.Y. Adventitious root culture of *Polygonum multiflorum* for phenolic compounds and its pilot-scale production in 500 L-tank. *Plant Cell Tiss. Organ Cult.* **2017**, *130*, 167–181. [CrossRef]
14. An, X.L.; Yu, Y.; Fan, M.Z.; Wu, X.H.; Li, X.F.; Piao, X.C.; Lian, M.L. A fungal mycelium elicitor efficiently improved ginsenoside synthesis during adventitious root culture of *Panax ginseng*. *J. Plant Biochem. Biotechnol.* **2022**, *31*, 657–664. [CrossRef]
15. Ho, T.T.; Lee, J.D.; Jeong, C.S.; Paek, K.; Park, S.Y. Improvement of biosynthesis and accumulation of bioactive compounds by elicitation in adventitious root cultures of *Polygonum multiflorum*. *Appl. Microbiol. Biotechnol.* **2018**, *102*, 199–209. [CrossRef]
16. Le, K.C.; Im, W.T.; Paek, K.; Park, S.Y. Biotic elicitation of ginsenoside metabolism of mutant adventitious root culture in *Panax ginseng*. *Appl. Microbiol. Biotechnol.* **2018**, *102*, 1687–1697. [CrossRef]
17. Huang, L.; Xiao, W.J.; Yang, G.; Mo, G.; Lin, S.F.; Wu, Z.G.; Guo, L.P. Mechanism exploration on synthesis of secondary metabolites in *Sorbus aucuparia* cell cultures treated with yeast extract. *Chin. Trad. Herb. Drugs.* **2014**, *39*, 2019–2023.
18. Goncharuk, E.A.; Saibel, O.L.; Zaitsev, G.P.; Zagoskina, N.V. The elicitor effect of yeast extract on the accumulation of phenolic compounds in *Linum grandiflorum* cells Cultured in vitro and their antiradical activity. *Biol Bull.* **2022**, *49*, 620–628. [CrossRef]
19. Maqsood, M.; Abdul, M. Yeast extract elicitation increases vinblastine and vincristine yield in protoplast derived tissues and plantlets in *Catharanthus roseus*. *Rev. Bras. Farmacogn.* **2017**, *27*, 549–556. [CrossRef]
20. Liu, Y.; Yuan, K.; Ji, C. New method of extraction on the chemical components of Chinese medicinal plants-extracting method by smashing of plant tissue (EMS). *Henan. Sci.* **1993**, *11*, 265–268.
21. Latif, A.; Maqbool, A.; Zhou, R.Z.; Arsalan, M.; Sun, K.; Si, Y.B. Optimized degradation of bisphenol A by immobilized laccase from *Trametes versicolor* using Box-Behnken design (BBD) and artificial neural network (ANN). *J. Environ. Chem. Eng.* **2022**, *10*, 107331. [CrossRef]
22. Hao, Y.J.; Zhang, K.X.; Jin, M.Y.; Piao, X.C.; Lian, M.L.; Jiang, J. Improving fed-batch culture efficiency of *Rhodiola sachalinensis* cells and optimizing flash extraction process of polysaccharides from the cultured cells by BBD-RSM. *Ind. Crop. Prod.* **2023**, *196*, 116513. [CrossRef]
23. Lee, E.J.; Park, S.Y.; Paek, K. Enhancement strategies of bioactive compound production in adventitious root cultures of *Eleutherococcus koreanum* Nakai subjected to methyl jasmonate and salicylic acid elicitation through airlift bioreactors. *Plant Cell Tiss. Organ Cult.* **2014**, *120*, 1–10. [CrossRef]
24. Li, B.; Wang, B.; Li, H.; Peng, L.; Ru, M.; Liang, Z.; Yan, X.; Zhu, Y. Establishment of *Salvia castanea* Diels f. tomentosa Stib. hairy root cultures and the promotion of tanshinone accumulation and gene expression with Ag^+, methyl jasmonate, and yeast extract elicitation. *Protoplasma* **2016**, *253*, 87–100. [CrossRef] [PubMed]
25. Liu, Q.; Kim, S.B.; Jo, Y.; Ahn, J.; Turk, A.; Kim, D.; Chang, B.; Kim, S.Y.; Jeong, C.S.; Hwang, B.Y.; et al. Curcubinoyl-conjugated flavonoids from methyl jasmonate-treated wild ginseng adventitious root cultures. *Sci. Rep.* **2021**, *11*, 12212. [CrossRef]
26. Wang, J.A.; Gao, W.Y.; Zhang, J.; Huang, T.; Wen, T.T.; Huang, L.Q. Combination effect of lactoalbumin hydrolysate and methyl jasmonate on ginsenoside and polysaccharide production in *Panax quinquefolium* L. cells cultures. *Acta Physiol. Plant.* **2011**, *33*, 861–866. [CrossRef]
27. Farjaminezhad, R.; Garoosi, G. Improvement and prediction of secondary metabolites production under yeast extract elicitation of *Azadirachta indica* cell suspension culture using response surface methodology. *AMB Express* **2021**, *11*, 1–16. [CrossRef]
28. Loc, N.H.; Giang, N.T. Effects of elicitors on the enhancement of asiaticoside biosynthesis in cell cultures of centella (*Centella asiatica* L. Urban). *Chem. Pap.* **2012**, *66*, 642–648. [CrossRef]
29. Vijayalakshmi, U.; Shourie, A. Yeast extract-mediated elicitation of anti-cancerous compounds licoisoflavone B, licochalcone A, and liquirtigenin in callus cultures of *Glycyrrhiza glabra*. *Biotechnologia* **2019**, *100*, 441–451. [CrossRef]
30. Mahood, H.E.; Sarropoulou, V.; Tzatzani, T.T. Effect of explant type (leaf, stem) and 2,4-D concentration on callus induction: Influence of elicitor type (biotic, abiotic), elicitor concentration and elicitation time on biomass growth rate and costunolide biosynthesis in gazania (*Gazania rigens*) cell suspension cultures. *Bioresour. Bioprocess.* **2022**, *9*, 100.
31. Abdul, M.N.A.; Kumar, I.S.; Nadarajah, K. Elicitor and receptor molecules: Orchestrators of plant defense and immunity. *Int. J. Mol. Sci.* **2020**, *21*, 963. [CrossRef] [PubMed]
32. Kochan, E.; Szymczyk, P.; Kuźma, Ł.; Lipert, A.; Szymańska, G.S. Yeast extract stimulates ginsenoside production in hairy root cultures of American ginseng cultivated in shake flasks and nutrient sprinkle bioreactors. *Molecules* **2017**, *22*, 880–895. [CrossRef]
33. Hao, Y.J.; An, X.L.; Sun, H.D.; Piao, X.C.; Gao, R.; Lian, M.L. Ginsenoside synthesis of adventitious roots in *Panax ginseng* is promoted by fungal suspension homogenate of *Alternaria panax* and regulated by several signaling molecules. *Ind. Crop. Prod.* **2020**, *150*, 112414. [CrossRef]
34. Chen, H.; Chen, F. Effects of yeast elicitor on the growth and secondary metabolism of a high-tanshinone-producing line of the ti transformed *Salvia miltiorrhiza* cells in suspension culture. *Process Biochem.* **2000**, *35*, 837–840. [CrossRef]

35. Yu, S.; Wu, X.H.; Wang, M.; Liu, L.L.; Ye, W.Q.; Jin, M.Y.; Piao, X.C.; Lian, M.L. Optimizing elicitation strategy of salicylic acid for flavonoid and phenolic production of fed-batch cultured *Oplopanax elatus* adventitious roots. *J. Biotechnol.* **2023**, *368*, 1–11. [CrossRef]
36. Kan, L.N.; Wang, L.; Ding, Q.Z.; Wu, Y.W.; Ouyang, J. Flash extraction and physicochemical characterization of oil from *Elaeagnus mollis* Diels seeds. *J. Oleo Sci.* **2017**, *66*, 345–352. [CrossRef] [PubMed]
37. Qin, D.Y.; Xi, J. Flash extraction: An ultra-rapid technique for acquiring bioactive compounds from plant materials. *Trends Food Sci. Tech.* **2021**, *112*, 581–591. [CrossRef]
38. Xu, L.; Xu, J.; Shi, G.H.; Xiao, S.N.; Dai, R.K.; Wu, S.; Sun, B.S.; Zhang, X.S.; Zhao, Y.Q. Optimization of flash extraction, separation of ginsenosides, identification by HPLC-FT-ICR-MS and determination of rare ginsenosides in mountain cultivated ginseng. *RSC Adv.* **2020**, *10*, 44050–44057. [CrossRef]
39. Zhang, Y.; Yang, C.; Huang, J.Q.; Xu, X.Q. Optimization of flash extraction of total polyphenols from leaves of *Eriobotrya japonica* lind by response surface methodology. *Guangzhou Chem. Ind.* **2019**, *47*, 112–115.
40. Li, Y.; Zhao, T.M.; Huang, L.R.; Zou, T.; Wu, X.L. Response surface optimized extraction of flavonoids from *Crotalaria ferruginea* and antioxidant activities. *Food Res. Dev.* **2019**, *40*, 79–84.
41. Murashige, T.; Skoog, F. A revised medium for rapid growth bioassays with tobacco tissues cultures. *Physiol. Plant.* **1962**, *15*, 473–497. [CrossRef]
42. Zhang, W.; Piao, X.C.; Li, J.R.; Jin, Y.H.; Lian, M.L. Optimized culture medium for the production of flavonoids from *Orostachys cartilaginea* V.N. Boriss. callus cultures. *In Vitro Cell. Dev. Biol.-Plant* **2017**, *53*, 1–11. [CrossRef]
43. Fu, R.; Zhang, Y.; Guo, Y.; Chen, F. Antioxidant and tyrosinase inhibition activities of the ethanol-insoluble fraction of water extract of *Sapium sebiferum* (L.) Roxb. leaves. *S. Afr. J. Bot.* **2014**, *93*, 98–104. [CrossRef]

Disclaimer/Publisher's Note: The statements, opinions and data contained in all publications are solely those of the individual author(s) and contributor(s) and not of MDPI and/or the editor(s). MDPI and/or the editor(s) disclaim responsibility for any injury to people or property resulting from any ideas, methods, instructions or products referred to in the content.

Article

Phytochemical Composition and Detection of Novel Bioactives in Anther Callus of *Catharanthus roseus* L.

Yashika Bansal [1], A. Mujib [1,*], Jyoti Mamgain [1], Yaser Hassan Dewir [2] and Hail Z. Rihan [3]

[1] Cellular Differentiation and Molecular Genetics Section, Department of Botany, Jamia Hamdard, New Delhi 110062, India; yashikab333@gmail.com (Y.B.); jyotimamgain93@gmail.com (J.M.)
[2] Plant Production Department, College of Food and Agriculture Sciences, King Saud University, Riyadh 11451, Saudi Arabia; ydewir@ksu.edu.sa
[3] School of Biological and Marine Sciences, Faculty of Science and Engineering, University of Plymouth, Drake Circus PL4 8AA, UK; hail.rihan@plymouth.ac.uk
* Correspondence: amujib3@yahoo.co.in

Abstract: *Catharanthus roseus* L. (G.) Don is the most widely studied plant because of its high pharmacological value. In vitro culture uses various plant parts such as leaves, nodes, internodes and roots for inducing callus and subsequent plant regeneration in *C. roseus*. However, till now, little work has been conducted on anther tissue using plant tissue culture techniques. Therefore, the aim of this work is to establish a protocol for in vitro induction of callus by utilizing anthers as explants in MS (Murashige and Skoog) medium fortified with different concentrations and combinations of PGRs. The best callusing medium contains high α-naphthalene acetic acid (NAA) and low kinetin (Kn) concentrations showing a callusing frequency of 86.6%. SEM–EDX analysis was carried out to compare the elemental distribution on the surfaces of anther and anther-derived calli, and the two were noted to be nearly identical in their elemental composition. Gas chromatography–mass spectrometry (GC–MS) analysis of methanol extracts of anther and anther-derived calli was conducted, which revealed the presence of a wide range of phytocompounds. Some of them are ajmalicine, vindolinine, coronaridine, squalene, pleiocarpamine, stigmasterol, etc. More importantly, about 17 compounds are exclusively present in anther-derived callus (not in anther) of *Catharanthus*. The ploidy status of anther-derived callus was examined via flow cytometry (FCM), and it was estimated to be 0.76 pg, showing the haploid nature of callus. The present work therefore represents an efficient way to produce high-value medicinal compounds from anther callus in a lesser period of time on a larger scale.

Keywords: anther culture; flow cytometry; GC–MS; phytochemical profiling; ploidy level; secondary metabolites; SEM–EDX

1. Introduction

Catharanthus roseus (L.) G. Don, a member of the Apocynaceae family, is a popular flowering plant. It is an indigenous species to Madagascar and is widely distributed throughout the African, American, Asian and southern European regions. In India, *C. roseus* has been spread across all the major parts of Gujarat, Madhya Pradesh, Assam, Bihar, Uttar Pradesh, Karnataka and Tamil Nadu [1]. The plant is well known for both its ornamental and medicinal value. It produces nearly 130 alkaloids, of which vincristine and vinblastine are the two major compounds that are used in the treatment of leukemia and Hodgkin's lymphoma [2]. For decades, this plant has been exploited for pharmaceutically active compounds from its native environments and thus is at risk of declining in the wild. Plant tissue culture proves to be an effective biotechnological tool for the rapid propagation of plants under aseptic conditions with a lesser risk of microbial infections [3]. Several in vitro studies using different explants have been successfully conducted for somatic embryogenesis [4] and organogenesis in *C. roseus* [5,6].

In recent times, double haploid (DH) production via anther is a promising option for developing improved plant varieties with high yields of medicinally important bioactive compounds [7]. In vitro anther culture has been attempted in various plants such as *Actinidia arguta* Planch [8] and *Triticum aestivum* L. [9]. Various factors such as stage of anther, culture conditions, plant growth regulators (PGRs) and genotypic and ploidy status determine the success of DH generation [10]. These factors necessitate ascertaining the ploidy status of anther-derived callus to generate true-to-type DH lines, which can be performed with a flow cytometric technique. The flow cytometry method (FCM) measures the genome size by examining the nuclei at a relatively faster rate and thus validates the ploidy levels of different plant tissues [11]. Recent investigations of genome size analysis using FCM have been reported for different plants [12,13]. Phytochemical profiling using gas chromatography coupled with mass spectrometry (GC–MS) has emerged as an important procedure for identifying and quantifying therapeutically significant compounds present in medicinal plants. This technique is relatively faster, accurate and needs a minimum volume of extracts to detect a wide range of bioactive compounds such as alkaloids, long-chain hydrocarbons, steroids, sugars, amino acids and nitro compounds [14]. Major bioactive compounds extracted from different plant parts of *C. roseus* such as stem, root and leaf include vincristine, vinblastine, reserpine, ajmalicine, vindolinine and catharine, which possess anti-cancerous, anti-diabetic, anti-fungal and anti-microbial activities [15]. GC–MS-based profiling has been recently reported for several plants including *Silybum marianum* L. [16] and *Chukrasia velutina* [17], but the information on tissue-culture-raised plants' phytocompound profiling is relatively much less. The present work, therefore, focuses on investigating the ploidy status of anther-derived callus of *C. roseus* using flow cytometry. The elemental composition of both anther and anther calli was studied using a scanning electron microscopy–energy-dispersive X-ray microanalysis (SEM–EDX) technique. The identification of the bioactive compounds present in methanolic extracts of anther and anther-derived calli was conducted for the first time in *C. roseus* using GC–MS analysis. This report will help to understand and improve the yield of the important pharmaceutical compounds synthesized from anther-derived callus.

2. Results

2.1. Callus Induction and Proliferation

In this study, the anthers were used as explants to induce callus on MS medium augmented with different concentrations and combinations of NAA and kinetin or TDZ alone (Figure 1A). The callusing response ranged from 13.3% to 86.6% on all the tested media (Table 1). Among the PGRs utilized, a combination of NAA and kinetin produced maximum callus (86.6%) at concentrations of 1.0 mg/L and 0.1 mg/L, respectively, followed by 0.75 mg/L TDZ with a frequency of 73.3%. On the other hand, TDZ alone at 0.5 mg/L showed the least incidence of callusing efficiency (13.3%). The highest callus fresh weight was noted to be 1.7 g on MS medium containing 1.0 mg/L NAA and 0.1 mg/L kinetin. The calli obtained were white to pale yellow in color and friable in nature (Figure 1B–D). The anther callus was noted to be recalcitrant, as plant regeneration (embryogenesis and organogenesis) was not achieved on any medium added with various PGR combinations.

Table 1. Effect of different concentrations and combinations of PGRs on callus induction and callus biomass (fresh weight) from anther explants of *C. roseus*.

PGRs	Concentration (mg/L)	Callusing Frequency (%)	Mean Fresh Weight (g)
Control	0	0 [e]	0 [c]
NAA + Kn	0.1 + 1.0	26.6 ± 12.4 [cde]	0.8 ± 0.3 [abc]
	0.5 + 0.75	33.3 ± 14.9 [cde]	0.9 ± 0.3 [ab]
	0.75 + 0.5	53.3 ± 16.9 [abc]	1.1 ± 0.3 [ab]
	1.0 + 0.1	86.6 ± 8.1 [a]	1.7 ± 1.7 [a]

Table 1. *Cont.*

PGRs	Concentration (mg/L)	Callusing Frequency (%)	Mean Fresh Weight (g)
TDZ	0.5	13.3 ± 8.1 de	0.5 ± 0.3 bc
	0.75	73.3 ± 27.8 ab	1.3 ± 0.2 ab
	1	46.6 ± 16.9 bcd	0.9 ± 0.2 ab

Mean values followed by the same superscripts within a column are not significantly different according to DMRT at $p \leq 0.05$ level.

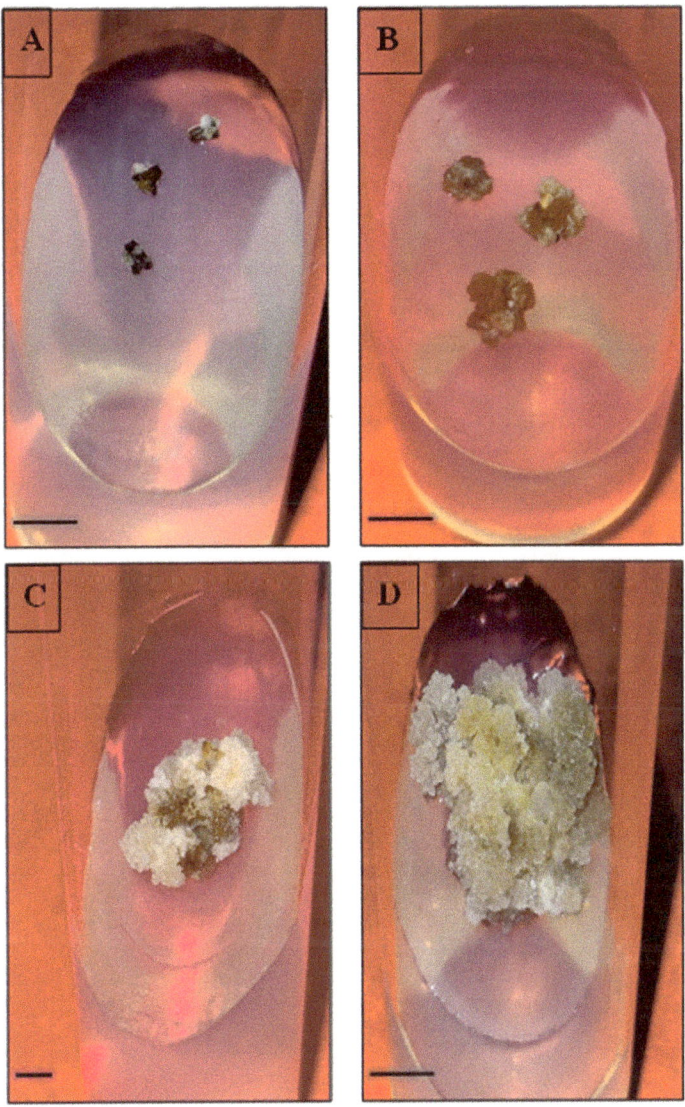

Figure 1. *Cont.*

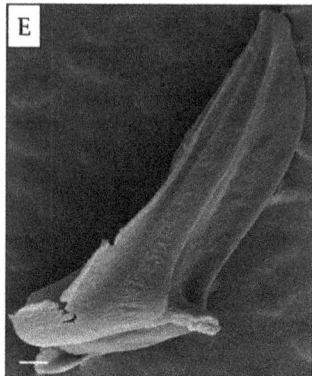
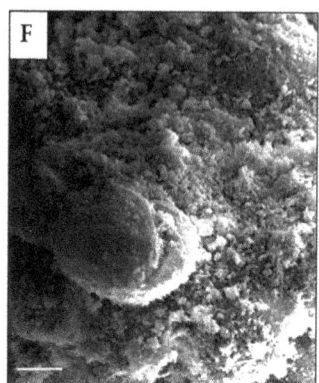

Figure 1. In vitro callus induction, proliferation and scanning electron microscopic (SEM) images of anther and anther-derived callus of *C. roseus*. (**A**,**B**): callus initiation (bars = 0.5 cm); (**C**,**D**): callus proliferation after 6 and 9 weeks, respectively (bars (**C**) = 1.0 cm, (**D**) = 0.5 cm); (**E**): side view of anther (bar = 200 μm); (**F**): a portion of anther-derived callus (bar = 20 μm).

2.2. Surface Morphology and Elemental Analysis

SEM–EDAX analysis was carried out to determine the elemental composition of anther as well as anther-derived callus. The SEM images and their respective spectra are shown in Figure 1E,F and Figure 2, respectively. The various peaks in both spectra reveal carbon, oxygen, sodium and phosphorous to be the major elements present on the surfaces of anther and anther-derived calli. In both the samples, the carbon and oxygen peaks are prominent and of high intensity, whereas those of sodium and phosphorous are of nearly equal intensity. The quantitative estimation of elements is presented in Table 2.

Table 2. Elemental composition of anther and anther-derived callus of *C. roseus* using SEM–EDX analysis.

S.No.	Element	Anther Explant		Anther-Derived Callus	
		Weight %	Atomic %	Weight %	Atomic %
1	Carbon	33.59	70.67	47.34	79.42
2	Oxygen	12.65	19.97	11.55	14.55
3	Sodium	1.93	2.12	1.63	1.42
4	Phosphorous	0.87	0.71	1.03	0.67

2.3. GC–MS Analysis

The bioactive compounds present in methanolic extracts of anthers (donor material) and anther-derived callus of *C. roseus* (Figure 3) were identified using the GC–MS technique. The active principles with their retention time (RT), peak area % (concentration), molecular formula and molecular weight from the NIST library are presented in Tables 3 and 4, and the GC–MS chromatograms are presented in Figure 4A,B. The chromatograms reveal more than 50 phytocompounds in both methanolic extracts belonging to various classes such as terpenoids, phenols, lignans, steroids, alkaloids and fatty acids.

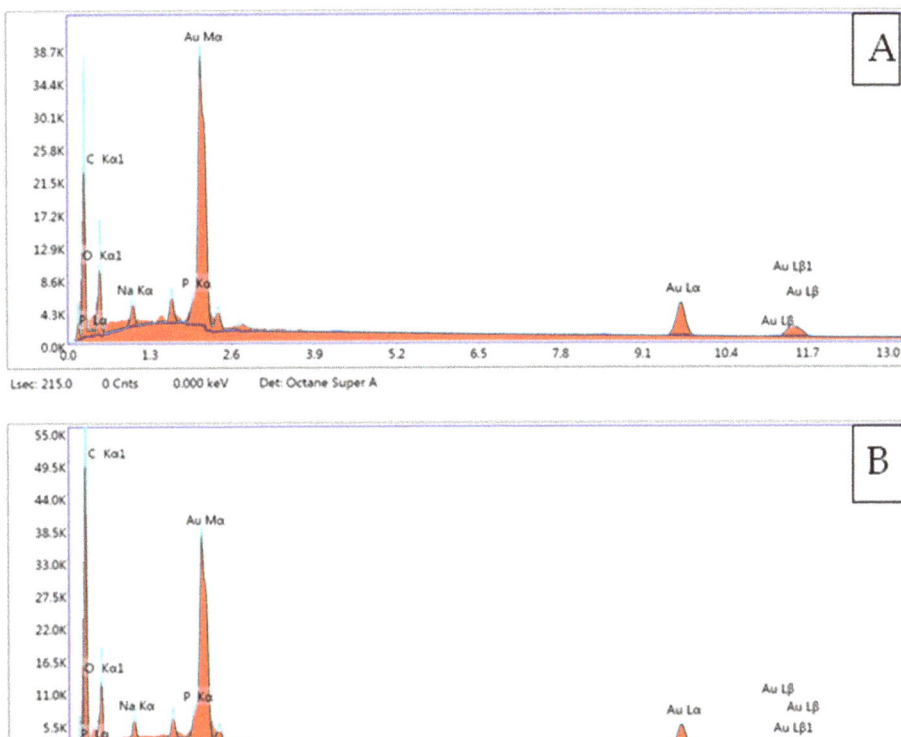

Figure 2. SEM–EDX analysis micrographs showing elemental composition of *C. roseus*. (**A**): field grown anther; (**B**): anther-derived callus.

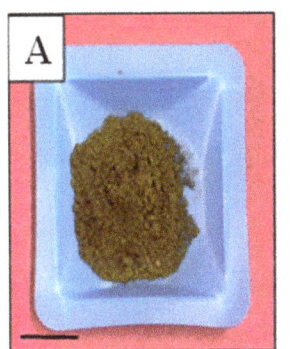
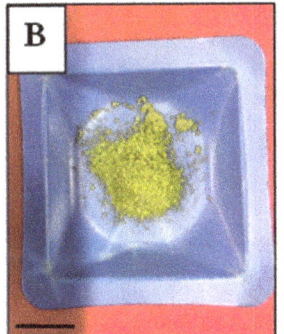
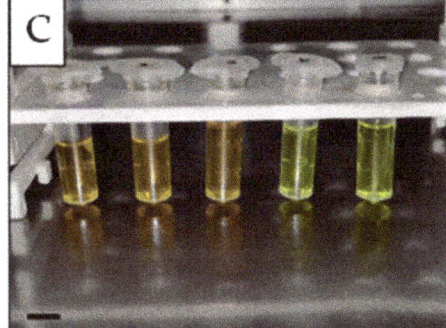

Figure 3. Extract preparation for GC–MS analysis of *C. roseus*. (**A**): dried powder of anther-derived callus; (**B**): dried powder of field-grown anther; (**C**): methanolic extracts of the samples (**A**,**B**).

Table 3. List of phytocompounds identified in the methanolic extract of field-grown anther of *C. roseus* using GC–MS analysis.

S.No.	RT (min)	Peak Area %	Name of the Compound	Molecular Formula	Molecular Weight
1	3.760	1.62	Ethylcyclopentenolone	$C_7H_{10}O_2$	126
2	4.436	1.12	Pyranone	$C_6H_8O_4$	144
3	5.484	1.17	Coumaran	C_8H_8O	120
4	5.739	0.42	1-monoacetin	$C_5H_{10}O_4$	134
5	6.287	0.26	6-oxoheptanoic acid	$C_7H_{12}O_3$	144
6	6.504	0.38	Indole	C_8H_7N	117
7	6.628	0.14	4-vinylguaiacol	$C_9H_{10}O_2$	150
8	7.616	2.12	1,2-octanediol	$C_8H_{18}O_2$	146
9	9.101	18.42	Guanosine	$C_{10}H_{13}N_5O_5$	283
10	9.816	0.45	2,6-dimethoxy-4-vinylphenol	$C_{10}H_{12}O_3$	180
11	10.086	0.78	1,2-benzenedicarboxylic acid, diethyl ester	$C_{12}H_{14}O_4$	222
12	10.473	0.09	Cedrol	$C_{15}H_{26}O$	222
13	10.796	0.17	Dihydromethyljasmonate	$C_{13}H_{22}O_3$	226
14	11.050	3.78	Quinic acid	$C_7H_{12}O_6$	192
15	11.914	0.08	2-benzylideneoctanal	$C_{15}H_{20}O$	216
16	12.061	2.86	Mome inositol	$C_7H_{14}O_6$	194
17	13.082	0.13	Diisobutyl phthalate	$C_{16}H_{22}O_4$	278
18	13.411	0.11	Heptadecane	$C_{17}H_{36}$	240
19	13.681	0.09	Methyl palmitate	$C_{17}H_{34}O_2$	270
20	14.117	0.18	n-hexadecanoic acid	$C_{16}H_{32}O_2$	256
21	14.403	0.12	Eicosane	$C_{20}H_{42}$	282
22	15.355	4.71	Hexacosane	$C_{26}H_{54}$	366
23	15.883	0.09	Docosanoic acid	$C_{22}H_{44}O_2$	340
24	16.259	0.79	Tetracosane	$C_{24}H_{50}$	338
25	16.910	0.35	9-tricosanol acetate	$C_{25}H_{50}O_2$	382
26	17.134	9.78	Hexatriacontane	$C_{36}H_{74}$	506
27	17.293	0.20	4,5-dihydro-2-[(8Z,11Z)-8,11-heptadecadienyl]oxazole	$C_{20}H_{35}NO$	305
28	17.398	0.19	4,8-cyclododecadien-1-one	$C_{12}H_{18}O$	178
29	17.963	1.13	Dotriacontane	$C_{32}H_{66}$	450
30	18.571	0.30	Octacosanol	$C_{28}H_{58}O$	410
31	18.765	2.82	n-tetracontane	$C_{40}H_{82}$	562
32	18.953	0.80	alpha-monostearin	$C_{21}H_{42}O_4$	358
33	19.536	0.22	1-bromotriacontane	$C_{30}H_{61}Br$	500
34	20.322	0.11	Linoleyl acetate	$C_{20}H_{36}O_2$	308
35	20.523	0.12	(-)-Coronaridine	$C_{21}H_{26}N_2O_2$	338
36	21.137	27.01	Squalene	$C_{30}H_{50}$	410
37	22.759	0.19	Arachidic acid, 3-methylbutyl ester	$C_{25}H_{50}O_2$	382
38	22.896	0.57	beta-tocopherol	$C_{28}H_{48}O_2$	416

Table 3. Cont.

S.No.	RT (min)	Peak Area %	Name of the Compound	Molecular Formula	Molecular Weight
39	23.543	0.30	Vitamin E	$C_{29}H_{50}O_2$	430
40	24.640	1.20	Campesterol	$C_{28}H_{48}O$	400
41	24.766	0.30	Ergostan-3-ol	$C_{28}H_{50}O$	402
42	25.105	0.08	Trans-24-ethylidenecholesterol	$C_{29}H_{48}O$	412
43	25.178	0.52	3-oxocholestane	$C_{27}H_{46}O$	386
44	25.440	0.22	p-coumaric acid, 2-methylpropyl ether, 2-methylpropyl ester	$C_{17}H_{24}O_3$	276
45	25.603	2.87	gamma-sitosterol	$C_{29}H_{50}O$	414
46	25.762	0.57	Stigmastanol	$C_{29}H_{52}O$	416
47	26.055	0.16	Ergosta-4,24(28)-dien-3-one	$C_{28}H_{44}O$	396
48	26.132	0.87	4-campestene-3-one	$C_{28}H_{46}O$	398
49	26.230	0.23	Cholestanone	$C_{27}H_{46}O$	386
50	27.321	2.72	Methyl commate C	$C_{31}H_{50}O_4$	486
51	28.021	5.54	alpha amyrin	$C_{30}H_{50}O$	426

Table 4. List of phytocompounds identified in the methanolic extract of anther-derived callus of *C. roseus* using GC–MS analysis.

S.No.	RT (min)	Peak Area %	Name of the Compound	Molecular Formula	Molecular Weight
1	3.598	0.58	1,3,5-triazine-2,4,6-triamine	$C_3H_6N_6$	126
2	4.320	0.10	Isopropylmethylnitrosamine	$C_4H_{10}N_2O$	102
3	4.498	5.49	1,2,3 propanetriol	$C_3H_8O_3$	92
4	5.040	0.23	3-cis-methoxy-5-trans-methyl-1R-cyclohexanol	$C_8H_{16}O_2$	144
5	5.270	0.35	Catechol	$C_6H_6O_2$	110
6	5.402	0.50	2,5,5-trimethylhepta-2,6-dien-4-ol	$C_{10}H_{18}O$	154
7	5.508	3.89	5-hydroxymethylfurfural	$C_6H_6O_3$	126
8	5.735	1.06	1-monoacetin	$C_5H_{10}O_4$	134
9	5.949	0.15	Decanoic acid	$C_{10}H_{20}O_2$	172
10	6.304	0.40	4-oxopentyl acetate	$C_7H_{12}O_3$	144
11	7.133	0.24	Eugenol acetate	$C_{12}H_{14}O_3$	206
12	8.022	0.07	Indan-1,3-diol monoacetate	$C_{11}H_{12}O_3$	192
13	8.728	6.16	Guanosine	$C_{10}H_{13}N_5O_5$	283
14	9.764	0.08	Dodecanoic acid	$C_{12}H_{24}O_2$	200
15	10.784	0.15	Dihydromethyljasmonate	$C_{13}H_{22}O_3$	226
16	10.986	0.10	1-(4-isopropylphenyl)-2-methylpropyl acetate	$C_{15}H_{22}O_2$	234
17	11.145	0.27	Benzoic acid, 2-hydroxy-, heptyl ester	$C_{14}H_{20}O_3$	236
18	11.555	0.19	Methyl myristate	$C_{15}H_{30}O_2$	242
19	11.934	0.61	4-((1E)-3-hydroxy-1-propenyl)-2-methoxyphenol	$C_{10}H_{12}O_3$	180

Table 4. Cont.

S.No.	RT (min)	Peak Area %	Name of the Compound	Molecular Formula	Molecular Weight
20	12.030	0.11	Tridecanoic acid	$C_{13}H_{26}O_2$	214
21	12.246	0.20	Stearic acid methyl ester	$C_{19}H_{38}O_2$	298
22	12.334	0.72	Octadecanoic acid, methyl ester	$C_{19}H_{38}O_2$	298
23	12.640	0.14	Pentadecanoic acid, methyl ester	$C_{16}H_{32}O_2$	256
24	13.075	0.29	Diisobutyl phthalate	$C_{16}H_{22}O_4$	278
25	13.255	0.03	1-hexadecanol	$C_{16}H_{34}O$	242
26	13.298	0.62	Hexadecanoic acid, methyl ester	$C_{17}H_{34}O_2$	270
27	13.467	0.19	Methyl palmitoleate	$C_{17}H_{32}O_2$	268
28	13.580	0.03	7,9-di-tert-butyl-1-oxaspiro(4,5)deca-6,9-diene-2,8-dione	$C_{17}H_{24}O_3$	276
29	13.676	3.96	Methyl palmitate	$C_{17}H_{34}O_2$	270
30	14.113	0.21	n-hexadecanoic acid	$C_{16}H_{32}O_2$	256
31	14.305	0.49	Decyl hexofuranoside	$C_{16}H_{32}O_6$	320
32	14.387	0.50	Eicosanoic acid, methyl ester	$C_{21}H_{42}O_2$	326
33	14.533	0.36	Cis-sinapyl alcohol	$C_{11}H_{14}O_4$	210
34	14.664	0.15	Heptadecanoic acid, methyl ester	$C_{18}H_{36}O_2$	284
35	14.925	0.12	Oxybenzone	$C_{14}H_{12}O_3$	228
36	15.316	3.31	Linoleic acid, methyl ester	$C_{19}H_{34}O_2$	294
37	15.374	1.97	Ethyl oleate	$C_{20}H_{38}O_2$	310
38	15.423	0.72	Oleic acid, methyl ester	$C_{19}H_{36}O_2$	296
39	15.607	0.79	Octadecanoic acid, methyl ester	$C_{19}H_{38}O_2$	298
40	16.399	0.70	cis-10-nonadecenoic acid, methyl ester	$C_{20}H_{38}O_2$	310
41	16.816	0.08	4,8,13-duvatriene-1,3-diol	$C_{20}H_{34}O_2$	306
42	17.290	0.08	4,5-dihydro-2-[(8Z,11Z)-8,11-heptadecadienyl]oxazole	$C_{20}H_{35}NO$	305
43	17.335	0.03	(Z)-2-(pentadec-8-en-1-yl)-4,5-dihydrooxazole	$C_{18}H_{33}NO$	279
44	17.379	0.16	Methyl arachidate	$C_{21}H_{42}O_2$	326
45	17.589	0.45	6-methyladenine, TMS derivative	$C_9H_{15}N_5Si$	221
46	17.888	0.09	Octadecanoic acid, 2,3-dihydroxypropyl ester	$C_{21}H_{42}O_4$	358
47	18.239	0.15	Henicosanal	$C_{21}H_{42}O$	310
48	18.746	0.12	Nonadecylpentafluoropropionate	$C_{22}H_{39}F_5O_2$	430
49	18.948	0.21	alpha-monostearin	$C_{21}H_{42}O_4$	358
50	19.006	0.28	Docosanoic acid, methyl ester	$C_{23}H_{46}O_2$	354
51	19.522	0.21	Vindolinine	$C_{21}H_{24}N_2O_2$	336
52	19.775	0.12	Methyl tricosanoate	$C_{24}H_{48}O_2$	368
53	20.046	0.09	Octocrylene	$C_{24}H_{27}NO_2$	361
54	20.317	0.25	n-propyl linoleate	$C_{21}H_{38}O_2$	322
55	20.593	0.10	Pleiocarpamine	$C_{20}H_{22}N_2O_2$	322
56	21.122	0.89	Squalene	$C_{30}H_{50}$	410

Table 4. Cont.

S.No.	RT (min)	Peak Area %	Name of the Compound	Molecular Formula	Molecular Weight
57	22.404	0.36	(+)-Pericyclivine	$C_{20}H_{22}N_2O_2$	322
58	22.793	0.47	Ajmalicine	$C_{21}H_{24}N_2O_3$	352
59	23.162	0.20	Cholesta-4,6-dien-3-ol	$C_{27}H_{44}O$	384
60	23.470	0.24	Ajmalicine oxindole	$C_{21}H_{24}N_2O_4$	368
61	24.641	1.69	Campesterol	$C_{28}H_{48}O$	400
62	24.901	1.23	Stigmasta-5,20(22)-dien-3-ol	$C_{29}H_{48}O$	412
63	25.035	0.84	19-epiajmalicine	$C_{21}H_{24}N_2O_3$	352
64	25.187	5.32	3-oxocholestane	$C_{27}H_{46}O$	386
65	25.476	2.09	beta-stigmasterol	$C_{29}H_{48}O$	412
66	25.600	2.42	gamma-sitosterol	$C_{29}H_{50}O$	414
67	25.790	1.76	(E)-1-(6,10-dimethylundec-5-en-2-yl)-4-methylbenzene	$C_{20}H_{32}$	272
68	25.990	0.30	(22E)-ergosta-4,7,22-trien-3-one	$C_{28}H_{42}O$	394
69	26.137	4.88	4-campestene-3-one	$C_{28}H_{46}O$	398
70	26.235	4.62	Cholestanone	$C_{27}H_{46}O$	386
71	26.459	4.47	Stigmasterone	$C_{29}H_{46}O$	410
72	26.547	0.22	6-dehydroprogesterone	$C_{21}H_{28}O_2$	312
73	26.640	0.56	Cycloartenol	$C_{30}H_{50}O$	426
74	26.776	0.20	3,5-cholestadien-7-one	$C_{27}H_{42}O$	382
75	26.869	0.61	Ergosta-4,6,22-trien-3-one	$C_{28}H_{42}O$	394
76	27.336	7.08	gamma-sitostenone	$C_{29}H_{48}O$	412
77	27.448	1.97	24-methylenecycloartanol	$C_{31}H_{52}O$	440
78	27.806	0.93	Stigmasta-3,5-dien-7-one	$C_{29}H_{46}O$	410
79	28.442	5.87	4,4-dimethylcholestan-3-one	$C_{29}H_{50}O$	414
80	28.846	3.78	(22E)-4-methylstigmast-22-en-3-one	$C_{30}H_{50}O$	426
81	30.011	5.55	3-acetylcholestan-2-one	$C_{29}H_{48}O_2$	428

Among the compounds identified, 1-monoacetin, guanosine, dihydromethyljasmonate, n-hexadecanoic acid, squalene, campesterol, cholestanone and gamma-sitosterol were the most prevalent present in both extracts. Only the methanolic extract of anthers contained bioactives such as cedrol (0.09%), (-)-coronaridine (0.12%), 4-vinylguaiacol (0.14%), vitamin E (0.30%), stigmastanol (0.57%), quinic acid (3.78%) and alpha amyrin (5.54%) (Table 3), and their respective mass spectra are shown in Figure S1A. The extract of anther-derived calli was found to have characteristic metabolites such as pleiocarpamine (0.10%), vindolinine (0.21%), cis-sinapyl alcohol (0.36%), (+)-pericyclivine (0.36%), ajmalicine (0.47%), cycloartenol (0.56%) and beta-stigmasterol (2.09%) (Tables 4 and 5) having specific mass spectra (Figure S1B).

2.4. Flow Cytometric Analysis

The ploidy status of callus obtained from anther was determined using a flow cytometric approach wherein good quality nuclei are a necessity. In this study, the leaves of field-grown *C. roseus* were utilized as an external standard reference (control). The flow cytometric histogram peak of callus reveals that its DNA content was nearly half to that of its diploid counterpart (control) (Figure 5A,B). The nuclear DNA content of anther-derived cell/callus was 0.76 pg compared to the diploid leaves' DNA (1.51 pg) with a DNA Index (DI) of 0.51 (Table 6). This estimation confirms the haploid DNA status of callus obtained from anther.

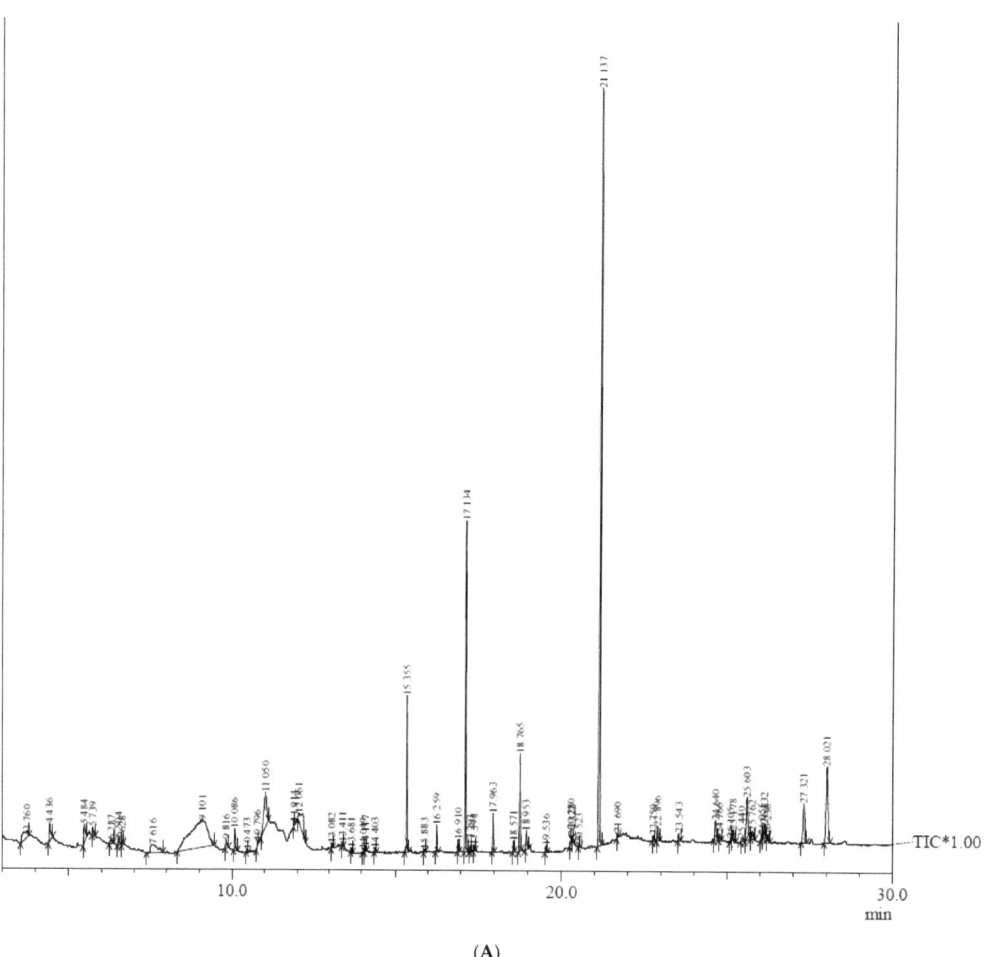

(A)

Figure 4. Cont.

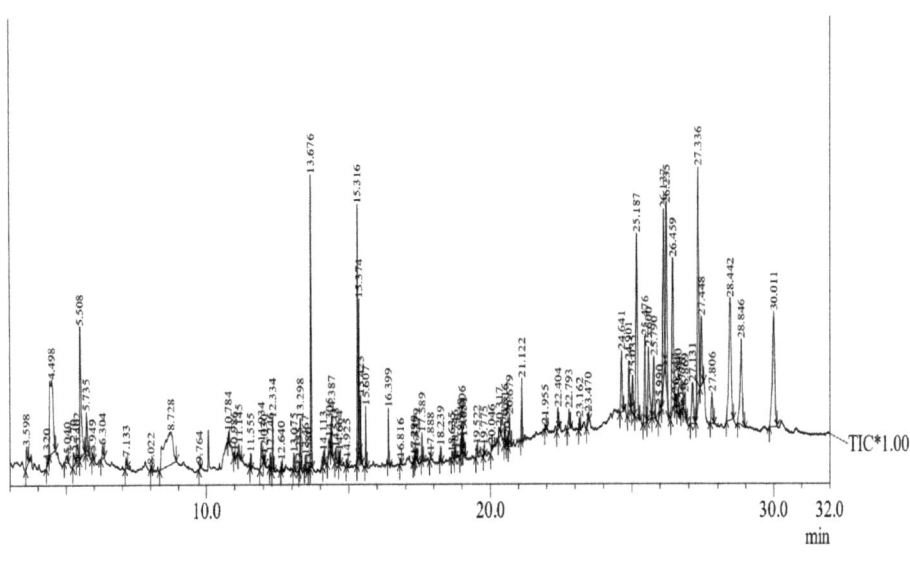

(**B**)

Figure 4. (**A**): GC–MS chromatogram (total ionic chromatogram) of methanolic extract of anthers of *C. roseus*; (**B**): GC–MS chromatogram (total ionic chromatogram) of methanolic extract of anther-derived callus of *C. roseus*.

Table 5. List of important phytocompounds identified exclusively in the methanolic extract of anther-derived callus of *C. roseus* using GC–MS analysis.

S.No.	RT (min)	Name of the Compound	Molecular Formula
1	7.133	Eugenol acetate	$C_{12}H_{14}O_3$
2	12.246	Stearic acid methyl ester	$C_{19}H_{38}O_2$
3	14.533	Cis-sinapyl alcohol	$C_{11}H_{14}O_4$
4	14.925	Oxybenzone	$C_{14}H_{12}O_3$
5	15.316	Linoleic acid, methyl ester	$C_{19}H_{34}O_2$
6	15.423	Oleic acid, methyl ester	$C_{19}H_{36}O_2$
7	19.522	Vindolinine	$C_{21}H_{24}N_2O_2$
8	20.046	Octocrylene	$C_{24}H_{27}NO_2$
9	20.593	Pleiocarpamine	$C_{20}H_{22}N_2O_2$
10	22.404	(+)-Pericyclivine	$C_{20}H_{22}N_2O_2$
11	22.793	Ajmalicine	$C_{21}H_{24}N_2O_3$
12	25.035	19-epiajmalicine	$C_{21}H_{24}N_2O_3$
13	25.476	beta-stigmasterol	$C_{29}H_{48}O$
14	26.459	Stigmasterone	$C_{29}H_{46}O$
15	26.547	6-dehydroprogesterone	$C_{21}H_{28}O_2$
16	26.640	Cycloartenol	$C_{30}H_{50}O$
17	27.336	gamma-sitostenone	$C_{29}H_{48}O$

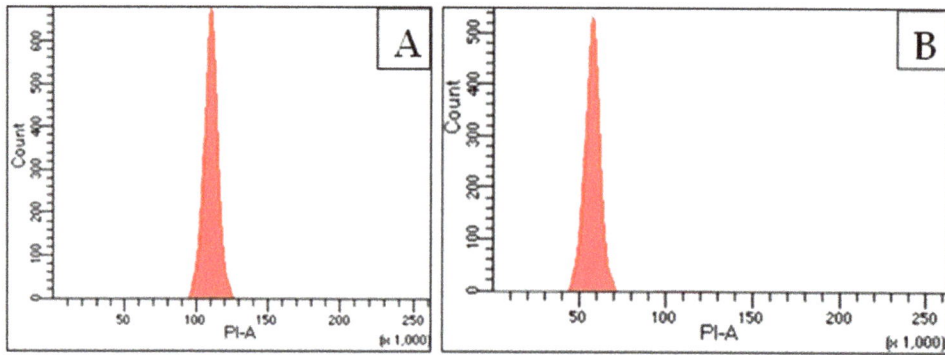

Figure 5. Flow cytometric histograms revealing ploidy level of (**A**) diploid leaves of *C. roseus* (standard) and (**B**) anther-derived callus of *C. roseus*.

Table 6. Estimation of nuclear DNA content, genome size and DNA index of anther-derived callus with respect to donor plant of *C. roseus* using flow cytometry technique.

Plant Sample Type	Nuclear DNA Content (pg)	Genome Size (Mbp) *	DNA Index (DI) **
Standard (leaves)	1.51	1476.7	-
Anther-derived callus	0.76	743.2	0.51

* 1 pg = 978 Mbp [18]. ** DNA Index = sample DNA content/standard DNA content.

3. Discussion

The present work was conducted to evaluate the callusing potentiality of anthers of *C. roseus* under in vitro culture conditions. The type and concentration of PGRs used in media strongly affect callusing ability and are different in different plant species. Initially, the anthers were subject to different concentrations and combinations of PGRs amended in MS medium. The results indicate that a high-to-low ratio of auxin: cytokinin concentrations was proven to be the best in inducing callus with a maximum mean fresh weight, which is very similar to Kou et al.'s [19] and Rout et al.'s [20] observations. Likewise, TDZ alone at different concentrations was found to be equally effective in producing callus and subsequent proliferation. Previous reports suggested that TDZ (a cytokinin-like PGR) alone may be used in improving callusing ability in different explants [21,22]. A comparison of the elemental distribution on the surfaces of anther and anther-derived callus was performed using SEM–EDX analysis, revealing a nearly similar elemental composition on both samples. EDX analyzes X-rays emitted from samples receiving a high-energy electron beam. This technique facilitates the qualitative and semi-quantitative detection of surface elements of samples and has been extensively used on various plant species such as sesame [23] and lemongrass [24].

Medicinal plants are an ingenious source of bioactive compounds that fight against several chronic diseases, and these phytocompounds can be identified and quantified using the GC–MS technique [25]. In the current study, phytochemical profiling with GC–MS of methanolic extracts (Figure 5) of anther and anther-derived callus of *C. roseus* has been conducted. The results obtained show the presence of various phytoconstituents, including carbohydrates, alkaloids, phenols, saponins, phytosterols, terpenoids, steroids, etc. A total of 14 bioactives are common in both the extracted samples. However, there are compounds that are exclusive to each sample that confer various biological properties to this plant. The presence of secondary metabolites in callus, which are otherwise not detected in anther tissue, may be due to the fact that certain bioactive compounds accumulate in specific cells or tissues or in a specific growth stage (mostly the stationary phase) of in vitro cultures [26]. Therefore, developing callus from different tissues to obtain therapeutically active compounds is of high significance.

The major compounds of medicinal value present in the methanolic extract of anthers were squalene (triterpene), alpha-amyrin (triterpene), coronaridine (alkaloid) and cedrol (essential oil), which possess anti-oxidant, gastroprotective and hepatoprotective, anti-cancerous and anti-inflammatory properties, respectively [27–30]. Similarly, in anther-derived calli exclusively, 17 compounds are present having diverse medicinal properties, and these compounds are listed in Table 5. These include stearic acid, linoleic acid, oleic acid, vindolinine, pleiocarpamine, pericyclivine, ajmalicine, 19-epiajmalicine, beta-stigmasterol, cycloartenol, etc. Ajmalicine and vindolinine are well-known alkaloids having anti-cancerous, anti-hypertensive and anti-oxidant properties [15,31]. Recently, an alkaloid named pleiocarpamine has been isolated from the stem bark of *Rauvolfia caffra* and is reported to possess anti-seizure activity [32]. Cycloartenol (a triterpenoid) and stigmasterol (a sterol) have also been detected in present studies and are associated with immunosuppressive, anti-hypercholestrolemic and anti-inflammatory activities, respectively [33,34]. Compounds such as cycloartenol, ajmalicine, vindolinine, pleiocarpamine and pericyclivine have been reported previously in leaf tissues of *C. roseus* [35,36]. Some reports of phytocompounds identified from different tissues using GC–MS were noted earlier [37,38], but till now, no information on the phytocompounds present in anther or anther-derived callus was available for *C. roseus*.

The ploidy status of anther-derived callus was checked using flow cytometry, and the results show that the ploidy of the calli was haploid in nature, confirming the involvement of microspores in developing callus. Similar observations have also been reported for other plant species [7,10,39]. FCM is the widely used approach for determining the ploidy of plants developed through callus, somatic embryos and other in vitro-regenerated pathways [40]. The origin of diploid plants from anthers may be due to the involvement of other somatic cells such as anther wall, filament or flower septum in developing callus. Spontaneous chromosomal doubling can also be a mechanism in the generation of polyploidy in anther-derived regenerants. In certain cases, mixoploids and aneuploids have also been noted in anther cultures of different plants [8,41], but these polyploids were not detected in this experiment. This is the first-ever report of GC–MS analysis of medically significant compounds from anther tissue of *C. roseus*, which enriches the phytocompound library of *Catharanthus* and may be utilized in the pharmaceutical and industrial sectors.

4. Materials and Methods

4.1. Anther Culture and Growth Conditions

The mature flowers of *C. roseus* were collected from the herbal garden, Jamia Hamdard, New Delhi, and the anthers were used as explants for experimentations. The surface sterilization of flowers was performed following the method of Bansal et al. [3] described earlier. The sterilized anthers were excised from the flowers and aseptically cultured onto agar-solidified basal Murashige and Skoog (MS) medium supplemented with various concentrations and combinations of plant growth regulators (PGRs) and sub-cultured every 3–4 weeks. The cultures were incubated at a temperature of 24 ± 2 °C with 48 $\mu mol/m^2/s^2$ illumination (white fluorescent light) for a 16 h photoperiod.

4.2. Callus Induction and Proliferation

The disinfected anthers were inoculated on MS augmented with different concentrations (alone or in combination) of α-naphthalene acetic acid (NAA), kinetin (Kn) and thidiazuron (TDZ) ranging from 0.1 to 1.0 mg/L for callus induction. Callus formation started within 14–16 days of culture and proliferated on the same medium with successive subculturing. The callus induction frequency and the callus fresh weight were recorded after 6 weeks of culture.

$$\text{Callus induction frequency (\%)} = \frac{\text{Number of explants showing callusing}}{\text{Total number of explants inoculated}} \times 100$$

4.3. Surface Morphology and Elemental Analysis

The surface morphology and elemental profile of anther and anther-derived callus were determined using energy-dispersive X-ray microanalysis (EDX) combined with scanning electron microscopy (SEM). For this purpose, the samples were primarily fixed with Karnovsky's fixative and washed with 0.1 M phosphate buffer at 4 °C. Afterward, a series of dehydrations with acetone (30%, 50%, 70%, 90% and 100%) were performed at 15 min intervals, and then critical-point drying was performed at 1100 p.s.i. These samples were then mounted on aluminum stubs and sputter-coated with gold having a 35 nm thick film. Finally, the coated samples were viewed at an accelerating voltage of 20 kV under a scanning electron microscope (Zeiss, Oberkochen, Germany) equipped with EDAX.

4.4. Preparation of Extracts

The methanolic extracts of both samples were prepared according to the protocol of Hussain et al. [42]. About 1.0 g of anther and anther callus were shade dried and crushed into fine powder using mortar and pestle (Figure 3A,B). Each sample was then extracted in 5.0 mL methanol in an orbital shaker for 48 h. Afterward, the extracts were filtered through Whatman filter paper no. 1 and evaporated to dryness. The obtained extracts were stored in an airtight container with proper labeling at 4 °C for further use (Figure 3C).

4.5. GC–MS Analysis

GC–MS analyses of these extracts were conducted on GC–MS QP-2010 equipment (Shimadzu, Japan) at Advanced Instrumentation Research Facility (AIRF), JNU, New Delhi. The program settings were as follows: Helium was used as a carrier gas (1 mL/min), and the initial and final temperatures were programmed at 100 °C and 260 °C, respectively, with a hold time of 18 min. Ion source temperature was 220 °C with an interface temperature of 270 °C and solvent cut time of 2.5 min. Other specifications included: detector gain mode relative to the tuning result, detector gain +0.00 kV, threshold of 1000, start time 3 min, end time 39.98 min, event time 0.3 s, scan speed of 2000, start m/z 40.00 and end m/z 600.00.

4.6. Metabolite Data Processing and Analysis

The bioactive compounds were identified using the mass spectral database of the NIST17 library. The unknown compounds' spectra were compared with the known phytocompound spectra available in the NIST library, and the name, molecular weight and structure of the compounds were determined.

4.7. Flow Cytometric Analysis

The ploidy status of anther-derived calli was examined using the flow cytometry method as described by Galbraith [43]. A total of 3 samples of anther-derived callus were randomly chosen, along with a reference standard of diploid leaves of *C. roseus* with a known 2C DNA content of 1.51 pg [44]. Approximately 50 mg of callus was added to a Petri plate having 1.0 mL ice-cold Galbraith's buffer (nuclei isolation buffer) and finely macerated with the help of a surgical blade. The homogenate was then filtered with a 100 µm nylon mesh to eliminate larger cellular remnants and was finally stained with 50 µg/mL PI RNase (propidium iodide RNase) (Sigma-Aldrich, St. Louis, MO, USA) for 8–10 min. The samples were incubated in the dark at 4 °C for about 40 min and eventually examined on a BD FACS(Calibur) flow cytometer (BD Biosciences, Franklin Lakes, NJ, USA). The relative nuclear DNA of anther-derived callus of *C. roseus* was estimated using the below formula [45]:

$$\text{Nuclear DNA content of sample (pg)} = 2C \text{ DNA content of standard (pg)} \times \frac{\text{mean position of G0/G1 peak of sample}}{\text{mean position of G0/G1 peak of standard}}$$

4.8. Statistical Analysis

In the tissue-culture experiment, three explants (anthers) per culture tube were inoculated with five replicates of every experimental treatment, and each experiment was repeated twice. The data are expressed as mean ± standard error, and the analysis was performed using one-way analysis of variance (ANOVA). The significance of mean difference was determined using Duncan's multiple range test (DMRT) at $p < 0.05$ using SPSS Ver. 26.0 (SPSS Inc., Chicago, IL, USA) [46]. The flow cytometric study was repeated thrice with randomly chosen standard (donor plant) and callus samples.

5. Conclusions

The in vitro culture technology was successfully employed to obtain callus from anther tissue of *C. roseus*, an important medicinal plant. The callus was checked for its ploidy status using flow cytometry and was found to be haploid in nature. The calli obtained from anther were then subjected to GC–MS analysis for phytocompound identification. Among the bioactive compounds identified, ajmalicine, vindolinine, pleiocarpamine, pericyclivine, stigmasterol, campesterol and squalenes were detected and have a wide range of biological activities. From this study, it can then be concluded that anther-derived calli are a potent source for developing new therapeutic drugs with larger-scale applicability in pharmaceutical sectors.

Supplementary Materials: The following supporting information can be downloaded at: https://www.mdpi.com/article/10.3390/plants12112186/s1, Figure S1A,B: Mass spectra of identified compounds from methanolic extract of anthers and anther derived callus of *C. roseus*.

Author Contributions: Conceptualization, A.M. and Y.B.; methodology, Y.B.; formal analysis, Y.B.; investigation, Y.B.; data curation, Y.B.; writing—original draft preparation, Y.B.; writing—review and editing, Y.B., A.M., J.M., Y.H.D. and H.Z.R.; validation, Y.H.D. and H.Z.R.; visualization, Y.H.D. and H.Z.R.; supervision, A.M.; project administration, A.M. All authors have read and agreed to the published version of the manuscript.

Funding: This research work is funded by the Department of Biotechnology (DBT), New Delhi, India (DBT/2020/JH/1336).

Institutional Review Board Statement: Not applicable.

Informed Consent Statement: Not applicable.

Data Availability Statement: All data are presented in the article.

Acknowledgments: The first author is thankful to the Department of Biotechnology (DBT) for the financial support given in the form of the Junior Research Fellowship and to AIRF, JNU, New Delhi, for providing the GC–MS facility. The authors are grateful for the laboratory facilities provided by the Department of Botany, Jamia Hamdard, New Delhi. The authors acknowledge Researchers Supporting Project number (RSP2023R375), King Saud University, Riyadh, Saudi Arabia.

Conflicts of Interest: The authors declare no conflict of interest.

Abbreviations

PGRs	Plant Growth Regulators
MS	Murashige and Skoog
NAA	α-Naphthaleneacetic acid
Kn	Kinetin
TDZ	Thidiazuron
DH	Double Haploid
SEM–EDX	Scanning Electron Microscopy–Energy-Dispersive X-ray
FCM	Flow Cytometric Method
DI	DNA Index
GC–MS	Gas Chromatography–Mass Spectrometry
DMRT	Duncan's Multiple Range Test

References

1. Das, A.; Sarkar, S.; Bhattacharyya, S.; Gantait, S. Biotechnological advancements in *Catharanthus roseus* (L.) G. Don. *Appl. Microbiol. Biotechnol.* **2020**, *10*, 44811–44835. [CrossRef]
2. Dhayanithy, G.; Subban, K.; Chelliah, J. Diversity and biological activities of endophytic fungi associated with *Catharanthus roseus*. *BMC Microbiol.* **2019**, *19*, 1–14. [CrossRef] [PubMed]
3. Bansal, Y.; Mujib, A.; Siddiqui, Z.H.; Mamgain, J.; Syeed, R.; Ejaz, B. Ploidy status, nuclear DNA content and start codon targeted (SCoT) genetic homogeneity assessment in *Digitalis purpurea* L., regenerated in vitro. *Genes* **2022**, *13*, 2335. [CrossRef] [PubMed]
4. Mujib, A.; Bansal, Y.; Malik, M.Q.; Syeed, R.; Mamgain, J.; Ejaz, B. Internal and external regulatory elements controlling somatic embryogenesis in *Catharanthus*: A model medicinal plant. In *Somatic Embryogenesis, Methods in Molecular Biology*; Ramirez-Mosqueda, M.A., Ed.; Humana: New York, NY, USA, 2022; Volume 2527, pp. 11–27. [CrossRef]
5. Dhandapani, M.; Kim, D.H.; Hong, S.B. Efficient plant regeneration via somatic embryogenesis and organogenesis from the explants of *Catharanthus roseus*. *In Vitro Cell. Dev. Biol. Plant* **2008**, *44*, 18–25. [CrossRef]
6. Verma, P.; Mathur, A.K. Direct shoot bud organogenesis and plant regeneration from pre-plasmolysed leaf explants in *Catharanthus roseus*. *Plant Cell Tissue Organ Cult.* **2011**, *106*, 401–408. [CrossRef]
7. Hoveida, Z.S.; Abdollahi, M.R.; Mirzaie-Asl, A.; Moosavi, S.S.; Seguí-Simarro, J.M. Production of doubled haploid plants from anther cultures of borage (*Borago officinalis* L.) by the application of chemical and physical stress. *Plant Cell Tissue Organ Cult.* **2017**, *130*, 369–378. [CrossRef]
8. Wang, G.F.; Qin, H.Y.; Sun, D.; Fan, S.T.; Yang, Y.M.; Wang, Z.X.; Xu, P.L.; Zhao, Y.; Liu, Y.X.; Ai, J. Haploid plant regeneration from hardy kiwifruit (*Actinidia arguta* Planch.) anther culture. *Plant Cell Tissue Organ Cult.* **2018**, *134*, 15–28. [CrossRef]
9. Abd El-Fatah, B.E.; Sayed, M.A.; El-Sanusy, S.A. Genetic analysis of anther culture response and identification of QTLs associated with response traits in wheat (*Triticum aestivum* L.). *Mol. Biol. Rep.* **2020**, *47*, 9289–9300. [CrossRef]
10. Sahoo, S.A.; Jha, Z.; Verulkar, S.B.; Srivastava, A.K.; Suprasanna, P. High-throughput cell analysis based protocol for ploidy determination in anther-derived rice callus. *Plant Cell Tissue Organ Cult.* **2019**, *137*, 187–192. [CrossRef]
11. Ejaz, B.; Mujib, A.; Mamgain, J.; Malik, M.Q.; Syeed, R.; Gulzar, B.; Bansal, Y. Comprehensive in vitro regeneration study with SCoT marker assisted clonal stability assessment and flow cytometric genome size analysis of *Carthamus tinctorius* L.: An important medicinal plant. *Plant Cell Tissue Organ Cult.* **2022**, *148*, 403–418. [CrossRef]
12. Bhusare, B.P.; John, C.K.; Bhatt, V.P.; Nikam, T.D. Induction of somatic embryogenesis in leaf and root explants of *Digitalis lanata* Ehrh.: Direct and indirect method. *S. Afr. J. Bot.* **2020**, *130*, 356–365. [CrossRef]
13. Mamgain, J.; Mujib, A.; Ejaz, B.; Gulzar, B.; Malik, M.Q.; Syeed, R. Flow cytometry and start codon targeted (SCoT) genetic fidelity assessment of regenerated plantlets in *Tylophora indica* (Burm. f.) Merrill. *Plant Cell Tissue Organ Cult.* **2022**, *150*, 129–140. [CrossRef]
14. Konappa, N.; Udayashankar, A.C.; Krishnamurthy, S.; Pradeep, C.K.; Chowdappa, S.; Jogaiah, S. GC–MS analysis of phytoconstituents from *Amomum nilgiricum* and molecular docking interactions of bioactive serverogenin acetate with target proteins. *Sci. Rep.* **2020**, *10*, 16438. [CrossRef]
15. Pham, H.N.; Vuong, Q.V.; Bowyer, M.C.; Scarlett, C.J. Phytochemicals derived from *Catharanthus roseus* and their health benefits. *Technology* **2020**, *8*, 80. [CrossRef]
16. Padma, M.; Ganesan, S.; Jayaseelan, T.; Azhagumadhavan, S.; Sasikala, P.; Senthilkumar, S.; Mani, P. Phytochemical screening and GC–MS analysis of bioactive compounds present in ethanolic leaves extract of *Silybum marianum* (L). *J. Drug. Deliv. Ther.* **2019**, *9*, 85–89. [CrossRef]
17. Jahan, I.; Tona, M.R.; Sharmin, S.; Sayeed, M.A.; Tania, F.Z.; Paul, A.; Chy, M.N.; Rakib, A.; Emran, T.B.; Simal-Gandara, J. GC-MS phytochemical profiling, pharmacological properties, and in silico studies of *Chukrasia velutina* leaves: A novel source for bioactive agents. *Molecules* **2020**, *25*, 3536. [CrossRef]
18. Dolezel, J. Nuclear DNA content and genome size of trout and human. *Cytom. Part A* **2003**, *51*, 127–128.
19. Kou, Y.; Ma, G.; Teixeira da Silva, J.A.; Liu, N. Callus induction and shoot organogenesis from anther cultures of *Curcuma attenuata* Wall. *Plant Cell Tissue Organ Cult.* **2013**, *112*, 1–7. [CrossRef]
20. Rout, P.; Naik, N.; Ngangkham, U.; Verma, R.L.; Katara, J.L.; Singh, O.N.; Samantaray, S. Doubled Haploids generated through anther culture from an elite long duration rice hybrid, CRHR32: Method optimization and molecular characterization. *Plant Biotechnol.* **2016**, *33*, 177–186. [CrossRef]
21. Cappelletti, R.; Sabbadini, S.; Mezzetti, B. The use of TDZ for the efficient in vitro regeneration and organogenesis of strawberry and blueberry cultivars. *Sci. Hortic.* **2016**, *207*, 117–124. [CrossRef]
22. Khan, T.; Abbasi, B.H.; Khan, M.A.; Shinwari, Z.K. Differential effects of thidiazuron on production of anticancer phenolic compounds in callus cultures of *Fagonia indica*. *Appl. Biochem. Biotechnol.* **2016**, *179*, 46–58. [CrossRef]
23. Nath, B.; Kalita, P.; Das, B.; Basumatary, S. Highly efficient renewable heterogeneous base catalyst derived from waste *Sesamum indicum* plant for synthesis of biodiesel. *Renew. Energy* **2020**, *151*, 295–310. [CrossRef]
24. Pandey, J.; Sarkar, S.; Verma, R.K.; Singh, S. Sub-cellular localization and quantitative estimation of heavy metals in lemongrass plants grown in multi-metal contaminated tannery sludge. *S. Afr. J. Bot.* **2020**, *131*, 74–83. [CrossRef]
25. Mamgain, J.; Mujib, A.; Syeed, R.; Ejaz, B.; Malik, M.Q.; Bansal, Y. Genome size and gas chromatography-mass spectrometry (GC–MS) analysis of field-grown and in vitro regenerated *Pluchea lanceolata* plants. *J. Appl. Genet.* **2022**, *64*, 1–21. [CrossRef]

26. Leng, T.C.; Ping, N.S.; Lim, B.P.; Keng, C.L. Detection of bioactive compounds from *Spilanthes acmella* (L.) plants and its various in vitro culture products. *J. Med. Plant Res.* **2011**, *5*, 371–378.
27. Diab, M.; Ibrahim, A.; Hadad, G. B: Review article on chemical constituents and biological activity of *Olea europaea*. *Rec. Pharm. Biomed. Sci.* **2020**, *4*, 36–45. [CrossRef]
28. Nogueira, A.O.; Oliveira, Y.I.S.; Adjafre, B.L.; de Moraes, M.E.A.; Aragão, G.F. Pharmacological effects of the isomeric mixture of alpha and beta amyrin from *Protium heptaphyllum*: A literature review. *Fundam. Clin. Pharmacol.* **2019**, *33*, 4–12. [CrossRef] [PubMed]
29. Naidoo, C.M.; Naidoo, Y.; Dewir, Y.H.; Murthy, H.N.; El-Hendawy, S.; Al-Suhaibani, N. Major bioactive alkaloids and biological activities of *Tabernaemontana* species (*Apocynaceae*). *Plants* **2021**, *10*, 313. [CrossRef] [PubMed]
30. Özek, G.; Schepetkin, I.A.; Yermagambetova, M.; Özek, T.; Kirpotina, L.N.; Almerekova, S.S.; Abugalieva, S.I.; Khlebnikov, A.I.; Quinn, M.T. Innate immunomodulatory activity of cedrol, a component of essential oils isolated from *Juniperus* species. *Molecules* **2021**, *26*, 7644. [CrossRef]
31. Hemmati, N.; Azizi, M.; Spina, R.; Dupire, F.; Arouei, H.; Saeedi, M.; Laurain-Mattar, D. Accumulation of ajmalicine and vinblastine in cell cultures is enhanced by endophytic fungi of *Catharanthus roseus* cv. Icy Pink. *Ind. Crops Prod.* **2020**, *158*, 112776. [CrossRef]
32. Chipiti, T.; Viljoen, A.M.; Cordero-Maldonado, M.L.; Veale, C.G.; Van Heerden, F.R.; Sandasi, M.; Chen, W.; Crawford, A.D.; Enslin, G.M. Anti-seizure activity of African medicinal plants: The identification of bioactive alkaloids from the stem bark of Rauvolfia caffra using an in vivo zebrafish model. *J. Ethnopharmacol.* **2021**, *279*, 114282. [CrossRef]
33. Bakrim, S.; Benkhaira, N.; Bourais, I.; Benali, T.; Lee, L.-H.; El Omari, N.; Sheikh, R.A.; Goh, K.W.; Ming, L.C.; Bouyahya, A. Health Benefits and Pharmacological Properties of Stigmasterol. *Antioxidants* **2022**, *11*, 1912. [CrossRef] [PubMed]
34. Lun, W.U.; Zhi-Li, C.H.; Yang, S.U.; Qiu-Hong, W.A.; Kuang, H.X. Cycloartenol triterpenoid saponins from *Cimicifuga simplex* (Ranunculaceae) and their biological effects. *China J. Nat. Med.* **2015**, *13*, 81–89. [CrossRef]
35. Murata, J.; Roepke, J.; Gordon, H.; De Luca, V. The leaf epidermome of *Catharanthus roseus* reveals its biochemical specialization. *Plant Cell* **2008**, *20*, 524–542. [CrossRef]
36. Wesołowska, A.; Grzeszczuk, M.; Wilas, J.; Kulpa, D. Gas chromatography-mass spectrometry (GC-MS) analysis of indole alkaloids isolated from *Catharanthus roseus* (L.) G. don cultivated conventionally and derived from in vitro cultures. *Not. Bot. Horti Cluj-Napoca* **2016**, *44*, 100–106. [CrossRef]
37. Syeda, A.M.; Riazunnisa, K. Data on GC-MS analysis, in vitro anti-oxidant and anti-microbial activity of the *Catharanthus roseus* and *Moringa oleifera* leaf extracts. *Data Brief* **2020**, *29*, 105258. [CrossRef] [PubMed]
38. Rani, S.; Singh, V.; Sharma, M.K.; Sisodia, R. GC-MS based metabolite profiling of medicinal plant-*Catharanthus roseus* under cadmium stress. *Plant Physiol. Rep.* **2021**, *26*, 491–502. [CrossRef]
39. Eshaghi, Z.C.; Abdollahi, M.R.; Moosavi, S.S.; Deljou, A.; Seguí-Simarro, J.M. Induction of androgenesis and production of haploid embryos in anther cultures of borage (*Borago officinalis* L.). *Plant Cell Tissue Organ Cult.* **2015**, *122*, 321–329. [CrossRef]
40. Gulzar, B.; Mujib, A.; Mushtaq, Z.; Malik, M.Q. Old *Catharanthus roseus* culture (14 years) produced somatic embryos and plants and showed normal genome size; demonstrated an increased antioxidant defense mechanism; and synthesized stress proteins as biochemical, proteomics, and flow-cytometry studies reveal. *J. Appl. Genet.* **2021**, *62*, 43–57. [CrossRef]
41. Jia, Y.; Zhang, Q.X.; Pan, H.T.; Wang, S.Q.; Liu, Q.L.; Sun, L.X. Callus induction and haploid plant regeneration from baby primrose (*Primula forbesii* Franch.) anther culture. *Sci. Hortic.* **2014**, *176*, 273–281. [CrossRef]
42. Hussain, S.A.; Ahmad, N.; Anis, M.; Alatar, A.A. Influence of meta-topolin on in vitro organogenesis in *Tecoma stans* L., assessment of genetic fidelity and phytochemical profiling of wild and regenerated plants. *Plant Cell Tissue Organ Cult.* **2019**, *138*, 339–351. [CrossRef]
43. Galbraith, D.W. Simultaneous flow cytometric quantification of plant nuclear DNA contents over the full range of described angiosperm 2C values. *Cytom. Part A J. Int. Soc. Adv. Cytom.* **2009**, *75*, 692–708. [CrossRef] [PubMed]
44. Mujib, A.; Malik, M.Q.; Bansal, Y.; Syeed, R.; Ejaz, B.; Mamgain, J. Somatic Embryogenesis in *Catharanthus roseus*: Proteomics of embryogenic and non-embryogenic tissues; and genome size analysis of regenerated plant. In *The Catharanthus Genome, Compendium of Plant Genomes*; Kole, C., Ed.; Springer: Cham, Switzerland, 2022; pp. 85–100. [CrossRef]
45. Doležel, J.; Greilhuber, J.; Suda, J. Estimation of nuclear DNA content in plants using flow cytometry. *Nat. Protoc.* **2007**, *2*, 2233–2244. [CrossRef] [PubMed]
46. Duncan, D.B. Multiple range and multiple F tests. *Biome* **1955**, *11*, 1478. [CrossRef]

Disclaimer/Publisher's Note: The statements, opinions and data contained in all publications are solely those of the individual author(s) and contributor(s) and not of MDPI and/or the editor(s). MDPI and/or the editor(s) disclaim responsibility for any injury to people or property resulting from any ideas, methods, instructions or products referred to in the content.

Article

Plant Growth Regulator- and Elicitor-Mediated Enhancement of Biomass and Andrographolide Production of Shoot Tip-Culture-Derived Plantlets of *Andrographis paniculata* (Burm.f.) Wall. (Hempedu Bumi)

Aicah Patuhai [1], Puteri Edaroyati Megat Wahab [1], Martini Mohammad Yusoff [1], Yaser Hassan Dewir [2], Ali Alsughayyir [3] and Mansor Hakiman [1,4,*]

1. Department of Crop Science, Faculty of Agriculture, Universiti Putra Malaysia (UPM), Serdang 43400, Selangor, Malaysia; aicahpatuhai@gmail.com (A.P.); putri@upm.edu.my (P.E.M.W.); martinimy@upm.edu.my (M.M.Y.)
2. Plant Production Department, College of Food and Agriculture Sciences, King Saud University, Riyadh 11451, Saudi Arabia; ydewir@ksu.edu.sa
3. Department of Plant and Soil Sciences, Mississippi State University, 75 B.S. Hood Rd, Starkville, MS 39762, USA; aa2942@msstate.edu
4. Laboratory of Sustainable Resources Management, Institute of Tropical Forestry and Forest Products, Universiti Putra Malaysia (UPM), Serdang 43400, Selangor, Malaysia
* Correspondence: mhakiman@upm.edu.my

Abstract: *Andrographis paniculata* (Burm.f.) Wall. (Acanthaceae) is revered for its medicinal properties. In vitro culture of medicinal plants has assisted in improving both the quantity and quality of their yield. The current study investigated the effects of different surface sterilization treatments, plant growth regulators (PGRs), and elicitors on culture establishment and axillary shoot multiplication of *A. paniculata*. Subsequently, the production of andrographolide in the in vitro plantlets was evaluated using high-performance liquid chromatography (HPLC) analysis. The shoot-tip explant was successfully sterilized using 60% commercial bleach for 5 min of immersion with a 90% survival rate and 96.67% aseptic culture. The optimal PGR for shoot growth was 6-benzylaminopurine (BAP) at 17.76 µM, supplemented into Murashige and Skoog (MS) media, producing 23.57 ± 0.48 leaves, 7.33 ± 0.10 shoots, and a 3.06 ± 0.02 cm length of shoots. Subsequently, MS medium supplemented with 5 mg/L chitosan produced 26.07 ± 0.14 leaves, 8.33 ± 0.07 shoots, and a 3.63 ± 0.02 cm length of shoots. The highest andrographolide content was obtained using the plantlets harvested from 5 mg/L chitosan with 2463.03 ± 0.398 µg/mL compared to the control (without elicitation) with 256.73 ± 0.341 µg/mL (859.39% increase). The results imply that the protocol for the shoot-tip culture of *A. paniculata* was developed, and that elicitation enhanced the herbage yield and the production of andrographolide.

Keywords: chitosan; contamination; elicitation; plant growth regulators; salicylic acid; surface sterilization

Citation: Patuhai, A.; Wahab, P.E.M.; Yusoff, M.M.; Dewir, Y.H.; Alsughayyir, A.; Hakiman, M. Plant Growth Regulator- and Elicitor-Mediated Enhancement of Biomass and Andrographolide Production of Shoot Tip-Culture-Derived Plantlets of *Andrographis paniculata* (Burm.f.) Wall. (Hempedu Bumi). *Plants* 2023, 12, 2953. https://doi.org/10.3390/plants12162953

Academic Editor: Sofia Caretto

Received: 19 June 2023
Revised: 8 July 2023
Accepted: 18 July 2023
Published: 15 August 2023

Copyright: © 2023 by the authors. Licensee MDPI, Basel, Switzerland. This article is an open access article distributed under the terms and conditions of the Creative Commons Attribution (CC BY) license (https://creativecommons.org/licenses/by/4.0/).

1. Introduction

Andrographis paniculata (Burm.f.) Wall., commonly known as the King of Bitters, belongs to the family Acanthaceae. It is widely used in East Asian, South Asian, and Southeast Asian traditional remedies. Leaves and aerial parts of this plant are usually employed in traditional medicine to treat hepatitis, bronchitis, colitis, cough, fever, mouth ulcers, sores, tuberculosis, bacillary dysentery, venomous snake bites, common cold, urinary tract infections, and diarrhea [1]. Lactone terpenoids, specifically andrographolide, dehydroandrographolide, neoandrographolide, and deoxyandrographolide, are the main bioactive compounds produced in *A. paniculata* [1]. Andrographolide is a promising candidate to remediate various diseases such as inflammation, colds, and cancer [2].

The initial step of in vitro culture is the obtaining of microorganism-free explants from the mother plant to be propagated by tissue culture. The contamination problem is the critical limiting factor that is frequently encountered at the early stage. The type of sterilization agent, its concentrations, and the time of immersion are all factors to consider during the sterilization process [3,4]. The most often-used sterilization agents include sodium hypochlorite (NaClO), mercuric chloride ($HgCl_2$), hydrogen peroxide (H_2O_2), ethanol (C_2H_5OH), and silver nitrate ($AgNO_3$). However, Suraya et al. [5] found that mercury chloride can be hazardous due to its volatile properties at room temperature and was excessively toxic to humans and the environment.

Biotechnological approaches, particularly those involving plant tissue culture, help produce desirable bioactive metabolites [6]. Higher production of bioactive and secondary metabolites has been reported in many tissue-cultured herbs compared to their mother plants. Plant tissue culture offers an alternative method that can overcome the limitations of extracting valuable bioactive and secondary metabolites, limits which arise from constrained natural resources [5].

Axillary shoot multiplication effectively addresses the problems related to conventional propagation techniques, such as limited seed viability and slow growth rate [7]. A previous study found that direct shoot regeneration from the nodal segment, seed and leaf-derived callus, and stem of *A. paniculata* was achieved when cultured on media supplemented with BAP and 1-naphthalene acetic acid (NAA) [8]. Another study reported that the multiplication of *Andrographis paniculata* shoots can be achieved using BAP with indole-3-acetic acid (IAA) [9]. Variables such as growth regulators, nutrient media, and light conditions can be optimized to enhance the biomass production and secondary metabolites of *A. paniculata* [10].

Elicitation using microbial, physical, and chemical agents is a practice commonly used to regulate plant growth and induce stress in the culture. Chitosan has also been reported to have enhanced natural defense responses in plants and has been used as a natural compound to control pre- and post-harvest pathogenic diseases [11]. Antimicrobial activities of chitosan-treated cultures against various phytopathogens have been reported [12]. Other than chitosan, salicylic acid (SA) can also enhance the bioactive metabolites from medicinal plants, which function in plant development and metabolism, crop production, and pest control [2]. Due to its hormone-like activity, SA has also synthesized and accumulated bioactive metabolites in various plant species via in vitro systems [6].

Enhancing bioactive metabolites through in vitro culture techniques is advantageous for developing models with scale-up potential. Hence, it can be widely employed as a novel technique to produce the desired bioactive metabolites [13]. The present study aimed to develop a reliable protocol for shoot-tip culture of *A. paniculata*. Surface sterilization, multiplication, elicitation, and the quantification of andrographolide production in in vitro plantlets of *A. paniculata* were conducted.

2. Results
2.1. Surface Sterilization

The results showed that different concentrations and immersion durations of commercial bleach affected the survival percentage of the explants (Figure 1a). The highest survival percentage (90%) was observed in Treatment 9 (60% commercial bleach; 5 min immersion), while the lowest survival percentage was recorded in Treatments 1 and 11 with 40% and 70% commercial bleach with immersion for 1 and 10 min, respectively, with 6.67% survival. On the other hand, based on the obtained results, Treatment 11 (70% of commercial bleach; 10 min immersion) showed the highest aseptic culture percentage (96.67%) (Figure 1b), while the lowest aseptic culture percentage was recorded in Treatment 1 (40% of commercial bleach; 1 min immersion), with 3.33% of aseptic culture.

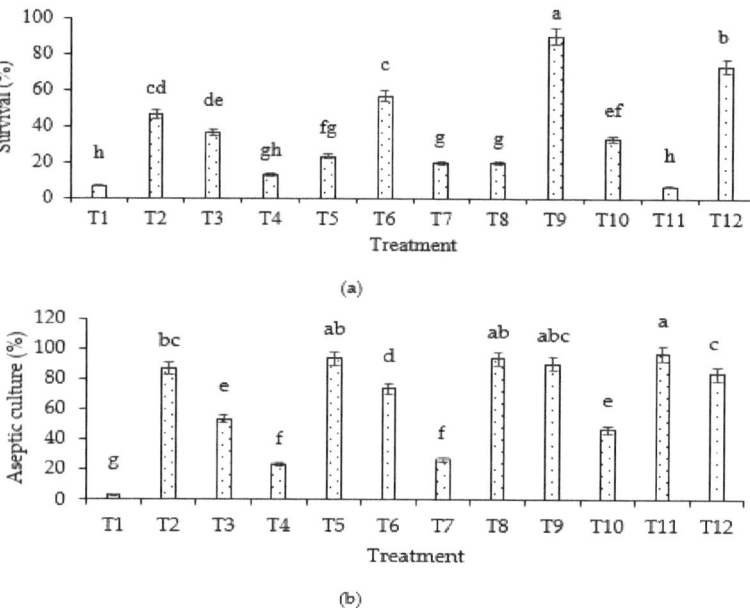

Figure 1. Effects of different commercial bleach concentrations and immersion durations on establishing sterile culture from shoot-tip explant of *A. paniculata*: (**a**) survival percentage; (**b**) aseptic culture percentage. T1 (40%, 1 min), T2 (40%, 10 min), T3 (40%, 5 min), T4 (50%, 1 min), T5 (50%, 10 min), T6 (50%, 5 min), T7 (60%, 1 min), T8 (60%, 10 min), T9 (60%, 5 min), T10 (70%, 1 min), T11 (70%, 10 min) and T12 (70%, 5 min), where values in percentage refer to commercial bleach concentration. Values are means of $n = 30$. T = treatment. Means followed by the same letters indicate that there is no significant difference ($p < 0.05$) using least significant means (LS-means).

2.2. Shoot-tip Culture

All the treatments of different PGRs, including the control, showed positive responses in the proliferation of buds when shoot-tip was used as explant. Parameters of the number of leaves, number of shoots, and length of shoots were increased from Treatments 1 to 6. The highest herbage yield for all parameters was observed in Treatment 6 (17.76 µM BAP), with 23.57 ± 0.48 leaves, 7.33 ± 0.10 shoots, and a 3.06 ± 0.02 cm length of shoot (Table 1, Figure 2), while Treatment 1 (control) showed the lowest herbage yield with 5.03 ± 0.08 leaves, 1.70 ± 0.04 shoots and a 2.25 ± 0.02 cm length of shoot, respectively.

Table 1. Effects of plant growth regulators and their concentrations on the number of leaves, shoots, and lengths of shoot from shoot-tip explants of *A. paniculata*.

Treatment	Number of Leaves	Number of Shoots	Length of Shoot (cm)
T1 (control)	5.03 ± 0.08 e	1.70 ± 0.04 d	2.25 ± 0.02 e
T2 (2.22 µM BAP + 0.49 µM IBA)	8.43 ± 0.17 d	2.00 ± 0.05 c	2.46 ± 0.01 d
T3 (2.22 µM BAP) + 5.37 µM NAA)	8.40 ± 0.18 d	1.80 ± 0.05 cd	2.28 ± 0.01 e
T4 (8.88 µM BAP + 2.69 µM NAA)	18.30 ± 0.21 c	3.60 ± 0.05 b	2.71 ± 0.02 c
T5 (8.88 µM BAP)	19.70 ± 0.38 b	3.8 ± 0.05 b	2.82 ± 0.01 b
T6 (17.76 µM BAP)	23.57 ± 0.48 a	7.33 ± 0.10 a	3.06 ± 0.02 a

Values are means ± standard error (SE) of $n = 30$. Means followed by the same letters within the same column indicate no significant difference ($p < 0.05$) using Tukey's honestly significant difference (HSD) analysis. BAP: 6-benzylaminopurine; IBA: indole-3-butyric acid; NAA: 1-naphthalene acetic acid.

Figure 2. Shoot-tip culture of *A. paniculata* cultured on MS media supplemented with 17.76 µM BAP on day 21. BAP: 6-benzylaminopurine.

2.3. Elicitation

From the previous experiment, an *A. paniculata* explant was excised from plantlets cultured on optimal medium for shoot-tip culture (MS + 17.76 µM BAP). MS medium supplemented with 17.76 µM BAP was subjected to different concentrations of chitosan and SA (1–5 mg/L, with 1 mg/L interval). The effect of the different chitosan- and SA-treated cultures were assessed in terms of the numbers of leaves and shoots, and the length of the shoots. The results showed an incremental trend in the parameters obtained when applied with different elicitors. The highest number of leaves and shoots and the highest length of shoots were recorded in MS media supplemented with 5.0 mg/L chitosan with 26.07 ± 0.14 leaves, 8.33 ± 0.07 shoots, and a 3.63 ± 0.02 cm length of shoots, respectively (Table 2). At the same time, the control treatment showed the lowest herbage yield with 23.57 ± 0.48 leaves, 7.33 ± 0.07 shoots, and a 3.06 ± 0.02 cm length of shoot, respectively. The same incremental trend could be observed when the shoot-tip explant was cultured onto MS media supplemented with SA until 4.0 mg/L SA, and then the trend starts to decline at the 5.0 mg/L SA treatment. This suggests that MS media supplemented with 4.0 mg/L SA showed the optimum treatment for SA-treated cultures, with 25.97 ± 0.06 leaves, 7.97 ± 0.10 shoots, and a 3.19 ± 0.00 cm length of shoot, respectively.

Table 2. The effect of different elicitors and their concentrations on the number of leaves, number of shoots, and the length of shoots (cm) using shoot-tip explant of *A. paniculata*.

Treatment (mg/L)	Number of Leaves	Number of Shoots	Length of Shoots (cm)
Control	23.57 ± 0.48 b	7.33 ± 0.07 c	3.06 ± 0.02 de
1.0 SA	23.60 ± 0.31 b	7.37 ± 0.10 c	2.69 ± 0.02 h
2.0 SA	23.63 ± 0.24 b	7.47 ± 0.04 c	2.88 ± 0.02 g
3.0 SA	24.70 ± 0.33 ab	7.53 ± 0.04 c	2.93 ± 0.02 gf
4.0 SA	25.97 ± 0.06 a	7.97 ± 0.10 b	3.19 ± 0.00 c
5.0 SA	24.77 ± 0.13 ab	7.90 ± 0.04 b	2.98 ± 0.01 ef
1.0 Chitosan	23.80 ± 0.50 b	7.37 ± 0.06 c	3.09 ± 0.02 d
2.0 Chitosan	24.03 ± 0.37 b	7.57 ± 0.03 c	3.10 ± 0.02 d
3.0 Chitosan	24.17 ± 0.30 b	7.90 ± 0.04 b	3.33 ± 0.02 b
4.0 Chitosan	24.53 ± 0.08 b	7.93 ± 0.04 b	3.34 ± 0.02 b
5.0 Chitosan	26.07 ± 0.14 a	8.33 ± 0.07 a	3.63 ± 0.02 a

Note: Values are means ± standard error (SE) of $n = 30$. Means followed by the same letters within the same column indicate that there is no significant difference ($p < 0.05$) using Tukey's honestly significant difference (HSD) analysis. SA: salicylic acid.

2.4. Quantification of Andrographolide in Shoot-Tip Extracts of Micropropagated A. paniculata

The present study investigates andrographolide production from plantlets obtained from shoot-tip explants cultured onto MS media supplemented with chitosan and SA. A total of 5 mg of standard andrographolide was dissolved in 5 mL of 100% methanol (HPLC grade) obtaining a stock of 1000 µg/mL. The stock solution was further diluted to obtain the dilution range of the calibration curve. The range of the calibration curve of andrographolide standards was 0, 20, 40, 60 and 100 ppm, with a correlation coefficient of 0.9989. This indicated that the linearity of the studied method complied with the regulatory requirement. Figure 3a shows the chromatogram standard at 100 ppm. The highest andrographolide production was found in MS media treated with 5.0 mg/L chitosan (2463.03 ± 0.398 µg/mL), while the lowest andrographolide production was obtained from the control treatment, with 256.73 ± 0.341 µg/mL (Figure 3b) at retention time 3.84 min and 254 nm.

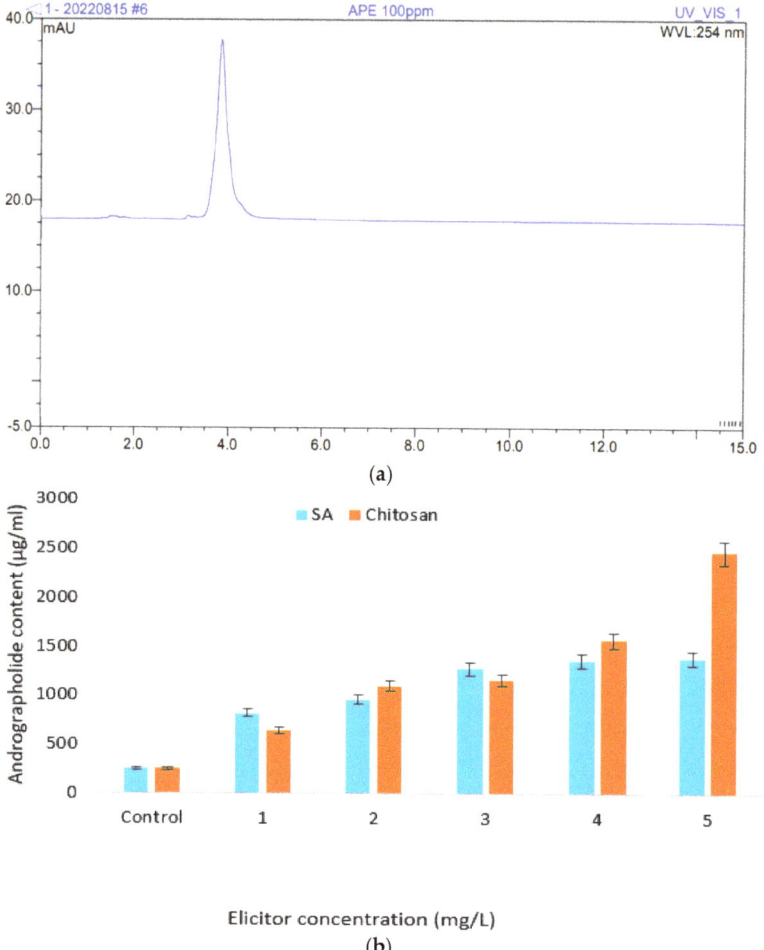

Figure 3. (**a**) Chromatogram of standard andrographolide; retention time 3.84 min at 254 nm and (**b**) andrographolide production of plantlets derived from shoot-tip culture under the influence of chitosan and salicylic acid at different concentrations.

3. Discussion

Surface sterilization is used to achieve sterility in order to prevent bacterial and fungal contamination. The surface sterilization protocol increased the aseptic percentage in the shoot-tip explants. However, it also led to necrosis of the explants, thus making it harder to rejuvenate them [14–16]. Contamination in cultured explants can occur within three to ten days after inoculation, as mentioned by Chen and David [17]. However, if the contamination occurs at a later stage, it can be concluded that the plant materials used might be infected. Hence, regardless of how optimal and effective the surface sterilization protocol is, contamination will occur as a result of exudation from the endogenous microbial culture of the planting materials. Although antibiotics can be used to remediate the problem, it is advisable to select a healthy mother plant rather than treat them with antibiotics, which would incur more costs. Clorox, a bleach containing 5.25% NaClO, is an effective sterilizing agent for various plant species, but their responses can vary, depending on explant types. Different species and plant parts can produce a variety of observations. Herbaceous plants are more easily surface sterilized and produce more positive responses as compared to woody plants due to the juvenility of the plant tissues. Different plant parts also need to be considered when choosing the concentration of the sterilant. Plant parts such as stems and embryos encapsulated in a seed pod will not become necrotic as frequently as would young shoots or petals when used as explants for surface sterilization, as these tissues are more fragile. As observed in many studies, different explant types have different concentration requirements for sterilants, including *Phyllanthus niruri*, *Gerbera hybrida*, *Aquilaria malaccensis*, and *Zingiber officinale* [5,18–20]. The concentration of commercial bleach affects its effectiveness in reducing contaminants and its potential to cause tissue injury. A higher concentration of bleach reduces contaminants but can cause tissue injury, resulting in browning, low percentages of explant survival, and the death of explants. Prolonged exposure to high concentrations of sterilants will produce phenolics, which will also lead to the browning and death of explants. Hence, a balance between commercial bleach concentration and the duration of immersion needs to be considered when conducting surface sterilization. Browning can be prevented by reducing sterilant exposure, increasing NaClO concentration, or keeping cultures in dark conditions. The browning of explants may also result from the stress conditions induced by the in vitro system, as observed in herbs such as *Cestrum nocturnum* [21], *Solanum tuberosum* [22], *Zingiber officinale* [23] and *Matricaria chamomilla* [1].

Axillary bud proliferation is one of the micropropagation pathways that utilize axil parts (e.g., meristem, shoot-tip, and nodal segment) to develop into plantlets [24,25]. Since this technology does not involve cell dedifferentiation of differentiated cells but rather the development and growth of new shoots from pre-existing meristems, it has usually been pointed out as the most commercially viable way of propagating plants in vitro [26]. Among the methods utilizing axils, shoot-tip culture is the most widely used and considered the most commercially viable method for guaranteeing the genetic stability of the plantlets obtained [25]. Not only that, but shoot-tip culture is also a rapid technique that does not require special equipment in the manner of meristem tip culture, in which the meristem part needs to be isolated under the microscope. The application of cytokinin is essential, as it helps to break the bud dormancy phase of the explants. Cytokinins control the size of the shoot meristem, the number of leaf primordia, and the growth of the leaf and shoot by promoting cell division [25]. Hence, it is important to choose juvenile plant tissues due to the abundance of natural phytohormones, compared to plant tissues that already aging. 6-benzylaminopurine (BAP) is more effective for shoot proliferation than are other cytokinins (kinetin, thidiazuron, picloram) [27,28]. In the present study, all shoot-tip explants showed positive growth. However, the growth was slower in treatments without PGRs. This observation concurred with the understanding of the roles of cytokinin in plant growth by stimulating cell division. According to Nor Mayati and Jamnah [29], slow growth is probably due to insufficient nutrients before the differentiation stage. These results differed from the findings by Suraya et al. [5], who stated that aseptic nodal segments of *Phyllanthus niruri* cultured on MS medium fortified with 1.0 mg/L BAP produced the highest number of shoots and the combination of kinetin and BAP induced multiple shoot formation in all nodal explants. MS medium supplemented with high BAP produced better results for *Dracocephalum kotschyi* shoot propagation [30]. Similar results were achieved for the bud proliferation of *Musa acuminata* [31], *Mentha piperita* [27] and *Scutellaria altissima* [6].

It was known that the explants' responses were different depending on the elicitors' types. The elicitors play a vital role in maintaining organisms' growth, but they are harmful to plants at high concentrations, as elicitors can act as stress agents to plants. When plants are under stress conditions, the herbage yield, biomass, and subsequently bioactive compounds will be altered. Using this knowledge, a number of research efforts have been conducted to enhance the herbage yield, biomass, and bioactive compounds in the treated plants. The increased biomass accumulation following chitosan application is due to the ability to boost the availability and absorption of water and essential nutrients by controlling the cells' osmotic pressure [32]. A similar observation of biomass production has been reported for various in vitro culture systems for different plant species, such as callus culture of *Fagonia indica* [13], as well as cell suspension of *Silybum marianum* [32]. Chitosan has been reported to have a symbiotic relationship with growth-promoting rhizobacteria, thus triggering the germination rate and improving plant nutrient uptake [33]. The application of chitosan for herbage yields can be observed in many species, including *Artemisia aucheri* [34], safflower [6], *Swertia paniculata* [35], and *Coffea arabica* [36]. Different SA concentrations promote or inhibit plant growth in different plant species by modulating cell division and expansion [2]. SA regulates plant growth via multiple pathways [37]. The altered endogenous SA levels in plants can result in abnormal growth phenotypes [38]. Further investigation by Wang et al. [2] found that SA accumulation in the cad1 mutant promotes the quiescent center of cell division through the accumulation of reactive oxygen species and downregulation of the transcription factor genes. The results of the current study on SA application align with a previous study reported by Golkar et al. [6] and Koo et al. [38], in which plant growth was inhibited by applying high levels of elicitors. The reduction of callus production after adding high amounts of SA might be attributed to the stress induced upon cell growth and cell division. Previous studies' results have shown the inhibitory effects of SA on plant growth development [39].

Plants are good sources for the discovery of pharmaceutical compounds and medicines. Natural products could be potential drugs for humans or livestock species, and these products and their analogs can act as intermediates for synthesizing useful drugs [40]. Phytochemical screening of medicinal plants is crucial in identifying new sources of therapeutically and industrially essential compounds [41,42]. In the current study, the production of andrographolide was higher when the MS media was supplemented with chitosan than it was with supplementation with SA. The andrographolide production was double when chitosan was applied in the MS media. The content of andrographolide and other diterpene lactones was higher than that reported by Jindal et al. [43] and Pawar et al. [44] when both studied different PGR combinations in the media. Another study on *Vitis vinifera* extract found that chitosan significantly enhanced the targeted bioactive compounds [45]. Moreover, the synthesis of polyphenolics, secoiridoid, glycosides, lignin, flavonoids, and phytoalexins was observed in *Fagonia indica* [13], *Swertia paniculata* [35] and *Silybum marianum* [32] after being treated with chitosan. SA-treated culture proved to increase the production of fatty acids from *Jatropha curcas* callus grown in vitro [46]. Additionally, SA regulates the production of swertiamarin and amarogentin glycosides in *Swertia paniculata* [35], and phenolic and flavonoid compounds in *Fagonia indica* [13]. Moreover, SA also acts as a self-protective agent in in vitro culture of *Nicotiana tabacum* [38] and *Arabidopsis thaliana* [47].

4. Materials and Methods

4.1. Plant Material

A. paniculata were collected from the nursery site at the Faculty of Agriculture, Universiti Putra Malaysia Serdang, Selangor. The plant sample was identified at the Biodiversity Unit, Institute of Bioscience, UPM, with reference number KM 0020/22.

4.2. Explant Surface Sterilization

The shoots of *A. paniculata* were chosen as explants in this experiment. The explant was prewashed with two drops of detergent for 15 min under tap water. The surface sterilization procedure was conducted under a laminar flow hood. The explants were sterilized for 1, 5, and 10 min of exposure using different concentrations (40, 50, 60, and 70%) of commercial bleach containing 5.2% sodium hypochlorite with one to two drops of Tween 20 as a wetting agent. After that, all the treatment solutions of T1 (40%, 1 min), T2 (40%, 10 min), T3 (40%, 5 min), T4 (50%, 1 min), T5 (50%, 10 min), T6 (50%, 5 min), T7 (60%, 1 min), T8 (60%, 10 min), T9 (60%, 5 min), T10 (70%, 1 min), T11 (70%, 10 min) and T12 (70%, 5 min) were discarded and the explants were rinsed at least three times with sterile distilled water.

4.3. Axillary Shoot Multiplication

Murashige and Skoog [48] (MS) medium supplemented with different concentrations of T1 (control), T2 (2.22 µM BAP + 0.49 µM IBA), T3 (2.22 µM BAP + 5.37 µM NAA), T4 (8.88 µM BAP + 2.69 µM NAA), T5 (8.88 µM BAP) and T6 (17.76 µM BAP) was mixed with 0.1 g/L of myo-inositol and 20 g/L of sucrose. A total of 3 g/L of Gelrite® as a gelling agent was added and stirred until completely dissolved before the pH was adjusted to pH 5.6–5.8. The solution was heated in the microwave before being placed in vials and covered with aluminum foil. All of the labeled vials were placed in an autoclave and sterilized at 121 °C and 1.05 kg/cm^2 for 20 min.

4.4. Elicitation

Chitosan ($C_{56}H_{103}N_9O_{39}$) and salicylic acid (SA) (Sigma-Aldrich, St. Louis, MO, USA) were used for elicitation. Different concentration levels (0, 1.0, 2.0, 3.0, 4.0, and 5.0 mg/L) of both chitosan and SA were introduced to the MS supplemented with 17.76 µM BAP.

4.5. Growing Conditions

Vials (25 × 95 mm; Phytotech®, Lenexa, KS, USA) containing cultured explants were incubated for four weeks at 25 ± 3 °C under a photoperiod of 16 h light and 8 h darkness supplied by white fluorescent light (45 µmol/m^2/s) in the incubation room.

4.6. Extraction of Andrographolide

Elicitated plantlets were harvested in the fourth week after inoculation. A total of 10 mg dried plantlets were powdered and dissolved in 5 mL of 100% methanol (HPLC grade). The sample was then extracted via sonication-assisted extraction (Fisherbrand® FB155055, Waltham, MA, USA) (40 °C for 30 min) and quantified for andrographolide content. Each extract was mixed using a vortex (ZX3 Advanced Vortex Mixer) for 3 min and filtered through 0.45 µm nylon syringe membranes (Macherey Nagel, Hoerdt, France), and stored at −4 °C before HPLC analysis.

4.7. Chromatographic Parameters

Quantitative analysis of andrographolide of *A. paniculata* was carried out as per the protocol of Masaenah et al. [49] with minor modification using HPLC (Thermo ScientificTM DionexTM UltiMate 3000 UHPLC, Waltham, MA, USA). Chromatographic separation was performed using XBridge® (Waters, Milford, MA, USA) C18 column (5 µm, 4.6 mm × 250 mm). The mobile phase consisted of a mixture of methanol and deionized water (50:50) and a 1 mL/min flow rate. The injection volume was 10 µL. The temperature of the column was controlled at 25 °C, and samples were detected using a PD-M20A photodiode array detector. The andrographolide (Merck Chemical, Saint-Quentin Fallavier, France) was identified by comparing the retention time of samples with reliable standard chromatographic peaks at 254 nm, and the andrographolide content was expressed as µg/mL.

4.8. Analysis of Data

The experiment was conducted in a completely randomized design and analyzed using one-way analysis of variance (ANOVA) through SAS software, version 9.4 (SAS Institute, Cary, NC, USA). Means comparisons were separated by the least significant mean (LS mean) and Tukey honestly significant difference (HSD) test at $p < 0.05$.

5. Conclusions

An in vitro culture of *A. paniculata* was successfully established using shoot-tip explants. BAP at 17.76 µM proved optimal for axillary shoot multiplication and growth. Chitosan at 5 mg/L further enhanced shoot multiplication and growth with 26.07 leaves, 8.33 shoots, and 3.63 cm of shoot-length per explant. An increase in andrographolide production was obtained in chitosan and SA-treated cultures. However, MS media supplemented with 5 mg/L chitosan produced higher andrographolide with 2463.034 ± 0.398 µg/mL, compared to that of the control with 256.73 ± 0.341 µg/mL.

Author Contributions: Conceptualization, A.P. and M.H.; methodology, A.P., M.H., P.E.M.W. and M.M.Y.; validation, A.P., Y.H.D., A.A. and M.H.; formal analysis, A.P.; investigation, A.P. and M.H.; data curation, A.P.; writing—original draft preparation, A.P. and M.H.; writing—review and editing, A.P., P.E.M.W., M.M.Y., Y.H.D., A.A. and M.H.; visualization, A.P., Y.H.D. and A.A.; project administration, M.H. All authors have read and agreed to the published version of the manuscript.

Funding: This research was funded by the Ministry of Higher Education Malaysia, grant number 5526700 and Researchers Supporting Project number RSP2023R375, King Saud University, Riyadh, Saudi Arabia. The APC was funded by University Putra Malaysia and King Saud University.

Data Availability Statement: All data are presented in the article.

Acknowledgments: The authors thank Universiti Putra Malaysia for the support and for providing Graduate Research Fellowships. The authors acknowledge Researchers Supporting Project number (RSP2023R375), King Saud University, Riyadh, Saudi Arabia.

Conflicts of Interest: The authors declare no conflict of interest.

References

1. Dawande, A.A.; Sanjay, S. Copper sulphate elicitation of optimized suspension culture of *Andrographis paniculata* Nees yields unprecedented level of andrographolide. *J. Microbiol. Biotechnol. Food Sci.* **2020**, *9*, 688–694. [CrossRef]
2. Wang, Z.; Rong, D.; Chen, D.; Xiao, Y.; Liu, R.; Wu, S. Salicylic acid promotes quiescent center cell division through ROS accumulation and down-regulation of PLT1, PLT2, and WOX5. *J. Integr. Plant Biol.* **2021**, *63*, 583–596. [CrossRef] [PubMed]
3. Daud, N.H.; Jayaraman, S.; Mohamed, R. An improved surface sterilization technique for introducing leaf, nodal and seed explants of *Aquilaria malaccensis* from field sources into tissue culture. *Asia-Pac. J. Mol. Biol. Biotechnol.* **2012**, *20*, 55–58.
4. Shukla, S.K.; Mishra, A.K.; Arotiba, O.A.; Mamba, B.B. Chitosan-based nanomaterials: A state-of-the-art review. *Int. J. Biol. Macromol.* **2013**, *59*, 46–58. [CrossRef] [PubMed]
5. Suraya, A.A.; Misran, A.; Hakiman, M. The efficient and easy micropropagation protocol of *Phyllanthus niruri*. *Plants* **2021**, *10*, 2141. [CrossRef] [PubMed]
6. Golkar, P.; Taghizadeh, M.; Yousefian, Z. The effects of chitosan and salicylic acid on elicitation of secondary metabolites and antioxidant activity of safflower under in vitro salinity stress. *Plant Cell Tissue Organ Cult.* **2019**, *137*, 575–585. [CrossRef]
7. Roy, P.K. *In vitro* propagation of *Andrographis paniculata* Nees.—A threatened medicinal plant of Bangladesh. *J. Biol. Sci.* **2014**, *3*, 67–73. [CrossRef]
8. Jindal, N.; Chaudhury, A.; Kajla, S. Shoot proliferation and multiplication from nodes of *Andrographis paniculata*. *Int. Res. J. Pharm.* **2015**, *6*, 654–657. [CrossRef]
9. Yadav, R.K.; Ram, L.; Mahala, K.R.; Maheshwari, R.K. Rapid in vitro multiplication, regeneration and rooting of Kalneg (*Andrographis paniculata* Nees.). *J. Emerg. Technol. Innov. Res.* **2021**, *8*, 5.

10. Benoy, G.K.; Animesh, D.K.; Aninda, M.; Priyanka, D.K.; Sandip, H. An overview on *A. paniculata* (Burm.f.) Nees. *Int. J. Res. Ayurveda Pharm.* **2012**, *3*, 752–760.
11. Matthias, E.; Daniel, J. Plant secondary metabolites as defenses, regulators, and primary metabolites: The blurred functional trichotomy. *Plant Physiol.* **2020**, *184*, 39–52.
12. Rahman, N.N.A.; Rosli, R.; Kadzimin, S.; Hakiman, M. Effects of auxin and cytokinin on callus induction in *Catharanthus roseus* (L.) G. Don. *Fundam. Appl. Agric.* **2019**, *4*, 928–932. [CrossRef]
13. Khan, T.; Khan, T.; Hano, C.; Abbasi, B.A. Effects of chitosan and salicylic acid on the production of pharmacologically attractive secondary metabolites in callus cultures of *Fagonia indica*. *Ind. Crops Prod.* **2019**, *129*, 525–535. [CrossRef]
14. Ahmad, A.; Qamar, M.T.; Shoukat, A.; Aslam, M.M.; Tariq, M.; Hakiman, M.; Joyia, F.A. The effects of genotypes and media composition on callogenesis, regeneration and cell suspension culture of chamomile (*Matricaria chamomilla* L.). *PeerJ* **2021**, *9*, e11464. [CrossRef] [PubMed]
15. Najhah, M.Y.; Jaafar, H.Z.; Nakasha, J.J.; Hakiman, M. Shoot multiplication and callus induction of *Labisia pumila* var. alata as influenced by different plant growth regulators treatments and its polyphenolic activities compared with the wild plant. *Molecules* **2021**, *26*, 3229. [CrossRef]
16. Ana, K.P.; Ana, P.S.; Joselita, C.S.; Silvio, L.T.; Juliana, M.R.; Ana, R.P.; Cristiane, D.P. Sodium hypochlorite sterilization of culture medium in micropropagation of *Gerbera hybrida* cv. Essandre. *Afr. J. Biotechnol.* **2016**, *15*, 1995–1998. [CrossRef]
17. Chen, Z.; David, A.E. General techniques in tissue culture in perennial crops, in hand. In *Book of Plant Cell Culture*; McGraw-Hill Publishing Company: New York, NY, USA, 2011; Volume 6, pp. 22–57.
18. Cesar, A.H.; Lucia, A.K.; Rafael, A.I.; Mario, L.A. Analysis of genetic variation in clones of rubber (*Hevea brasiliensis*) from Asian, South and Central American origin using RAPDs markers. *Rev. Colomb. Biotecnol.* **2009**, *2*, 29–34.
19. Ong, C.W.; Shamsul, B.A.R. The introduction of rubber planting recommendations by The Rubber Research Institute of Malaysia since 1925. *J. Biol.* **2011**, *4*, 2224–3208.
20. Zahid, N.A.; Jaafar, H.Z.E.; Hakiman, M. Micropropagation of ginger (*Zingiber officinale* Roscoe) 'Bentong' and evaluation of its secondary metabolites and antioxidant activities compared with the conventionally propagated plant. *Plants* **2021**, *10*, 630. [CrossRef]
21. Sameer, N.M.; Nabeel, K.A. Effect of different sterilization methods on contamination and viability of nodal segments of *Cestrum nocturnum* L. *Int. J. Res. Stud. Biosci.* **2016**, *4*, 4–9.
22. Anoop, B.; Chauhan, J.S. In vitro sterilization protocol for in vitro regeneration of *Solanum tuberosum* cv. 'Kufri Himalini'. *Sci. Publ. J.* **2010**, *1*, 24–27.
23. Zahid, N.A.; Jaafar, H.Z.E.; Hakiman, M. Alterations in microrhizome induction, shoot multiplication and rooting of ginger (*Zingiber officinale* Roscoe) var. Bentong with regards to sucrose and plant growth regulators application. *Agronomy* **2021**, *11*, 320. [CrossRef]
24. Chawla, H.S. *Introduction to Plant Biotechnology*, 2nd ed.; Science Press: Beijing, China, 2002.
25. Nowakowska, K.; Pińkowska, A.; Siedlecka, E. The effect of cytokinins on shoot proliferation, biochemical changes and genetic stability of *Rhododendron* 'Kazimierz Odnowiciel' in the in vitro cultures. *Plant Cell Tissue Organ Cult.* **2022**, *149*, 675–684. [CrossRef]
26. Sharma, S.K.; Bryan, G.J.; Winfield, M.O.; Millam, S. Stability of potato (*Solanum tuberosum* L.) plants regenerated via somatic embryos, axillary bud proliferated shoots, microtubers and true potato seeds: A comparative phenotypic, cytogenetic and molecular assessment. *Planta* **2007**, *226*, 1449–1458. [CrossRef] [PubMed]
27. Mehta, J.; Naruka, R.; Sain, M.; Dwiredi, A.; Sharma, D.; Mirza, J. An efficient protocol for clonal micropropagation of *Mentha piperita* L. (Pippermint). *Asian J. Plant Sci. Res.* **2012**, *2*, 518–523.
28. Grzegorczyk-Karolak, I.; Kuźma, Ł.; Wysokińska, H. The influence of cytokinins of proliferation and polyphenol accumulation in shoot cultures of *Scutellaria altissima* L. *Phytochem. Lett.* **2017**, *20*, 449–455. [CrossRef]
29. Nor Mayati, C.H.; Jamnah, A.R. Induction of shoots and roots from vegetative tissue culture of *Hevea brasiliensis* RRIM 2020. *J. Trop. Plant Physiol.* **2014**, *6*, 1–9.
30. Otroshy, M.; Moradi, K. Rapid regeneration of *Dracocephalum kotschyi* Boiss. from nodal explants. *Int. Life Sci. Med. Res.* **2013**, *3*, 11–14.
31. Jafari, N.; Rofina Yasmin, O.; Norzulaani, K. Effect of benzylaminopurine (BAP) pulsing on in vitro shoot multiplication of *Musa acuminata* (banana) cv. Berangan. *Afr. J. Biotechnol.* **2010**, *10*, 2446–2450.
32. Shah, M.; Jan, H.; Drouet, S.; Tungmunnithum, D.; Shirazi, J.H.; Hano, C.; Abbasi, B.H. Chitosan elicitation impacts flavonolignan biosynsthesis in *Silybum marianum* (L.) Gaertn cell suspension and enhances antioxidant and anti-inflammatory activities of cell extracts. *Molecules* **2021**, *26*, 791. [CrossRef]
33. Sharif, R.; Mujtaba, M.; Ur Rahman, M.; Shalmani, A.; Ahmad, H.; Anwar, T.; Tianchan, D.; Wang, X. The multifunctional role of chitosan in horticultural crops; A review. *Molecules* **2018**, *23*, 872. [CrossRef] [PubMed]
34. Asghari, G.R.; Ghasemi, R.; Yosefi, M.; Mehdinezhad, N. Effect of hormones, salicylic acid, chitosan on phenolic compounds in *Artemisia aucheri* in vitro. *J. Plant Process Funct.* **2015**, *3*, 93–100.
35. Kaur, P.; Gupta, R.C.; Dey, A.; Malik, T.; Pandey, D.K. Optimization of salicylic acid and chitosan treatment for bitter secoiridoid and xanthone glycosides production in shoot cultures of *Swertia paniculata* using response surface methodology and artificial neural network. *BMC Plant Biol.* **2020**, *20*, 225. [CrossRef] [PubMed]

36. Dzung, N.A. Enhancing crop production with chitosan and its derivatives. In *Chitin, Chitosan, Oligosaccharides and their Derivatives: Biological Activity and Application*; Kim, S.K., Ed.; CRC Press: Boca Raton, FL, USA; Taylor & Francis: Abingdon, UK, 2010; pp. 619–632.
37. Li, A.; Sun, X.; Liu, L. Action of salicylic acid on plant growth. *Front. Plant Sci.* **2022**, *13*, 878076. [CrossRef]
38. Koo, Y.M.; Heo, A.Y.; Choi, H.W. Salicylic acid as a safe plant protector and growth regulator. *Plant Pathol. J.* **2020**, *36*, 1–10. [CrossRef]
39. Lee, C.W.; Mahendra, S.; Zodrow, K.; Li, D.; Tsai, Y.C.; Braam, J.; Alvarez, P.J.J. Developmental phytotoxicity of metal oxide nanoparticles to *Arabidopsis thaliana*. *Environ. Toxicol. Chem.* **2010**, *29*, 669–675. [CrossRef]
40. Haida, Z.; Ab Ghani, S.; Nakasha, J.J.; Hakiman, M. Determination of experimental domain factors of polyphenols, phenolic acids and flavonoids of lemon (*Citrus limon*) peel using two-level factorial design. *Saudi J. Biol. Sci.* **2022**, *29*, 574–582. [CrossRef] [PubMed]
41. Savithramma, N.; Linga Rao, M.; Suhrulatha, D. Screening of medicinal plants for secondary metabolites. *Middle-East J. Sci. Res.* **2011**, *8*, 579–584.
42. Ranjha, M.M.A.N.; Shafeeqa, I.; José, M.L.; Bakhtawar, S.; Rabia, K.; Mirian, P.; Rai, N.A.; Lufeng, W.; Gulzar, A.N.; Roobab, U.; et al. Sonication, a potential technique for extraction of phytoconstituents: A systematic review. *Processes* **2021**, *9*, 1406. [CrossRef]
43. Jindal, N.; Kajla, S.; Chaudhury, A. Establishment of callus cultures of *Andrographis paniculata* for the assessment of andrographolide content. *Int. J. Res. Ayurveda Pharm.* **2016**, *7*, 197–201. [CrossRef]
44. Pawar, S.D.; Yeole, P.T.; Bhadane, P.V.; Kadam, S.R. Standardization of callus induction protocol and effect of hormone concentration on synthesis of andrographolide from *Andrographis paniculata*. *Int. J. Chem. Stud. J.* **2018**, *6*, 1384–1387.
45. Sae-Lee, N.; Kerdchoechuen, O.; Laohakunjit, N.; Thumthanaruk, B.; Sarkar, D.; Shetty, K. Improvement of phenolic antioxidant-linked cancer cell cytotoxicity of grape cell culture elicited by chitosan and chemical treatments. *Hortic. Sci.* **2017**, *52*, 1577–1584. [CrossRef]
46. Eganathan, P.; Ravi Mahalakshmi, R.; Kumar, A.P. Salicylic acid elicitation on production of secondary metabolite by cell cultures of *Jatropha curcas* L. *Int. J. Pharm. Pharm. Sci.* **2013**, *5*, 655–659.
47. Lovelock, D.A.; Šola, I.; Marschollek, S.; Donald, C.E.; Rusak, G.; van Pée, K.H.; Ludwig-Müller, J.; Cahill, D.M. Analysis of salicylic acid-dependent pathways in *Arabidopsis thaliana* following infection with *Plasmodiophora brassicae* and the influence of salicylic acid on disease. *Mol. Plant Pathol.* **2016**, *17*, 1237–1251. [CrossRef]
48. Murashige, T.; Skoog, F. A revised medium for rapid growth and bio assays with tobacco tissue cultures. *Physiol. Plant.* **1962**, *15*, 473–497. [CrossRef]
49. Masaenah, E.; Elya, B.; Setiawan, H.; Fadhilah, Z.; Arianti, V. Quantification of andrographolide in *Andrographis paniculata* (Burm.f.) Nees, myricetin in *Syzygium cumini* (L.) Skeels, and Brazilin in *Caesalpinia sappan* L. by HPLC method. *Pharmacogn. J.* **2021**, *13*, 1437–1444. [CrossRef]

Disclaimer/Publisher's Note: The statements, opinions and data contained in all publications are solely those of the individual author(s) and contributor(s) and not of MDPI and/or the editor(s). MDPI and/or the editor(s) disclaim responsibility for any injury to people or property resulting from any ideas, methods, instructions or products referred to in the content.

Article

Fungal Elicitation Enhances Vincristine and Vinblastine Yield in the Embryogenic Tissues of *Catharanthus roseus*

Dipti Tonk [1], Abdul Mujib [1,*], Mehpara Maqsood [2], Mir Khusrau [3], Ali Alsughayyir [4] and Yaser Hassan Dewir [5]

[1] Cellular Differentiation and Molecular Genetics Section, Department of Botany, Jamia Hamdard, New Delhi 110062, India; dipti10785@gmail.com
[2] Department of Botany, Government College for Women, M.A. Road, Srinagar 190001, India; meh_heem@yahoo.co.in
[3] Department of Botany, Government Degree College (Boys), Anantnag 231213, India; mirkhusrau@gmail.com
[4] Department of Plant and Soil Sciences, Mississippi State University, 75 B.S. Hood Rd, Starkville, MS 39762, USA; aa2942@msstate.edu
[5] Plant Production Department, College of Food and Agriculture Sciences, King Saud University, Riyadh 11451, Saudi Arabia; ydewir@ksu.edu.sa
* Correspondence: amujib3@yahoo.co.in

Citation: Tonk, D.; Mujib, A.; Maqsood, M.; Khusrau, M.; Alsughayyir, A.; Dewir, Y.H. Fungal Elicitation Enhances Vincristine and Vinblastine Yield in the Embryogenic Tissues of *Catharanthus roseus*. *Plants* 2023, 12, 3373. https://doi.org/10.3390/plants12193373

Academic Editor: Iyyakkannu Sivanesan

Received: 29 August 2023
Revised: 17 September 2023
Accepted: 20 September 2023
Published: 25 September 2023

Copyright: © 2023 by the authors. Licensee MDPI, Basel, Switzerland. This article is an open access article distributed under the terms and conditions of the Creative Commons Attribution (CC BY) license (https://creativecommons.org/licenses/by/4.0/).

Abstract: Fungal elicitation could improve the secondary metabolite contents of in vitro cultures. Herein, we report the effect of *Fusarium oxysporum* on vinblastine and vincristine alkaloid yields in *Catharanthus roseus* embryos. The study revealed increased yields of vinblastine and vincristine in *Catharanthus* tissues. Different concentrations, i.e., 0.05% (T1), 0.15% (T2), 0.25% (T3), and 0.35% (T4), of an *F. oxysporum* extract were applied to a solid MS medium in addition to a control (T0). Embryogenic calli were formed from the hypocotyl explants of germinating seedlings, and the tissues were exposed to *Fusarium* extract elicitation. The administration of the *F. oxysporum* extract improved the growth of the callus biomass, which later differentiated into embryos, and the maximum induction of somatic embryos was noted T2 concentration (102.69/callus mass). A biochemical analysis revealed extra accumulations of sugar, protein, and proline in the fungus-elicitated cultivating tissues. The somatic embryos germinated into plantlets on full-strength MS medium supplemented with 2.24 µM of BA. The germination rate of the embryos and the shoot and root lengths of the embryos were high at low doses of the *Fusarium* treatment. The yields of vinblastine and vincristine were measured in different treated tissues via high-pressure thin-layer chromatography (HPTLC). The yield of vinblastine was high in mature (45-day old) embryos (1.229 µg g^{-1} dry weight), which were further enriched (1.267 µg g^{-1} dry weight) via the *F. oxysporum*-elicitated treatment, especially at the T2 concentration. Compared to vinblastine, the vincristine content was low, with a maximum of 0.307 µg g^{-1} dry weight following the addition of the *F. oxysporum* treatment. The highest and increased yields of vinblastine and vincristine, 7.88 and 15.50%, were noted in *F. oxysporum*-amended tissues. The maturated and germinating somatic embryos had high levels of SOD activity, and upon the addition of the fungal extracts, the enzyme's activity was further elevated, indicating that the tissues experienced cellular stress which yielded increased levels of vinblastine and vincristine following the T2/T1 treatments. The improvement in the yields of these alkaloids could augment cancer healthcare treatments, making them easy, accessible, and inexpensive.

Keywords: callus induction; elicitation; Madagascar periwinkle; somatic embryogenesis; vincristine; vinblastine

1. Introduction

Catharanthus roseus, commonly known as Madagascar periwinkle, is a tropical, perennial, medicinal plant belonging to the family Apocynaceae. It is a source of several important indole alkaloids of medicinal importance such as vinblastine, vincristine, ajmalicine, vindoline, catharanthine, and serpentine [1]. Due to its immense pharmaceutical importance and low (0.0005%) contents of vinblastine and vincristine, *C. roseus* has been regarded as an important model for secondary metabolism studies. In recent decades, an inclusive, multidimensional research study has attempted to improve the alkaloid contents in *C. roseus* [2,3]. Birat et al. [4] recently reported that the fungus *Nigrospora zimmermanii*, which is present within the leaves of *Catharanthus roseus*, also produced vincristine successfully. The strategies frequently used for enriching the levels of alkaloids are the optimization of media, plant growth regulators, and cultural practices; the culture of high-yielding cell lines; the use of precursors; the incorporation of elicitors; and improving the expression of the regulatory enzymes of metabolic pathways [5–7].

In recent years, researchers have attempted to influence the production of secondary metabolites from diverse tissue sources through the use of different biotic and abiotic elicitors [8]. The techniques used drastically reduced the processing times needed to obtain active compounds [9,10]. Elicitors are a large target group of compounds which have been added to media at various stages of cultural growth for improving secondary compounds. Traditionally, an 'elicitor' is a molecule which is introduced into a medium in small levels to improve the biosynthesis of compounds by triggering cellular defense response genes [11,12]. The process also refers to compounds of various sources which stimulate physiological and morphological responses in inducing compounds of a defensive nature [13]. It is well established that the application of an elicitor or the invasion of a pathogen produces an array of defensive secondary reactions in plant cells. Singh et al. [14] categorized diverse types of elicitors diverse types: (a) biotic elicitors such as bacterial and fungal cell walls or glycoproteins, (b) abiotic elicitors like UV irradiation, salt, and various non-constitutive compounds, and (c) endogenous elicitors, which are signaling compounds of plant-cell origins. A large number of biotic elicitors have been recognized to be very efficient at enriching secondary metabolites and are exploited in a variety of cultures [15]. Yeast extract was used as a biotic elicitor in cultures which induced the synthesis of a variety of phytocompounds in several investigations into plant–microbe interactions [16,17]. Endophytic fungi (used as fungal elicitors) isolated from *C. roseus* could also be used to enrich indole alkaloid production in culture [18]. The culture filtrate of *Fusarium sporotrichioides* Sherbakoff, isolated from *Narcissus tazetta var. italicus* rhizosphere and grown on a potato dextrose broth, stimulated the production of alkaloids in cultivated tissues [19]. A marked increase in vasicine content in *Adhatoda vasica* was observed via the amendment of select elicitors like methyl jasmonate (MeJA), chitosan, yeast extract, ascorbic acid, and sodium salicylate at optimized concentrations [20]. Arbuscular mycorrhizal fungi (a group of beneficial microorganisms) were reported to play a major role in enhancing alkaloid production in root organ cultures [21]. In *Centella asiatica*, the influence of various elicitors, like the use of *Trichoderma harzianum*, *Colletotrichum lindemuthianum*, and *Fusarium oxysporum* to improve the accumulation of secondary metabolites, was reviewed and discussed [22,23].

In addition, a number of abiotic factors have been widely incorporated to augment product synthesis in cultured tissues, such as elevated temperature, excess salinity, osmotic stress, ultra-violet (UV) rays, and heavy metal stress [24]. In this specific plant, *C. roseus*, a variety of abiotic compounds such as NaCl, cerium (CeO_2 and $CeCl_3$), yttrium (Y_2O_3), and neodymium ($NdCl_3$) were used successfully to enhance alkaloid yield [25]. $CaCl_2$ was used as an elicitor for the enhancement of vinblastine in a *C. roseus* embryogenic cell suspension [26]. When used, these elicitors caused stresses and improved the synthesis of secondary compounds in several investigated genera. Elicitor-induced cellular stress is measured by monitoring antioxidant enzymes, which ameliorate stresses in cultures [27,28]. Various enzymes such as superoxide dismutase (SOD), catalase (CAT), ascorbate peroxidase (APX), and glutathione reductase (GR) are assayed to ascertain the level of stress in cultured tissues and were studied in different plant genera [29]. Although the enhancement of alkaloids is noted to be treatment-specific, the use of elicitors could be a valuable strategy for enriching phytocompounds. In this study, the fungus *Fusarium oxysporum* was used as biotic elicitor, and the yields of vinblastine and vincristine were measured in cultures. This is perhaps one of the first fungal (biotic) elicitation studies on alkaloid yield mediated via embryogenesis. The growth of the callus biomass and the biochemical alterations/associations during the course of its growth and morphogenesis were monitored.

2. Results

2.1. Callus Induction and Biomass Growth in a Medium Containing a Fusarium oxysporum Extract

On MS medium supplemented with 4.52 µM of 2,4-D, the hypocotyls of seedlings grown in vitro produced profuse calli. The calli were friable, light-yellow, and fast-growing; they later turned into embryogenic calli (Figure 1a). These hypocotyl calli were subjected to various levels of *F. oxysporum* elicitation and routinely subcultured at regular intervals. Growth is an indicator of cell division with the rapid multiplication of the callus; therefore, the growth the biomass of each callus was measured in response to the elicitor treatments. We observed that with the *F. oxysporum* treatment, the growth of the embryogenic calli was faster compared to the control. The biomass of each calli increased up to T2, and for this treatment, maximum fresh, dry, and absolute dry mass % values were observed (1.55, 0.183 g, and 11.803%, respectively). Upon elicitation, the calli appeared friable and white, especially those that received the T1 and T2 treatments. The calli that received higher concentrations, i.e., T3 and T4, were less responsive; the calli turned light-brown, were compact, and showed poor growth.

Figure 1. (**a**) Embryogenic callus grown in MS medium containing 4.52 µM of 2,4-D and the T2 fungal elicitor (bar 2 mm); (**b**) embryo on a maturation medium containing 2.60 µM of GA$_3$ and an elicitor (bar 2 mm); (**c**) germinated embryos at early stage with a root (bar 0.5 cm).

2.2. F. oxysporum Treatments and the Number of Embryos

The embryogenic calli were cultured on MS medium supplemented with 5.37 µM of NAA and 6.72 µM of BA, and different concentrations of the *F. oxysporum* extract were added in order to monitor the influence of the fungus elicitor on the number of embryos and their growth. The maximum fresh, dry, and absolute dry weight values were observed for T2 (2.066, 0.237 g, and 11.442%, respectively) compared to the other treatments and the control, T0. Under all the tested conditions, the embryogenic calli differentiated into embryos, and for the T2 concentration of the *F. oxysporum* treatment, the maximum number of embryos was formed (102.69/culture) (Table 1). The next important treatment was T1 (94.36/culture), which induced a good number of embryos; the embryo numbers declined gradually at higher elicitor levels.

Table 1. Number of somatic embryos for various *Fusarium oxysporum* treatments.

Treatment	Number of Somatic Embryos/Culture
T0	82.53 ± 1.074 d
T1	94.36 ± 0.899 b
T2	102.69 ± 0.835 a
T3	84.78 ± 0.868 c
T4	85.14 ± 0.945 c

The different *F. oxysporum* levels used were a control (T0), 0.05% (T1), 0.15% (T2), 0.25% (T3), and 0.35% (T4). The MS was added with the addition of 6.72 μM of BA and 5.37 μM of NAA. The data were scored after four weeks of culture, and the values are the means ± standard errors of three replicates. Means with the same letters are not significantly different at $p \leq 0.05$, according to DMRT.

2.3. The Maturation and Germination of Somatic Embryos in a Medium Containing F. oxysporum

The cotyledonary embryos were cultured on MS medium supplemented with 2.60 μM of GA_3 for maturation; the medium was additionally supplemented with the fungal elicitors (Figure 1b). For the concentrations T1 and T2, the embryos were elongated, coiled, and turned green, and they later germinated into plantlets (Figure 1c). For the concentrations T3 and T4, however, embryo development was poor; a few remained in an advanced cotyledonary stage, while the others turned brown. The embryos that reached maturity were thin and showed poor growth. The embryos germinated into plantlets on the MS medium containing 2.24 μM of BA. The percent germination and the shoot and root lengths of the germinated somatic embryos were higher under the *F. oxysporum*-elicitated conditions compared to the control (Table 2).

Table 2. The germination of somatic embryos in *Fusarium oxysporum*-elicitated treatments.

Treatment	Germination (%)	Shoot Length (mm)	Root Length (mm)
T0	38.56 ±1.87 c	3.36 ± 0.30 c	4.34 ± 0.29 b
T1	47.58 ±2.60 b	8.96 ± 0.39 b	4.90 ± 0.23 b
T2	56.63 ±1.88 a	11.16 ± 0.29 a	6.67 ± 0.30 a
T3	16.76 ±1.92 d	1.94 ± 0.22 d	2.12 ± 0.28 c
T4	12.7 ±1.91 e	1.24 ± 0.21 d	2.07 ± 0.31 c

The different *F. oxysporum* levels used were a control (T0), 0.05% (T1), 0.15% (T2), 0.25% (T3), and 0.35% (T4). The MS was supplemented with 2.24 μM of BA. The data were scored after four weeks of culture, and the values are the means ± standard errors of three replicates. Within each column, the means with the letters are not significantly different at $p \leq 0.05$, according to DMRT.

2.4. Vinblastine and Vincristine Yields

The yields of vinblastine and vincristine were quantified in different in vitro-cultivated tissues. The mobile phase showed sharp standard vinblastine and vincristine peaks. A regression analysis also showed a good linearity, with r = 0.999 and 0.993 for vinblastine and vincristine, respectively. It is evident from Table 3 that the maximum yields of vinblastine were achieved in the embryos' maturation (0.788 μg g^{-1} dry weight; Figure 2a,c) and germination stages (0.835 μg g^{-1} dry weight; Figure 2b,d) compared to the other two stages, i.e., the induction and proliferation stages of the embryo tissues. With *F. oxysporum* elicitation at T2, the vinblastine yield was further improved (0.886 μg g^{-1} dry weight), and the T1 treatment was equally efficient in promoting its yield. Compared to vinblastine, the yield of vincristine was low, and the maximum content was achieved in germinating embryos compared to the other stages. Upon the addition of *F. oxysporum*, an improved vincristine yield was noted in the cultured tissues (Table 4), with the maximum identified for the T2 treatment (0.307 μg g^{-1} dry weight), followed by the T1 treatment (0.275 μg g^{-1} dry weight). The maximum increased yields of vinblastine and vincristine, 7.88 and 15.50%, respectively, were noted for the *F. oxysporum*-elicitated treatment T2 over the control tissues.

Table 3. Vinblastine contents (μg g^{-1} DW) for the different stages of the embryos in *Fusarium oxysporum*-elicitated treatments.

Treatment	Induction	Proliferation	Maturation	Germination
T0	0.422 ± 0.010 c	0.401 ± 0.001 b	0.788 ± 0.005 c	0.835 ± 0.012 c
T1	0.429 ± 0.011 b	0.406 ± 0.0006 b	0.813 ± 0.0007 b	0.861 ± 0.009 b
T2	0.451 ± 0.007 a	0.417 ± 0.0009 a	0.839 ± 0.002 a	0.886 ± 0.011 a
T3	0.408 ± 0.013 d	0.395 ± 0.001 c	0.775 ± 0.004 d	0.827 ± 0.010 d
T4	0.415 ± 0.011 d	0.392 ± 0.003 c	0.771 ± 0.001 d	0.823 ± 0.018 d

The different *F. oxysporum* levels used were a control (T0), 0.05% (T1), 0.15% (T2), 0.25% (T3), and 0.35% (T4). The MS was supplemented with 2.60 μM of gibberellic acid (GA$_3$). The data were scored after 45 days of culture. The values are the means ± standard errors of three replicates. Within each column, the means with the letters are not significantly different at $p \leq 0.05$, according to DMRT.

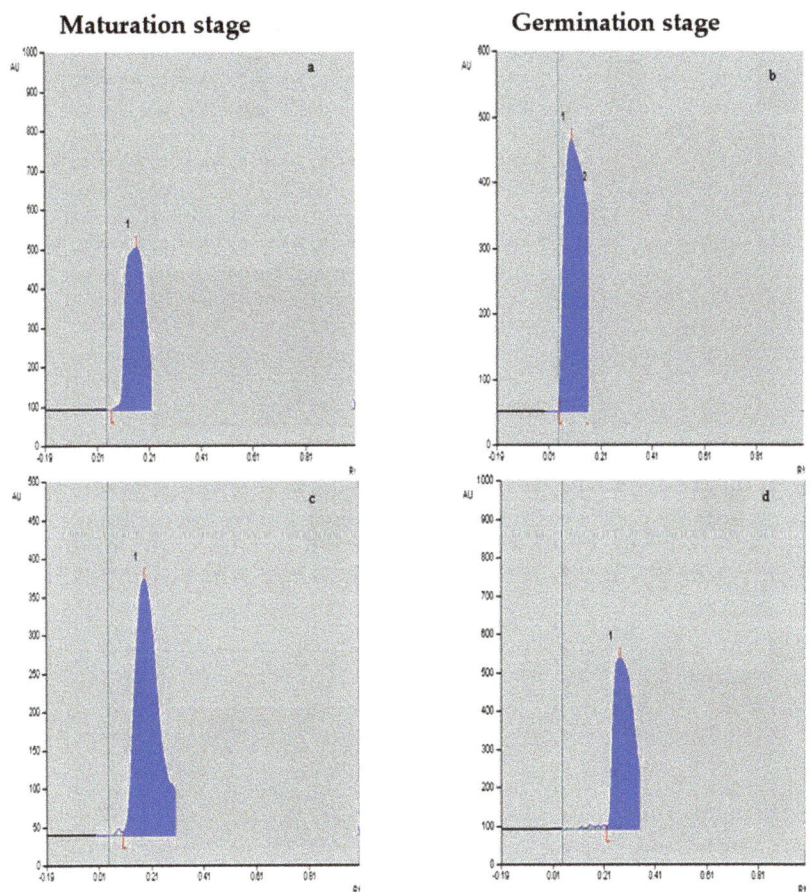

Figure 2. Vinblastine (**a**,**b**) and vincristine (**c**,**d**) peaks/levels at the maturation stage and germination stage, respectively, in response to *F. oxysporum* elicitation treatment at 0.25%.

Table 4. Vincristine contents (µg g^{-1} DW) in the different stages of the embryos in *Fusarium oxysporum*-elicitated treatments.

Treatment	Induction	Proliferation	Maturation	Germination
T0	0.083 ± 0.014 c	0.185 ± 0.011 c	0.181 ± 0.002 b	0.254 ± 0.007 c
T1	0.088 ± 0.011 b	0.191 ± 0.008 b	0.184 ± 0.001 b	0.275 ± 0.011 b
T2	0.095 ± 0.010 a	0.199 ± 0.012 a	0.192 ± 0.0007 a	0.307 ± 0.016 a
T3	0.076 ± 0.011 d	0.182 ± 0.013 c	0.170 ± 0.002 c	0.242 ± 0.013 d
T4	0.074 ± 0.009 d	0.177 ± 0.011 d	0.168 ± 0.0006 c	0.239 ± 0.008 d

The different *F. oxysporum* levels used were the control (T0), 0.05% (T1), 0.15% (T2), 0.25% (T3), and 0.35% (T4). The MS was supplemented with 2.24 µM of BA. The data were scored after 45 days of culture. The values are the means ± standard errors of three replicates. Within each column, the means with the letters are not significantly different at $p \leq 0.05$, according to DMRT.

2.5. Fusarium Oxysporum Elicitation and Biochemical Attributes

2.5.1. Sugar, Proline, and Protein Contents

As the *F. oxysporum* elicitation, especially at low levels, improved alkaloid yields, we attempted to monitor various non-enzymatic stress markers for different tissues. The sugar content was noted to be high during the embryos' maturation stage compared to the germination stage. Upon the addition of increased levels of elicitors, the sugar level increased further, reaching a maximum in T2 (21.663 mg g^{-1}). The proline level was also high in the maturation stage (8.255 mg g^{-1}), but the proline accumulation declined with the growth and maturation of the embryos (7.254 mg g^{-1}) at T2. The total soluble protein, on the other hand, was found to be more or less the same at these two advanced stages, i.e., maturation and germination,; comparative details of the elicitation doses and biochemical attributes are presented in Tables 5 and 6.

Table 5. Sugar, protein, and proline contents (mg g^{-1} FW) during the maturation stage of the embryos in a *Fusarium oxysporum*-treated culture.

Treatment	Sugar	Protein	Proline
T0	16.475 ± 0.009 d	4.517 ± 0.018 d	6.692 ± 0.010 d
T1	18.957 ± 0.011 b	5.084 ± 0.011 b	7.745 ± 0.011 b
T2	21.663 ± 0.010 a	5.378 ± 0.013 a	8.255 ± 0.009 a
T3	17.434 ± 0.009 c	4.657 ± 0.019 c	6.947 ± 0.010 c
T4	17.785 ± 0.006 c	4.695 ± 0.016 c	7.016 ± 0.008 c

The different *F. oxysporum* levels used were a control (T0), 0.05% (T1), 0.15% (T2), 0.25% (T3), and 0.35% (T4). The MS medium was supplemented with 2.60 µM of gibberellic acid (GA$_3$). The data were scored after 30 days of culture. The values are the means ± standard errors of three replicates. Within each column, the means with the letters are not significantly different at $p \leq 0.05$, according to DMRT.

Table 6. Sugar, protein, and proline contents (mg g^{-1} FW) during the germination stage of the embryos in a *Fusarium oxysporum*-treated culture.

Treatment	Sugar	Protein	Proline
T0	12.355 ± 0.011 d	4.675 ± 0.019 d	5.847 ± 0.010 d
T1	13.282 ± 0.008 b	5.116 ± 0.017 b	6.696 ± 0.008 b
T2	14.967 ± 0.010 a	5.457 ± 0.011 a	7.254 ± 0.009 a
T3	12.674 ± 0.011 c	4.817 ± 0.014 c	6.065 ± 0.008 c
T4	12.742 ± 0.009 c	4.863 ± 0.018 c	6.146 ± 0.010 c

The different *F. oxysporum* levels used were a control (T0), 0.05% (T1), 0.15% (T2), 0.25% (T3), and 0.35% (T4). The MS medium was supplemented with 2.24 µM of BA. The data were scored after 30 days of culture. The values are the means ± standard errors of three replicates. Within each column, the means with the letters are not significantly different at $p \leq 0.05$, according to DMRT.

2.5.2. SOD, CAT, and APX Activities

The germinating and maturated somatic embryos showed enhanced levels of alkaloids, especially on the *F. oxysporum*-treated culture. The addition of an elicitor might cause stress for tissues. To better understand the impact of the elicitor treatments on plant defense and later on secondary metabolism, the antioxidant activities of various enzymes were investigated as stress markers. The maturated and germinating somatic embryos had higher levels of antioxidant enzyme activities than the early embryogenic tissues. The antioxidant enzyme activities were higher upon the addition of the *F. oxysporum* treatments, which indicated extra cellular stress on the cultivated tissues. It is evident from Figure 3 that at T_2, the levels of SOD activity were high in the maturation (4.115 EU min^{-1} mg^{-1} proteins) and germination (3.693 EU min^{-1} mg^{-1} proteins) stages of the embryos compared to the control (3.785 and 3.415 EU min^{-1} mg^{-1} proteins respectively), which yielded the highest levels of vinblastine and vincristine. Compared to SOD, the activities of CAT and APX were, however, low, i.e., 2.355 and 1.075 min^{-1} mg^{-1} protein, respectively, in the embryos' maturation stage. The germinating somatic embryos also had similarly low levels of CAT and APX enzyme activity (Figure 4).

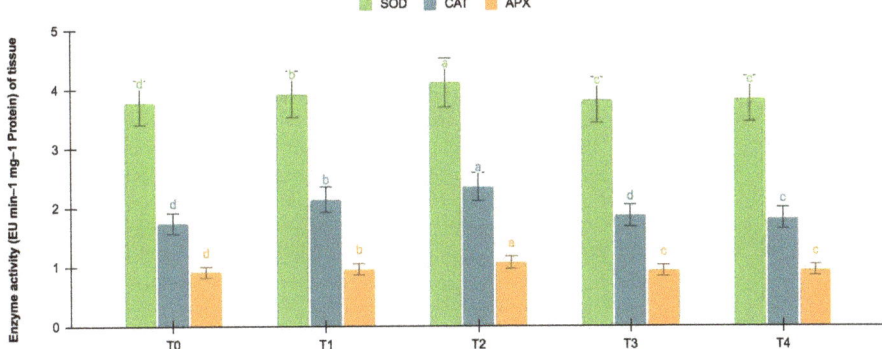

Figure 3. SOD, CAT, and APX activities (EU mg^{-1} protein min^{-1}) in the maturation stage of the embryos for different *Fusarium oxysporum* treatments. The different *F. oxysporum* levels used were a control (T0), 0.05% (T1), 0.15% (T2), 0.25% (T3), and 0.35% (T4). The data were scored after 30 days of culture. The values are the means ± standard errors of three replicates. Within each column, the means with the letters are not significantly different at $p \leq 0.05$, according to DMRT.

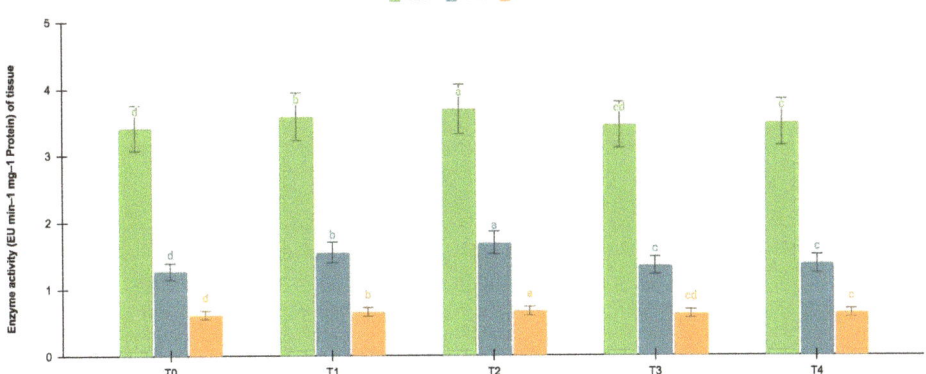

Figure 4. SOD, CAT and APX activities (EU mg^{-1} protein min^{-1}) of the germination stage of the embryos for different *Fusarium oxysporum* treatments. The different *Fusarium oxysporum* levels used were a control (T0), 0.05% (T1), 0.15% (T2), 0.25% (T3), and 0.35% (T4). The data were scored after 30 days of culture. The values are the means ± standard errors of three replicates. Within each column, the means with the letters are not significantly different at $p \leq 0.05$, according to DMRT.

3. Discussion

In the present study, the yields of vinblastine and vincristine were quantified following *F. oxysporum* elicitation in embryogenic cultures of *C. roseus*. The callus was induced from hypocotyls on MS medium supplemented with 2,4-D in which-high frequency somatic embryos were formed; other auxins used induced embryos at a slower rate. Here, embryo differentiation was noted on the embryogenic callus, i.e., indirectly, but in other observed cases, embryos were also formed directly on explants without an intervening callus [30]. In both embryo-forming developmental pathways, the use of exogenous auxins/auxin analogues like 2,4-D efficiently promoted embryogenesis. These synthetic auxin analogs play a central signaling role in the acquisition of embryogenic competence from a somatic state [31,32]. In our study, an *F. oxysporum* extract was used at varying concentrations, of which T2 (0.15%) was observed to be more efficient at promoting callus biomass growth compared to T1, T3, and T4. We also observed that the callus biomass and the number of embryos increased significantly in T2 with *F. oxysporum* elicitation. The induced embryos were distinct and showed fast growth and development under the elicitated condition. The results of the present study indicate that the high concentrations (T3 and T4) of elicitation decreased the growth of the callus biomass by inhibiting cell division, and this reduction may have been due to the toxicity of the fungal extract or the excessive availability of stress ions [33]. In the present study, the addition of a low level of the *F. oxysporum* extract improved the number of somatic embryos in the culture. Similar responses, i.e., stress-induced embryogenesis, were described earlier in a number of previous observations [34,35]. Once an embryo is induced, the presence of 2,4-D in the medium inhibits the embryo's development; therefore, other PGR combinations were tested and suggested to be necessary [35]. The involvement of cytokinins alone or with low doses of a weak auxin like NAA successfully influenced in vitro embryogenesis and plant morphogenesis [36,37].

The cultivation of plant cells and tissues or complex, organized structures is practiced in vitro as an efficient renewable source for the production of a variety of phytochemicals, and the importance of these methods were reviewed in recent years [38,39]. Calli and suspensions are cultivated more frequently because of their ease of cultivation and the possibility of scaling up their production in bioreactors. Aside from bioreactors, a number of other important strategies such as liquid culture, the use of mist, and liquid overlaying are used to improve biomass/embryogenesis to generate raw materials for alkaloid synthesis [40]. Liquid overlaying is a technique in which a thin film of a liquid nutrient is added on a solid medium to improve somatic embryogenesis in cultures [41]. The yields of active compounds are often high in complex, differentiated structures like shoots, roots, and leaves [14,42]. The method of extracting metabolites synthesized and accumulated in specialized cells or tissues is difficult, but genetically constructed biosensors can detect the precise locations of specialized metabolites at the tissue or cell level [43]. Different techniques have recently been adopted for the collection of alkaloids from specialized tissues. In the present study, we noted that compact embryo structures like maturated and germinating embryos synthesized higher yields of vinblastine and vincristine compared to embryos in early stages. Upon receiving *F. oxysporum* elicitation treatment, a 7.88% increased yield of vinblastine and a 15.50% increased yield of vincristine were noted. The same low level (T1/T2) of elicitation was noted earlier to be very efficient for improving the callus biomass. This rapid growth of the embryogenic callus may have been due to fast cell mitosis triggered by cell-cycle genes which were strongly upregulated in the dividing cells [44,45]. The influence of *F. oxysporum* on biochemical attributes was investigated as the addition of the elicitor improved the alkaloid yield. In the present study, extra sugar, protein, and proline accumulations were noted; however, these declined with increased levels of elicitation. Similar increases in protein, phenolics, hydrogen peroxide, and carbohydrates in response to stress were noted in several investigated plant genera, and these enhancements are considered good adaptation mechanisms in tolerant genotypes [46,47]. The protein level also increases gradually with the progress of tissues, and a change in protein with a progressing developmental stage was reported earlier in other investigated plant materials [42,48]. In tomato, enriched proline and lysine and glutamine accumulation were noted at an early stage of embryonic development, and this probably confers tolerance to drought [49]. Here, in the *F. oxysporum*-elicited tissues, the increased accumulation of proline may have been due to the up-regulation of a proline synthesis gene which produced P5C reductase (PYCR) and proline dehydrogenase/oxidase (PRODH/POX) enzymes participating in the interconversion of intermediates in proline biosynthesis pathways [50,51]. Transcriptome data reveal that rice universally downregulates photosynthesis in response to abiotic and biotic stresses. At the same time, it also upregulates the hormone-responsive genes of the abscisic acid, jasmonic acid and salicylic acid pathways during stress [52]. In transgenic tobacco, the overexpression of AhCytb6 regulates the expression of various genes to enhance plant growth under a N_2 deficit and abiotic stress conditions by modulating the plant's physiology [53]. Enzymes like Cipk6, a Calcineurin B-like interacting protein kinase (CIPK) of tomato, regulates programmed cell death in immunity, transforming Ca^{2+} signaling in the formation of reactive oxygen species [54,55].

As the yields of alkaloids were high in the advanced-staged embryos, we tried to investigate the level of stress by measuring the activities of antioxidant enzymes in these cultivated tissues. The level of SOD activity was high in both of these two tissues, and upon the addition of the elicitor, the activity was further elevated. Increased SOD activity under various stresses was observed in several investigated plant genera [56,57]. CAT and APX also showed similar trends with added levels of elicitors, although tissue- and dose-specific variations were not uncommon [58,59]. In addition to the increases in the activities of stress marker enzymes and the alteration of physiological reserves, a molecular analysis indicated that the expression of the *Salt Overly Sensitive 1* (*SOS1*) gene is an important event in response to adaptive stress caused by biotic and abiotic factors [60]. It is very evident from the present study that the *F. oxysporum* elicitor promoted cultural growth in *C. roseus* and later stimulated enriched levels of alkaloids; however, the underlying mechanism is still not fully understood. It was reported earlier that the fungus extract in general contained compounds like sugars and proteins [61]. A chemical analysis showed that the hyphal walls of *F. oxysporum* are primarily composed of N-acetyl-glucosamine, glucose, mannose, galactose, uronic acid, and proteins or peptides [62,63]. The roles of various sugars, sugar alcohols, and related energy sources in improving synthesis were indicated earlier in several studies [64,65]. But the roles of protein or truncated proteins like small, moderate, or large peptides in triggering the synthesis of alkaloids have not been determined in a major way. Although the best mechanism of improving synthesis is not fully known, the process may be due to the formation of an 'elicitor-receptor complex' [66,67] which stimulates a cascade of defense genes in promoting alkaloid synthesis [68,69]. Thus, experimentations on elicitation through the use of various agents are immensely valuable as the technique promises to promote alkaloid biosynthesis in cultivated tissues.

4. Materials and Methods

The fruits/seeds of *Catharanthus roseus* (L.) G. Don were procured from the herbal garden of Jamia Hamdard (Hamdard University). The material was identified earlier, and a voucher specimen (JH-002-98) was maintained.

4.1. In Vitro Seed Germination and Culture Conditions

Seed germination and the process of establishing a culture of *C. roseus* L. (G). Don were carried out using the protocol established in our laboratory by [64]. In a nutshell, from twenty to twenty-five surface-disinfected seeds were placed in a 250 mL conical flask (Borosil, Mumbai, India) containing 50 mL of solid MS medium without any plant growth regulator (PGR). The germinated seedlings were maintained until the shoots attained a height of 2–4 cm. Various parts (the nodal stem, leaf, and hypocotyl) were used and inoculated in test tubes (Borosil, India) as explants. For the induction of an embryogenic callus, the MS medium was supplemented with 4.52 lM of 2,4-Dichlorophenoxyacetic acid (2,4-D). For the fast proliferation of embryos, the medium was fortified with 6.72 μM of

N^6-Benzyladenine (BA) and 5.37 µM of naphthalene acetic acid (NAA). All the above PGRs were procured from Sigma-Aldrich, St. Louis, MO, USA. The medium was solidified with 8 g L^{-1} of agar (Hi-media, Mumbai, India), and each tube contained 20 mL of medium. The pH of the medium was adjusted to 5.7 before it was autoclaved at 121 °C. All the cultures were incubated at 25 ± 2 °C under a 16 h photoperiod provided by cool-white fluorescent tubes at a photosynthetic photon flux density (PPFD) of 100 µmol m^{-2} s^{-1}.

4.2. The Procurement and Culture of Fungi and the Preparation of the Elicitor

Fusarium oxysporum (Figure 5) was obtained from the Department of Pathology, Indian Agricultural Research Institute (IARI), Pusa, New Delhi, India. The fungus was grown in 100 mL conical flasks containing potato dextrose agar (Hi-media, India). After 7 d, the conical flasks containing fungal growth were sterilized and filtrated using Whatman no. 1 filter paper. The mycelium was washed several times with sterilized, distilled water and stored at 4 °C after being suspended in 100 mL water; this was designated as the culture media filtrate. The fungal mat was washed several times with sterilized, distilled water, and an aqueous extract was prepared [70] via homogenization with a mortar and pestle. This extract was filtered through centrifugation at 5000 rpm, and the supernatant was taken. It was later sterilized (designated as the mat extract) and kept at 4 °C for future investigations. Four different fungal elicitor treatments, i.e., 0.05% (T1), 0.15% (T2), 0.25% (T3), and 0.35% (T4), were prepared and added to the culture medium. A control (T0), i.e., a culture medium without the fungal filtrate, was also used for comparative evaluations of the elicitor's influence. Morphogenetic and biochemical studies were conducted at periodic intervals.

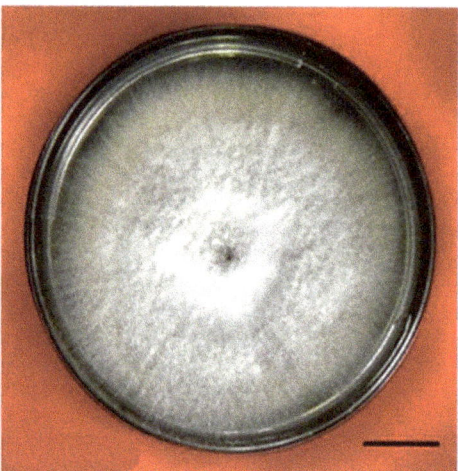

Figure 5. *Fusarium oxysporum* culture grown on potato dextrose medium (bar 0.5 cm).

4.3. Callus Induction under Fungus-Treated and Non-Treated Conditions

Hypocotyls of 5–6 d old seedlings were placed on MS and supplemented with an optimized 2,4-D concentration (4.52 µM). Four different treatments containing the *Fusarium oxysporum* fungal elicitor were added in order to assess the effect of the elicitors on callus induction and growth. A control, i.e., a medium without fungal filtrate, was also used for comparison. For a growth index analysis, callus biomass samples, i.e., the fresh and dry weights of calli at various growth stages, were taken and investigated. For the determination of the fresh weight, the calli (with or without elicitor treatment) were weighed immediately after isolation at regular intervals (15, 30, and 45 d). To determine the dry weight, the calli were dried at 60 °C for 18 h and measured, and the absolute dry mass was finally calculated using the method and formula of Winkelmann et al. (2004): Absolute dry mass (%) = Dry weight/fresh weight × 100.

4.4. The Proliferation, Maturation, and Germination of Embryos under the Influence of Biotic Elicitors

The embryogenic callus (40–50 mg) was cultured on MS supplemented with optimized concentrations of BAP (6.62 µM) and NAA (5.36 µM) for embryo proliferation. The medium was additionally amended with the above-mentioned fungus for the treatments indicated earlier treatments. The somatic embryos were induced in masses and were counted; this stage was called the proliferation stage. Vincristine and vinblastine alkaloids were extracted from the proliferation-stage embryos, and some of the proliferated embryos were cultured in medium for embryo maturation. The somatic embryos on MS supplemented with 2.89 µM of GA_3 became larger and turned green, which is a good morphological indicator of matured embryos. The green, matured embryos were later placed on the same MS, supplemented with 2.22 µM of BAP for germination. The above two stages (maturation and germination) of the embryo development media were additionally supplemented with the *Fusarium oxysporum* extract for the above-indicated treatments. The somatic embryos started to germinate within a week or so, and the germination percentage and shoot and root lengths were measured and compared to assess the impact of the elicitor on the embryos. Matured and germinating embryos were harvested and oven-dried for the extraction of vincristine and vinblastine.

4.5. Vinblastine and Vincristine Quantification through HPTLC

Vinblastine and vincristine were extracted following methods described earlier methods [71,72] and their contents were measured in different in vitro-grown tissues and compared with standard vinblastine and vincristine obtained from Sigma-Aldrich (St. Luis, MO, USA). The selected tissues/embryos were collected from optimized media with their best growth. A total of 1 gm (dry weight) of tissues/embryos was refluxed in 30 mL of methanol for 5 h; later the supernatant was warmed at 60 °C, and the volume was finally reduced to 1–2 mL. Then, 1 mg of vinblastine and vincristine each was dissolved in 1.0 mL of methanol to make a stock solution concentration of 1.0 mg mL^{-1}. Various concentrations were prepared from the stock solutions to obtain 200, 400, 600, 800, and 1000 µg per band of the standard and were assessed separately via HPTLC. A standard curve was plotted between the peak area (y-axis) and concentration (x-axis), which showed good linearity. For the stationary phase, thin-layer chromatography (TLC) aluminum sheets which measured 20 × 10 cm and were coated with silica gel (60 F 254, Merck, Bengaluru, India) were used. The freshly prepared mobile solution (phase) contained toluene, carbinol, acetone, and ammonia in a ratio of 40:20:80:2. The samples were applied using a 100 µL micro-syringe via a Linomat 3 (CAMAG) applicator. The silica plates were air-dried for 10–15 min and kept in a chamber (Twin Through Chamber CAMAG, 20 × 10 cm) filled with mobile solution. The solvent system was allowed to move up to about 85 mm. The plates were later removed from the chamber and air-dried again for about 10–20 min. The silica gel plates were documented using a CAMAG Reprostar under UV light without any chemical spray applied. The vinblastine- and vincristine-containing stationary phase was scanned via a CAMAG Scanner 3. The vinblastine and vincristine were scanned at 280 and 300 nm, respectively. The peaks of vinblastine and vincristine were fixed, and the identification of the alkaloids in the tissue samples was achieved by comparing the peaks of standard alkaloids. Finally, the alkaloid yields were measured in µg gm^{-1} of dry weight.

4.6. Estimation of Total Sugar, Proline, and Protein Contents

The estimation of the total sugar content was carried out according to the Dey method [73]. Tissues at different stages (0.5 g) were extracted twice with 90% ethanol (AR, New Delhi, India), and the extracts were pooled. The final volume of the pooled extract was increased to 25 mL via the addition of double-distilled water. To an aliquot of 1.0, 1.0 mL of 5.0% phenol and 5.0 mL of concentrated analytical-grade sulfuric acid were added and cooled in air. The optical density was measured at 485 nm. A solution containing 1.5 mL of 55% glycerol (AR, India), 0.5 mL of ninhydrin (AR, India), and 4.0 mL of double-distilled water was used as a calibration standard. For the measurement of proline, 0.2 g of specific stages of tissues were homogenized in 5.0 mL of 3% aqueous sulfosalicylic acid and filtered through Whatman filter paper (No. 1). To 1.0 mL of the extract, 1.0 mL of acid ninhydrin and 1.0 mL of glacial acetic acid (AR, India) were added, and the reaction mixture was incubated at 100 °C for 1 h. The reaction mixture was placed on ice and extracted using 2.0 mL of toluene. The proline content in the extract was subject to the spectrophotometric assay of Bates et al. [74]. The protein content was estimated via the Bradford method [75]; 0.5 g of tissue was ground in a pre-cooler mortar and pestle with 1.5 mL (0.1 M) of phosphate buffer (pH 7.0), placed on ice, and centrifuged at 5000 rpm for 10 min. With 0.5 mL of trichloroacetic acid (TCA), the sample was again centrifuged at 5000 rpm for 10 min. The supernatant was discarded, and the pellet was washed with chilled acetone and dissolved in 1.0 mL of 0.1 N sodium hydroxide (NaOH). Later, a 0.5 mL aliquot was added to 5.0 mL of Bradford reagent, and the optical density was measured at 595 nm.

4.7. Assay of Antioxidant Enzyme Activity

The catalase (CAT) activity was measured following the Aebi method [76]. It was measured by observing the decay in H_2O_2, and a decrease was measured at an absorbance of 240 nm in a reaction mixture containing 1.0 mL of a 0.5 M phosphate buffer (Na-phosphates, pH 7.5, AR, India), 0.1 mL of EDTA (AR, India), 0.2 mL of enzyme extract, and 0.1 mL of H_2O_2. The chemical reaction was continued for 3 min. The enzyme activity was represented as EU mg^{-1} protein min^{-1}. A single unit of enzyme represents the amount used to decompose 1.0 µmol of H_2O_2/min. The enzyme activity was registered using the coefficient of absorbance at 0.036 mM^{-1} cm^{-1}. The superoxide dismutase (SOD) activity was measured following the method of Dhindsa et al. [77]. Different stages of tissues/embryos (0.1 g) were homogenized in 2.0 mL of extraction solution (0.5 M of sodium phosphate buffer, pH 7.3, + 3.0 mM of EDTA + 1.0% (w/v) polyvinylpyrollidone (PVP, AR, India) + 1.0% (v/v) + Triton X100, AR, India), and the mixture was centrifuged (10,000 rpm) at 4 °C. The enzyme activity was measured by the ability to inhibit photochemical reduction. The assay mixture contained 1.5 mL of reaction buffer, 0.2 mL of methionine, 0.1 mL of enzyme extract, an equal amount of 1.0 M $NaCO_3$ and 2.25 mM Nitro Blue Tetrazolium (NBT) solution, 3.0 mM of EDTA, riboflavin, and 1.0 mL of Millipore H_2O. The whole mixture was kept in test tubes and incubated at 25 °C for 10 min under light. A 50% loss in color was considered 1.0 unit, and the enzyme content was expressed as EU mg^{-1} protein min^{-1}. For ascorbate peroxidase (APX), the Nakano and Asada [78] method was used. The assay mixture contained 1.0 mL of 0.1 M sodium buffer, pH 7.2, + 0.1 mL pf EDTA + 0.1 mL of enzyme extract. The ascorbate was added to the solution and the reaction mixture was run for 3 min at 25 °C. The APX activity was measured by observing the reduction of absorbance by ascorbate mediated breakdown of APX. Enzyme activity was measured by using co-efficient of absorbance 2.81 mM^{-1} cm^{-1}. Similar to other enzymes, the activity was expressed in EU mg^{-1} protein min^{-1} i.e., one unit of enzyme determines the amount necessary in decomposing 1.0 µm of ascorbate/min.

4.8. Statistical Analysis

The data on the effect of *Fusarium oxysporum* elicitor on callus growth and embryogenesis and differences in biochemical attributes, antioxidant enzyme activity, the alkaloid yield, and other parameters were analyzed via a one-way analysis of variance (ANOVA). The data or the values are the means of three replicates from two experiments and the presented mean values were separated using Duncan's multiple range test (DMRT) at $p \leq 0.05$.

5. Conclusions

Low doses of an *F. oxysporum* extract proved effective for improving callus biomass growth, embryogenesis, plant regeneration, and alkaloid yield in *C. roseus*. The percent germination and shoot and root lengths of somatic embryos were high at a low level (from 0.05% to 0.15%). Maturated and germinating somatic embryos had high levels of vinblastine and vincristine, which were further improved (to 7.8 and 15.5%) via elicitation. The addition of the elicitor caused cellular stress, which was evidenced by the biochemical attributes and high levels of antioxidant enzyme activities. We therefore recommend low doses of the fungal extract for enhancing the synthesis of alkaloids in *C. roseus*. The improvement in the yields of alkaloids could augment cancer healthcare in an easy and inexpensive manner.

Author Contributions: Conceptualization, D.T. and A.M.; methodology, formal analysis, D.T.; investigation, D.T.; data curation, M.M. and M.K.; writing—original draft preparation, D.T., A.M. and M.K.; writing—review and editing, M.K., A.M. and Y.H.D.; validation, Y.H.D. and A.A.; visualization, A.M., Y.H.D. and A.A.; project administration, A.M. All authors have read and agreed to the published version of the manuscript.

Funding: The authors are highly thankful to the University Grant Commission (UGC) and Department of Botany, Jamia Hamdard, for help, in addition to other research facilities and the Researchers Supporting Project (number RSP2023R375), King Saud University, Riyadh, Saudi Arabia.

Data Availability Statement: All data are presented in the article.

Acknowledgments: The authors acknowledge the Researchers Supporting Project, number RSP2023R375, King Saud University, Riyadh, Saudi Arabia.

Conflicts of Interest: The authors declare no conflict of interest.

References

1. Almagro, L.; Fernández-Pérez, F.; Pedreño, M.A. Indole alkaloids from *Catharanthus roseus*: Bioproduction and their effect on human health. *Molecules* **2015**, *20*, 2973–3000. [CrossRef] [PubMed]
2. Mujib, A.; Ilah, A.; Aslam, J.; Fatima, S.; Siddiqui, Z.H. *Catharanthus roseus* alkaloids: Application of biotechnology for improving yield. *Plant Growth Regul.* **2012**, *68*, 111–127. [CrossRef]
3. Hemmati, N.; Azizi, M.; Spina, R.; Dupire, F.; Arouei, H.; Saeedi, M.; Laurain-Mattar, D. Accumulation of ajmalicine and vinblastine in cell cultures is enhanced by endophytic fungi of *Catharanthus roseus* cv. Icy Pink. *Ind. Crops Prod.* **2020**, *158*, 112776. [CrossRef]
4. Birat, K.; Binsuwaidan, R.; Siddiqi, T.O.; Mir, S.R.; Alshammari, N.; Adnan, M.; Nazir, R.; Ejaz, B.; Malik, M.Q.; Dewangan, R.P.; et al. Report on vincristine producing endophytic fungus *Nigrospora zimmermanii* from leaves of *Catharanthus roseus*. *Metabolites* **2022**, *12*, 1119. [CrossRef]
5. Singh, N.R.; Rath, S.K.; Behera, S.; Naik, S.K. In vitro secondary metabolite production through fungal elicitation: An approach for sustainability. In *Fungal Nanobionics: Principles and Applications*; Springer: Singapore, 2018; pp. 215–242.
6. Satheesan, J.; Sabu, K.K. Endophytic fungi for a sustainable production of major plant bioactive compounds. In *Plant-Derived Bioactives: Production, Properties and Therapeutic Applications*; Springer: Berlin/Heidelberg, Germany, 2020; pp. 195–207.
7. Mujib, A.; Tonk, D.; Gulzar, B.; Maqsood, M.; Ali, M. Quantification of taxol by high performance thin layer chromatography (HPTLC) in *Taxus wallichiana* callus cultivated in vitro. *BioTechnologia* **2020**, *101*, 337–347. [CrossRef]
8. Naik, P.M.; Al-Khayri, J.M. Impact of abiotic elicitors on in vitro production of plant secondary metabolites: A review. *J. Adv. Res. Biotechnol.* **2016**, *1*, 1–7.
9. Coste, A.; Vlase, L.; Halmagyi, A.; Deliu, C.; Coldea, G. Effects of plant growth regulators and elicitors on production of secondary metabolites in shoot cultures of *Hypericum hirsutum* and *Hypericum maculatum*. *Plant Cell Tissue Organ Cult. PCTOC* **2011**, *106*, 279–288. [CrossRef]
10. Halder, M.; Sarkar, S.; Jha, S. Elicitation: A biotechnological tool for enhanced production of secondary metabolites in hairy root cultures. *Eng. Life Sci.* **2019**, *19*, 880–895. [CrossRef]
11. Siddiqui, Z.H.; Mujib, A. Accumulation of vincristine in calcium chloride elicitated *Catharanthus roseus* cultures. *Nat. Prod. J.* **2012**, *2*, 307–315. [CrossRef]
12. Syeed, R.; Mujib, A.; Dewir, Y.H.; Malik, M.Q.; Bansal, Y.; Ejaz, B.; Alsughayyir, A. Methyl Jasmonate elicitation for in vitro lycorine accumulation in three *Zephyranthes* species and comparative analysis of tissue-cultured and field grown plants. *Horticulturae* **2023**, *9*, 832. [CrossRef]

13. Birat, K.; Siddiqi, T.O.; Mir, S.R.; Aslan, J.; Bansal, R.; Khan, W.; Dewangan, R.P.; Panda, B.P. Enhancement of vincristine under in vitro culture of *Catharanthus roseus* supplemented with *Alternaria sesami* endophytic fungal extract as a biotic elicitor. *Int. Microbiol.* **2022**, *25*, 275–284. [CrossRef] [PubMed]
14. Singh, S.; Pandey, S.S.; Shanker, K.; Kalra, A. Endophytes enhance the production of root alkaloids ajmalicine and serpentine by modulating the terpenoid indole alkaloid pathway in *Catharanthus roseus* roots. *J. Appl. Microbiol.* **2020**, *128*, 1128–1142. [CrossRef] [PubMed]
15. Alcalde, M.A.; Perez-Matas, E.; Escrich, A.; Cusido, R.M.; Palazon, J.; Bonfill, M. Biotic elicitors in adve in adventitious and hairy root cultures: A review from 2010 to 2022. *Molecules* **2022**, *27*, 5253. [CrossRef] [PubMed]
16. Wilczańska-Barska, A.; Królicka, A.; Głód, D.; Majdan, M.; Kawiak, A.; Krauze-Baranowska, M. Enhanced accumulation of secondary metabolites in hairy root cultures of *Scutellaria lateriflora* following elicitation. *Biotechnol. Lett.* **2012**, *34*, 1757–1763. [CrossRef]
17. Hedayati, A.; Hemmaty, S.; Nourozi, E.; Amirsadeghi, A. Effect of yeast extract on H6H gene expression and tropane alkaloids production in *Atropa belladonna* L. hairy roots. *Russ. J. Plant Physiol.* **2021**, *68*, 102–109. [CrossRef]
18. Pandey, S.S.; Singh, S.; Babu, C.V.; Shanker, K.; Srivastava, N.K.; Shukla, A.K.; Kalra, A. Fungal endophytes of *Catharanthus roseus* enhance vindoline content by modulating structural and regulatory genes related to terpenoid indole alkaloid biosynthesis. *Sci. Rep.* **2016**, *6*, 26583. [CrossRef]
19. Taleb, A.M.A.; Ramadan, E.H.; Zaki, S.A.; Salama, A.B.; Kapiel, T.Y. Impact of fungal elicitor and culture conditions on induction of calli and alkaloids production in *Narcissus tazetta* var. italicus tissue cultures. *Plant Cell. Biotechnol. Mol. Biol.* **2021**, *22*, 88–98.
20. Bhambhani, S.; Karwasara, V.S.; Dixit, V.K.; Banerjee, S. Enhanced production of vasicine in *Adhatoda vasica* (L.) Nees. cell culture by elicitation. *Acta Physiol. Plant* **2012**, *34*, 1571–1578. [CrossRef]
21. Kumar, S.; Arora, N.; Upadhyay, H. Arbuscular mycorrhizal fungi: Source of secondary metabolite production in medicinal plants. In *New and Future Developments in Microbial Biotechnology and Bioengineering*; Singh, J., Gehlot, P., Eds.; Elsevier: Amsterdam, The Netherlands, 2021; pp. 155–164.
22. Prasad, A.; Mathur, A.; Kalra, A.; Gupta, M.M.; Lal, R.K.; Mathur, A.K. Fungal elicitor- mediated enhancement in growth and asiaticoside content of *Centella asiatica* L. shoot cultures. *Plant Growth Regul.* **2013**, *69*, 265–273. [CrossRef]
23. Ganie, I.B.; Ahmad, Z.; Shahzad, A.; Zaushintsena, A.; Neverova, O.; Ivanova, S.; Wasi, A.; Tahseen, S. Biotechnological intervention and secondary metabolite production in *Centella asiatica* L. *Plants* **2022**, *30*, 2928. [CrossRef]
24. Espinosa-Leal, C.A.; Mora-Vásquez, S.; Puente-Garza, C.A.; Alvarez-Sosa, D.S.; García-Lara, S. Recent advances on the use of abiotic stress (water, UV radiation, atmospheric gases, and temperature stress) for the enhanced production of secondary metabolites on in vitro plant tissue culture. *Plant Growth Regul.* **2022**, *97*, 1–20. [CrossRef]
25. Das, A.; Sarkar, S.; Bhattacharyya, S.; Gantait, S. Biotechnological advancements in *Catharanthus roseus* (L.) G. Don. *Appl. Microbiol. Biotechnol.* **2020**, *104*, 4811–4835. [CrossRef]
26. Siddiqui, Z.H.; Mujib, A.; Abbas, Z.K.; Noorani, M.S.; Khan, S. Vinblastine synthesis under the influence of CaCl2 elicitation in embryogenic cell suspension culture of *Catharanthus roseus*. *S. Afr. J. Bot.* **2023**, *154*, 319–329. [CrossRef]
27. Cai, Z.; Kastell, A.; Mewis, I.; Knorr, D.; Smetanska, I. Polysaccharide elicitors enhance anthocyanin and phenolic acid accumulation in cell suspension cultures of *Vitis vinifera*. *Plant Cell Tissue Org. Cult.* **2012**, *108*, 401–409. [CrossRef]
28. Ben Romdhane, A.; Chtourou, Y.; Sebii, H.; Baklouti, E.; Nasri, A.; Drira, R.; Maalej, M.; Drira, N.; Rival, A.; Fki, L. Methyl jasmonate induces oxidative/nitrosative stress and the accumulation of antioxidant metabolites in *Phoenix dactylifera* L. *Biotechnol. Lett.* **2022**, *44*, 1323–1336. [CrossRef] [PubMed]
29. Fatima, S.; Mujib, A.; Tonk, D. NaCl amendment improves vinblastine and vincristine synthesis in *Catharanthus roseus*: A case of stress signalling as evidenced by antioxidant enzymes activities. *Plant Cell Tiss. Organ. Cult.* **2015**, *121*, 445–458. [CrossRef]
30. Mujib, A. *Somatic Embryogenesis in Ornamentals and Its Applications*; Springer: Berlin/Heidelberg, Germany, 2016; p. 267.
31. Gulzar, B.; Mujib, A.; Malik, M.Q.; Mamgain, J.; Syeed, R.; Zafar, N. Plant tissue culture: Agriculture and industrial applications. In *Transgenic Technology Based Value Addition in Plant Biotechnology*; Academic Press: Cambridge, MA, USA, 2020; pp. 25–49.
32. Cordeiro, D.; Pérez-Pérez, Y.; Canhoto, J.; Testillano, P.S.; Correia, S. H3K9 methylation patterns during somatic embryogenic competence expression in tamarillo (*Solanum betaceum* Cav.). *Sci. Hortic.* **2023**, *321*, 112259. [CrossRef]
33. Robinson, J.R.; Isikhuemhen, O.S.; Anike, F.N. Fungal-Metal Interactions: A review of toxicity and homeostasis. *J. Fungi* **2021**, *7*, 225. [CrossRef]
34. Karami, O.; Saidi, A. The molecular basis for stress-induced acquisition of somatic embryogenesis. *Mol. Biol. Rep.* **2010**, *37*, 2493–2507. [CrossRef]
35. Feher, A. Somatic embryogenesis—Stress-induced remodeling of plant cell fate. *Biochim. Biophys. Acta* **2015**, *1849*, 385–402. [CrossRef]
36. Hazubska-Przybył, T.; Ratajczak, E.; Obarska, A.; Pers-Kamczyc, E. Different roles of auxins in somatic embryogenesis efficiency in two *Picea* species. *Int. J. Mol. Sci.* **2020**, *21*, 3394. [CrossRef] [PubMed]
37. Sosnowski, J.; Truba, M.; Vasileva, V. The impact of auxin and cytokinin on the growth and development of selected crops. *Agriculture* **2023**, *13*, 724. [CrossRef]
38. Isah, T.; Umar, S.; Mujib, A.; Sharma, M.P.; Rajasekharan, P.E.; Zafar, N.; Frukh, A. Secondary metabolism of pharmaceuticals in the plant in vitro cultures: Strategies, approaches, and limitations to achieving higher yield. *Plant Cell Tissue Organ. Cult.* **2018**, *132*, 239–265.

39. Espinosa-Leal, C.A.; Puente-Garza, C.A.; García-Lara, S. In vitro plant tissue culture: Means for production of biological active compounds. *Planta* **2018**, *248*, 1–18. [CrossRef] [PubMed]
40. Tonk, D.; Mujib, A.; Maqsood, M.; Ali, M.; Zafar, N. Aspergillus flavus fungus elicitation improves vincristine and vinblastine yield by augmenting callus biomass growth in *Catharanthus roseus*. *Plant Cell Tissue Organ Cult. PCTOC* **2016**, *126*, 291–303. [CrossRef]
41. Siddiqui, Z.H.; Mujib, A.; Maqsood, M. Liquid overlaying improves somatic embryogenesis in *Catharanthus roseus*. *Plant Cell Tiss Org Cult* **2011**, *104*, 247–256. [CrossRef]
42. Li, Y.; Kong, D.; Fu, Y.; Sussman, M.R.; Wu, H. The effect of developmental and environmental factors on secondary metabolites in medicinal plants. *Plant Physiol. Biochem.* **2020**, *148*, 80–89. [CrossRef]
43. Garagounis, C.; Delkis, N.; Papadopoulou, K.K. Unraveling the roles of plant specialized metabolites: Using synthetic biology to design molecular biosensors. *New Phytol.* **2021**, *231*, 1338–1352. [CrossRef]
44. Ikeuchi, M.; Sugimoto, K.; Iwase, A. Plant callus: Mechanisms of induction and repression. *Plant Cell* **2013**, *25*, 3159–3173. [CrossRef]
45. Guo, F.; Wang, H.; Lian, G.; Cai, G.; Liu, W.; Zhang, H.; Li, D.; Zhou, C.; Han, N.; Zhu, M.; et al. Initiation of scutellum-derived callus is regulated by an embryo-like developmental pathway in rice. *Commun. Biol.* **2023**, *6*, 457. [CrossRef]
46. Hosseinifard, M.; Stefaniak, S.; Ghorbani Javid, M.; Soltani, E.; Wojtyla, Ł.; Garnczarska, M. Contribution of exogenous proline to abiotic stresses tolerance in plants: A Review. *Int. J. Mol. Sci.* **2022**, *23*, 5186. [CrossRef]
47. Dvojković, K.; Plavšin, I.; Novoselović, D.; Šimić, G.; Lalić, A.; Čupić, T.; Horvat, D.; Viljevac Vuletić, M. Early antioxidative response to desiccant-stimulated drought stress in field-grown traditional wheat varieties. *Plants* **2023**, *12*, 249. [CrossRef]
48. Lim, C.Y.; Lee, K.J.; Oh, D.B.; Ko, K. Effect of the developmental stage and tissue position on the expression and glycosylation of recombinant glycoprotein GA733-FcK in transgenic plants. *Front. Plant Sci.* **2015**, *5*, 778. [CrossRef] [PubMed]
49. Li, J.; Wang, Y.; Wei, J.; Pan, Y.; Su, C.; Zhang, X. A tomato proline-, lysine-, and glutamic-rich type gene SpPKE1 positively regulates drought stress tolerance. *Biochem. Biophys. Res. Comm.* **2018**, *499*, 777–782. [CrossRef] [PubMed]
50. Kononczuk, J.; Czyzewska, U.; Moczydlowska, J.; Surażyński, A.; Palka, J.; Miltyk, W. Proline oxidase (pox) as a target for cancer therapy. *Curr. Drug Targets* **2015**, *16*, 1464–1469. [CrossRef] [PubMed]
51. Chalecka, M.; Kazberuk, A.; Palka, J.; Surazynski, A. P5C as an interface of proline interconvertible amino acids and its role in regulation of cell survival and apoptosis. *Int. J. Mol. Sci.* **2021**, *22*, 11763. [CrossRef]
52. Cohen, S.P.; Leach, J.E. Abiotic and biotic stresses induce a core transcriptome response in rice. *Sci. Rep.* **2019**, *9*, 6273. [CrossRef]
53. Alexander, A.; Singh, V.K.; Mishra, A. Overexpression of differentially expressed AhCytb6 gene during plant-microbe interaction improves tolerance to N_2 deficit and salt stress in transgenic tobacco. *Sci. Rep.* **2021**, *11*, 1–20. [CrossRef]
54. Gutiérrez-Beltrán, E.; Personat, J.M.; de la Torre, F.; Del Pozo, O. A universal stress protein involved in oxidative stress is a phosphorylation target for protein kinase CIPK6. *Plant Physiol.* **2017**, *173*, 836–852. [CrossRef]
55. Ren, H.; Zhao, X.; Li, W.; Hussain, J.; Qi, G.; Liu, S. Calcium signaling in plant programmed cell death. *Cells* **2021**, *10*, 1089. [CrossRef]
56. Leonowicz, G.; Trzebuniak, K.F.; Zimak-Piekarczyk, P.; Ślesak, I.; Mysliwa-Kurdziel, B. The activity of superoxide dismutases (SODs) at the early stages of wheat deetiolation. *PLoS ONE* **2018**, *13*, e0194678. [CrossRef] [PubMed]
57. Zafar, N.; Mujib, A.; Ali, M.; Tonk, D.; Gulzar, B. Aluminum chloride elicitation (amendment) improves callus biomass growth and reserpine yield in *Rauvolfia serpentina* leaf callus. *Plant Cell Tiss. Organ. Cult.* **2017**, *130*, 357–368. [CrossRef]
58. Sofo, A.; Scopa, A.; Nuzzaci, M.; Vitti, A. Ascorbate peroxidase and catalase activities and their genetic regulation in plants subjected to drought and salinity stresses. *Int. J. Mol. Sci.* **2015**, *16*, 13561–13578. [CrossRef] [PubMed]
59. Gutiérrez-Martínez, P.B.; Torres-Morán, M.I.; Romero-Puertas, M.C.; Casas-Solís, J.; Zarazúa-Villaseñor, P.; Sandoval-Pinto, E.; Ramírez-Hernández, B.C. Assessment of antioxidant enzymes in leaves and roots of Phaseolus vulgaris plants under cadmium stress. *Biotecnia Hermosillo* **2020**, *22*, 110–118. [CrossRef]
60. Rolly, N.K.; Imran, Q.M.; Lee, I.-J.; Yun, B.-W. Salinity stress-mediated suppression of expression of salt overly sensitive signaling pathway genes suggests negative regulation by AtbZIP62 transcription factor in Arabidopsis thaliana. *Int. J. Mol. Sci.* **2020**, *21*, 1726. [CrossRef] [PubMed]
61. Scholtmeijer, K.; van den Broek, L.A.M.; Fischer, A.R.H.; van Peer, A. Potential protein production from lignocellulosic materials using edible mushroom forming fungi. *J. Agric. Food Chem* **2023**, *71*, 4450–4457. [CrossRef]
62. Barbosa, L.P.; Kemmelmeier, C. Chemical composition of the hyphal wall from *Fusarium graminearum*. *Exp. Mycol.* **1993**, *17*, 274–283. [CrossRef]
63. Garcia-Rubio, R.; de Oliveira, H.C.; Rivera, J.; Trevijano-Contador, N. The Fungal Cell Wall: Candida, Cryptococcus, and Aspergillus Species. *Front. Microbiol.* **2020**, *10*, 2993. [CrossRef]
64. Junaid, A.; Mujib, A.; Bhat, M.A.; Sharma, M.P. Somatic embryo proliferation, maturation and germination in *Catharanthus roseus*. *Plant Cell Tissue Org. Cult.* **2006**, *84*, 325–332. [CrossRef]
65. Thakur, M.; Bhattacharya, S.; Khosla, P.K.; Puri, S. Improving production of plant secondary metabolites through biotic and abiotic elicitation. *J. Appl. Res. Med. Aromat. Plants* **2019**, *12*, 1–12. [CrossRef]
66. Ramirez-Estrada, K.; Vidal-Limon, H.; Hidalgo, D.; Moyano, E.; Golenioswki, M.; Cusidó, R.M.; Palazon, J. Elicitation, an effective strategy for the biotechnological production of bioactive high-added value compounds in plant cell factories. *Molecules* **2016**, *21*, 182. [CrossRef] [PubMed]

67. Jeyasri, R.; Muthuramalingam, P.; Karthick, K.; Shin, H.; Choi, S.H.; Ramesh, M. Methyl jasmonate and salicylic acid as powerful elicitors for enhancing the production of secondary metabolites in medicinal plants: An updated review. *Plant Cell Tiss. Organ. Cult.* **2023**, *153*, 447–458. [CrossRef]
68. Abdul Malik, N.A.; Kumar, I.S.; Nadarajah, K. Elicitor and receptor molecules: Orchestrators of plant defense and immunity. *Int. J. Mol. Sci.* **2020**, *21*, 963. [CrossRef]
69. Guo, J.; Cheng, Y. Advances in fungal elicitor-triggered plant immunity. *Int. J. Mol. Sci.* **2022**, *23*, 12003. [CrossRef] [PubMed]
70. Staniszewska, I.; Krolicka, A.; Malinski, E.; Jojkowska, E.; Szafranek, J.Z. Elicitation of secondary metabolites is in vitro cultures of *Ammi majus* L. *Enzym. Microb. Technol.* **2003**, *33*, 565–568. [CrossRef]
71. Miura, Y.; Hirata, K.; Miyamoto, K.; Uchida, K. Formation of vinblastine from multiple shoot culture of *Catharanthus roseus*. *Planta Med.* **1988**, *54*, 8–20. [CrossRef] [PubMed]
72. Junaid, A.; Mujib, A.; Fatima, Z.; Sharma, M.P. Variations in vinblastine production at different stages of somatic embryogenesis, embryo, and field-grown plantlets of *Catharanthus roseus* L. (G) Don, as revealed by HPLC. In Vitro Cell. *Dev. Biol.-Plant* **2010**, *46*, 348–353. [CrossRef]
73. Dey, P.M. *Methods in Plant Biochemistry. Carbohydarates, Volume 2*; Academic Press: London, UK, 1990.
74. Bates, L.; Waldren, P.P.; Teare, J.D. Rapid determination of free proline of water stress studies. *Plant Soil.* **1973**, *39*, 205–207. [CrossRef]
75. Bradford, M.M. A rapid and sensitive method for quantification of microgram quantities of protein, utilizing the principle of protein dye binding. *Anal. Biochem.* **1976**, *72*, 248–541. [CrossRef]
76. Aebi, H. Catalase in vitro. *Methods Enzymol.* **1984**, *105*, 121–126.
77. Dhindsa, R.H.; Plumb-Dhindsa, R.; Thorpe, T.A. Leaf senescence correlated with increased level of membrane permeability, lipid peroxidation and decreased level of SOD and CAT. *J. Exp. Bot.* **1981**, *32*, 93–101. [CrossRef]
78. Nakano, Y.; Asada, K. Hydrogen peroxide is scavenged by ascorbate specific peroxidase in spinach chloroplasts. *Plant Cell Physiol.* **1981**, *22*, 867–880.

Disclaimer/Publisher's Note: The statements, opinions and data contained in all publications are solely those of the individual author(s) and contributor(s) and not of MDPI and/or the editor(s). MDPI and/or the editor(s) disclaim responsibility for any injury to people or property resulting from any ideas, methods, instructions or products referred to in the content.

Article

Suspension Cell Culture of *Polyscias fruticosa* (L.) Harms in Bubble-Type Bioreactors—Growth Characteristics, Triterpene Glycosides Accumulation and Biological Activity

Maria V. Titova [1,*], Dmitry V. Kochkin [1,2,*], Elena S. Sukhanova [1], Elena N. Gorshkova [1], Tatiana M. Tyurina [1], Igor M. Ivanov [1], Maria K. Lunkova [1], Elena V. Tsvetkova [3,4], Anastasia Orlova [1], Elena V. Popova [1] and Alexander M. Nosov [1,2]

[1] K.A. Timiryazev Institute of Plant Physiology, Russian Academy of Sciences, 127276 Moscow, Russia; orlova@ifr.moscow (A.O.); popova@ifr.moscow (E.V.P.); al_nosov@mail.ru (A.M.N.)
[2] Biology Faculty, M.V. Lomonosov Moscow State University, 119234 Moscow, Russia
[3] Department of Biochemistry, Saint Petersburg State University, 199034 Saint Petersburg, Russia; e.v.tsvetkova@spbu.ru
[4] Department of General Pathology and Pathological Physiology, Institute of Experimental Medicine, 197022 Saint Petersburg, Russia
* Correspondence: ifr@ippras.ru (M.V.T.); dmitry-kochkin.m@lecturer.msu.ru (D.V.K.)

Abstract: *Polyscias fruticosa* (L.) Harms, or Ming aralia, is a medicinal plant of the Araliaceae family, which is highly valued for its antitoxic, anti-inflammatory, analgesic, antibacterial, anti-asthmatic, adaptogenic, and other properties. The plant can be potentially used to treat diabetes and its complications, ischemic brain damage, and Parkinson's disease. Triterpene glycosides of the oleanane type, such as 3-O-[β-D-glucopyranosyl-(1→4)-β-D-glucuronopyranosyl] oleanolic acid 28-O-β-D-glucopyranosyl ester (PFS), ladyginoside A, and polysciosides A-H, are mainly responsible for biological activities of this species. In this study, cultivation of the cell suspension of *P. fruticosa* in 20 L bubble-type bioreactors was attempted as a sustainable method for cell biomass production of this valuable species and an alternative to overexploitation of wild plant resources. Cell suspension cultivated in bioreactors under a semi-continuous regime demonstrated satisfactory growth with a specific growth rate of 0.11 day^{-1}, productivity of 0.32 g (L · day)$^{-1}$, and an economic coefficient of 0.16 but slightly lower maximum biomass accumulation (~6.8 g L^{-1}) compared to flask culture (~8.2 g L^{-1}). Triterpene glycosides PFS (0.91 mg gDW^{-1}) and ladyginoside A (0.77 mg gDW^{-1}) were detected in bioreactor-produced cell biomass in higher concentrations compared to cells grown in flasks (0.50 and 0.22 mg gDW^{-1}, respectively). In antibacterial tests, the minimum inhibitory concentrations (MICs) of cell biomass extracts against the most common pathogens *Staphylococcus aureus*, methicillin-resistant strain MRSA, *Pseudomonas aeruginosa*, and *Escherichia coli* varied within 250–2000 µg mL^{-1} which was higher compared to extracts of greenhouse plant leaves (MIC = 4000 µg mL^{-1}). Cell biomass extracts also exhibited antioxidant activity, as confirmed by DPPH and TEAC assays. Our results suggest that bioreactor cultivation of *P. fruticosa* suspension cell culture may be a perspective method for the sustainable biomass production of this species.

Keywords: *Ming aralia*; PFS; ladyginoside A; antimicrobial activity; antioxidant activity; cell aggregation; cell farming; plant cell biotechnology

Citation: Titova, M.V.; Kochkin, D.V.; Sukhanova, E.S.; Gorshkova, E.N.; Tyurina, T.M.; Ivanov, I.M.; Lunkova, M.K.; Tsvetkova, E.V.; Orlova, A.; Popova, E.V.; et al. Suspension Cell Culture of *Polyscias fruticosa* (L.) Harms in Bubble-Type Bioreactors—Growth Characteristics, Triterpene Glycosides Accumulation and Biological Activity. *Plants* **2023**, *12*, 3641. https://doi.org/10.3390/plants12203641

Academic Editor: Iyyakkannu Sivanesan

Received: 11 September 2023
Revised: 12 October 2023
Accepted: 19 October 2023
Published: 22 October 2023

Copyright: © 2023 by the authors. Licensee MDPI, Basel, Switzerland. This article is an open access article distributed under the terms and conditions of the Creative Commons Attribution (CC BY) license (https://creativecommons.org/licenses/by/4.0/).

1. Introduction

Polyscias fruticosa (L.) Harms (Syns = *Panax fruticosa* = *Nothopanax fruticosum* = *Parsley panax* = *Ming aralia*) belonging to family Araliaceae, is a medicinal plant widespread in the Polynesian islands of the Pacific region and Southeast Asia (India, Malaysia, Indonesia, [1]. The name *Polyscias* is composed of two Greek words: 'poly' meaning many, and 'skia' meaning shade, indicating the dense foliage typical for this genus [1]. The plant grows relatively slowly with stem reaching one or two meters tall; the roots resemble parsley in smell and taste [2]. *P. fruticosa* is sensitive to stresses, particularly temperature stress, and has an optimal temperature range of 19–29 °C [3].

Roots and leaves of *P. fruticosa* contain triterpene glycosides, alkaloids, vitamins, amino acids, cyanogenic glycosides, polyacetylenes, sterols, tannins, essential oils, essential micro- and macroelements, and sugars [1,4–6]. Triterpene glycosides are represented mainly by the oleanane-type compounds, such as 3-*O*-[β-*D*-glucopyranosyl-(1→4)-β-*D*-glucuronopyranosyl] oleanolic acid 28-*O*-β-*D*-glucopyranosyl ester (PFS), ladyginoside A (LadA), and polysciosides A-H [1,7,8]. According to Do et al. [9], in 2019, methylglucuronate glycosides polyscioside J and polyscioside K and chikusetsusaponin IVa were isolated and identified in the leaves of *P. fruticosa* for the first time.

P. fruticosa is actively used in traditional medicine due to a wide range of pharmacological effects and safety: according to Koffuor et al. [10], the No-Observable-Adverse Effect-level (NOAEL) of the ethanolic leaf extract is below 1000 mg kg^{-1}. Over the past 20 years, studies have demonstrated antitoxic, anti-inflammatory, analgesic, and molluscicidal properties of *P. fruticosa* extracts [10–13] and their antibacterial activity against Gram-positive and Gram-negative bacteria [13,14]. The high potential of using *P. fruticosa* for treating diabetes and its complications has been reported in the in vivo models [11,15–17]. *P. fruticosa* leaf extract also exhibited anti-asthmatic and antihistamine activities increasing protection against histamine-induced bronchospasm and reducing recovery time [10,18]. Roots are used to treat neuralgia and rheumatic pain, to improve brain function, and as diuretic and anti-dysentery agent [19]. In old male rats, root extract stimulated sexual activity and increased fertility, and significantly increased memory function and lifespan [20–22]. *P. fruticosa* leaf and root extracts stimulated the immune system to produce antibodies or interferons, thus enhancing the body's defense against diseases, and exhibited adaptogenic activity due to modulation of the body's endocrine system to counteract stress [1,14]. Several studies reported the neuroprotective effect and therapeutic potential of *P. fruticosa* against ischemic brain damage and Parkinson's disease [23,24].

In addition to medicinal purposes, *P. fruticosa* is widely used as a spice and as a fresh vegetation for salads, as well as in tonic drinks and tea [25,26]. Several studies investigated changes in the phytochemical characteristics of *P. fruticosa* during drying and storage to select optimal conditions for its processing and preservation [25,27,28].

Despite the high potential value of *P. fruticosa* for functional foods and pharmacology, information on in vitro cultivation of this species is scarce. Several protocols have been published on in vitro micropropagation using shoot tips, and stem and leaf explants [2,3,29]. Vinh et al. [30] worked on the optimization of culture medium to enhance the growth and biosynthetic characteristics of *P. fruticosa* callus. Kim et al. [31] investigated the effects of carbon sources and plant growth regulators on growth and oleanolic acid accumulation in *P. fruticosa* cell suspension cultures. Phuong et al. [32] examined the effects of two auxins, 2,4-dichlorophenoxyacetic acid (2,4-D) and α-naphthaleneacetic acid (NAA), on biomass accumulation, rhizogenesis, and somatic embryogenesis of *P. fruticosa* suspension-cultured cells. Hau et al. [33] observed the accumulation of oleanolic acid (40.1 μg g^{-1}) and saponins (396.2 μg g^{-1}) in proliferated *P. fruticosa* hairy roots. Adventitious root induction was also reported for this species [34].

Production of suspension cell biomass in bioreactors could potentially provide a sustainable, renewable source of plant material compared to quantitatively limited and chemically variable wild plants [35–39]. However, to our knowledge, there are no publications on the cultivation of *P. fruticosa* cells in bioreactors.

In the previous studies, we reported the development of callus and suspension cell cultures of *P. fruticosa* in flasks followed by biochemical analysis of their triterpene glycosides (TG) content [40–44]. The present study aimed to establish a suspension cell culture of *P. fruticosa* in 20 L laboratory bubble-type bioreactors to explore the feasibility of upscaling the cell biomass production and its potential use as a novel food supplement and a source of TG. The content of two main TG in *P. fruticosa* cell biomass, as well as the antioxidant and antimicrobial activities of the cell biomass extracts, were evaluated for quality control.

2. Results

2.1. Suspension Cell Culture Growth in Flasks

The suspension cell culture of *Polyscias fruticosa* has been maintained as a flask culture since 2005, and its growth characteristics were monitored periodically [40–44]. The growth dynamics of the cell culture in 250 mL flasks was recorded before [43] and in parallel with bioreactor cultivation. The representative growth curves in normal and semi-logarithmic coordinates are shown in Figure 1a,b, respectively. Particularly for this part of the research, the cultivation was extended to over 30 days to record the complete growth cycle, including the degradation phase.

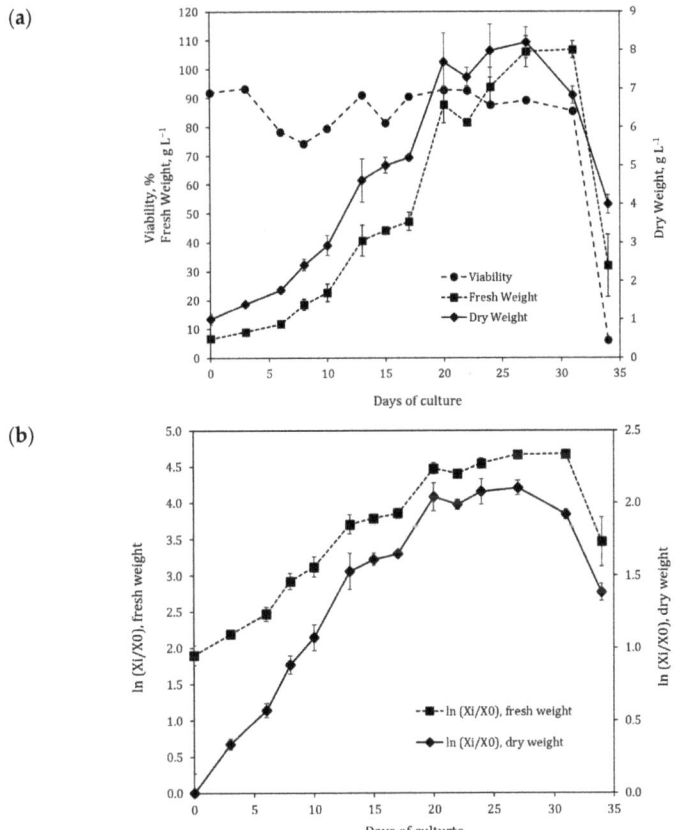

Figure 1. Representative growth curves of the suspension cell culture of *Polyscias fruticosa* during cultivation in 250 mL flasks: (**a**) fresh and dry weights and cell viability plotted in normal coordinates; (**b**) fresh and dry weights plotted in semi-logarithmic coordinates.

At the initial inoculum density near 1.1 g L^{-1}, the lag phase (~2 days) and acceleration phase (~4 days) were observed. The duration of the exponential growth phase was 7–9 days followed by the 7-day stationary phase, and then the degradation phase. Cell viability during the cultivation cycle varied within the 75–95% range, followed by a sharp dropdown after 31 days during the degradation phase.

The main growth parameters for the suspension cell culture in flasks are presented in Table 1. The culture grew well, with average maximum biomass accumulation of 8.2 g L^{-1}, specific growth rate of 0.21 day^{-1}, and productivity of 0.45 g (L · day)$^{-1}$ calculated based on dry weight (DW).

Table 1. Growth parameters of the suspension cell culture of *Polyscias fruticosa* during cultivation in 250 mL flasks and 20 L bioreactors (calculated based on dry weight).

Parameter	Cultivation System				
	250 mL Flasks	20 L Bioreactors *			
		MC 1	MC 2	MC 3	Average of MC 1–3
X_0, g L^{-1}	1.10 ± 0.45	2.40 ± 0.37	1.97 ± 0.10	1.83 ± 0.27	2.07 ± 0.30
Δt_{max}, days	22.5 ± 3.5 [a]	15.0 ± 2.5 [b]	16.5 ± 4.0 [ab]	18.5 ± 4.5 [ab]	16.7 ± 2.0
X_{max}, g L^{-1}	8.19 ± 0.81 [a]	7.31 ± 1.43 [a]	6.31 ± 0.34 [a]	6.76 ± 0.81 [a]	6.79 ± 0.50
μ, day^{-1}	0.21 ± 0.03 [a]	0.10 ± 0.02 [b]	0.13 ± 0.03 [b]	0.11 ± 0.02 [b]	0.11 ± 0.03
τ, day	3.3 ± 0.4 [b]	6.9 ± 1.2 [a]	5.3 ± 1.6 [ab]	6.3 ± 1.3 [a]	6.3 ± 1.4
Y	0.27 ± 0.05 [a]	0.16 ± 0.05 [b]	0.14 ± 0.01 [b]	0.16 ± 0.03 [b]	0.16 ± 0.01
P_{max}, g (L · day)$^{-1}$	0.45 ± 0.12 [a]	0.33 ± 0.06 [ab]	0.30 ± 0.05 [b]	0.35 ± 0.15 [ab]	0.32 ± 0.13

MC—multi-cycle (continuous cultivation in bioreactors under semi-continuous mode consisting of three (MC 1 and 2) or 11 (MC 3) subcultivation cycles), X_0—inoculum density, Δt_{max}—time from the beginning of the subculture cycle to achieving maximum biomass accumulation, X_{max}—maximum dry cell weight, μ—the maximum value of the specific growth rate calculated for the analyzed growing cycle, τ—doubling time, Y—economic coefficient, P_{max}—productivity. * For each MC, parameters were calculated as an average of all subcultivation cycles within the respective MC. Mean values in rows followed by the same letter are not significantly different at $p < 0.05$, according to Duncan's multiple range test.

The cytological analysis revealed that cells in the suspension were, on average, relatively small (20–35 μm in diameter) and round-shaped. Typical microphotographs of cells in the exponential growth phase (day 10) are presented in Figure 2. Cell aggregates were predominantly round with uneven edges. The aggregation level (percentage of aggregates consisting of a certain number of cells) estimated at the early exponential phase (day 10) and beginning of the stationary phase (day 20) is shown in Table 2. The culture was relatively small-aggregated: about 70% of the aggregates in the suspension consisted of less than 50 cells, and the aggregates were less than 1 mm in diameter. The number of viable single cells and small aggregates (up to 5 cells) was insignificant (Table 2).

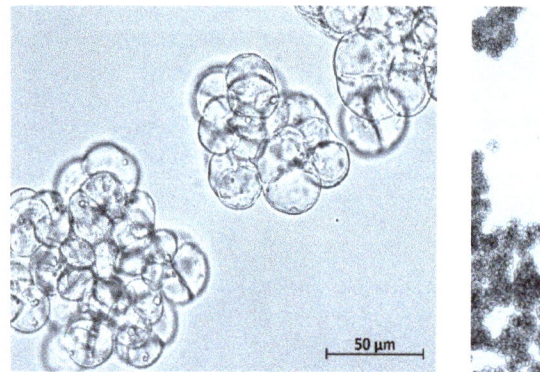
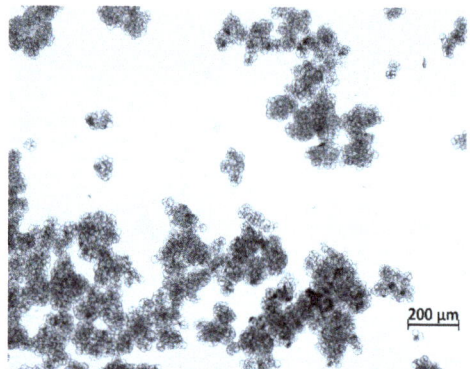

Figure 2. Representative photographs of cells and cell aggregates of *Polyscias fruticosa* suspension cell culture grown in flasks at different magnifications.

Table 2. Aggregation level of *Polyscias fruticosa* suspension cell culture during cultivation in 250 mL flasks and 20 L bioreactor estimated at the early exponential phase (day 10) and the beginning of the stationary phase (day 20, only for flasks). Aggregation level is presented as the percentage of aggregates consisting of a certain number of cells out of the total number of cell aggregates counted.

Day of the Growth Cycle	Number of Cells in the Aggregate					
	1–5	6–10	11–20	21–30	31–50	>50
	% of Aggregates Consisted of a Given Number of Cells					
	250 mL flasks					
Day 10	4.3 ± 2.1 [a]	12.0 ± 2.0 [a]	17.0 ± 1.0 [a]	15.0 ± 6.2 [a]	24.3 ± 3.2 [a]	27.6 ± 5.5 [b]
Day 20	2.3 ± 0.6 [a]	5.7 ± 1.5 [b]	13.0 ± 1.7 [ab]	17.7 ± 8.4 [a]	29.0 ± 4.3 [a]	33.0 ± 6.9 [ab]
	20 L bioreactors					
Day 10	4.0 ± 1.7 [a]	7.7 ± 1.5 [b]	10.0 ± 3.6 [b]	15.4 ± 7.4 [a]	25.3 ± 7.6 [a]	39.3 ± 2.9 [a]

Mean values in columns followed by the same letter are not significantly different at $p < 0.05$, according to Duncan's multiple range test.

Based on growth and the cytological data, the suspension cell culture was considered potentially suitable for the cultivation in bioreactors.

2.2. Suspension Cell Culture Growth in 20 L Bioreactor under a Semi-Continuous Cultivation Regime

To establish the bioreactor cultivation for cell suspension of *P. fruticosa* in bubble-type 20 L glass bioreactors, three sequential multi-cycle (MC) runs were performed under a semi-continuous regime. The first two MC runs consisted of three subculture cycles, and the third MC run had 11 subculture cycles (Figure 3). In each subculture cycle, a portion of the suspension was harvested at the end of the exponential—beginning of the stationary growth phase, and the remaining suspension was diluted by adding fresh medium to reach the suspension concentration around 1.5–2.0 gDW L^{-1} that allowed further growth without a lag phase. The total duration of multicycles 1, 2, and 3 was 42, 53, and 204 days, respectively. The growth curves and main growth characteristics of the cell culture during MC 1, 2, and 3 are shown in Figure 3 and Tables 1 and 2.

Growth parameters of the cell suspension in bioreactors changed compared to flasks (Table 1). An increase in inoculum density to ~2 g L^{-1} allowed to omit the lag phase and thus shorten the cultivation time until reaching a maximum biomass production point from 22 to 16–17 days on average (Table 1) while keeping a nearly similar maximum biomass concentration level (X_{max} = 6.7 g L^{-1} in bioreactors vs. 8.2^{-1} in flasks). However, the specific growth rate of the culture in bioreactors decreased almost twice compared to flasks. Some reductions in productivity (by 22%) and economic coefficient (by 41%) were also recorded. Despite the overall good cell viability (above 70%) during cultivation in bioreactors, a steady decrease in maximum fresh weight (FW) and DW concentrations was observed within each MC run, and this tendency persisted through all three MC runs performed in this study (Figure 3). However, growth parameters were not significantly different between three independent MC bioreactor runs.

Compared to flask culture, a minor shift towards higher aggregation was observed for the cell suspension in bioreactors, as the percentage of aggregates with over 50 cells increased from 27.6 to 39.3% (Table 2). However, this change had no negative impact on the technological aspects of the cultivation process.

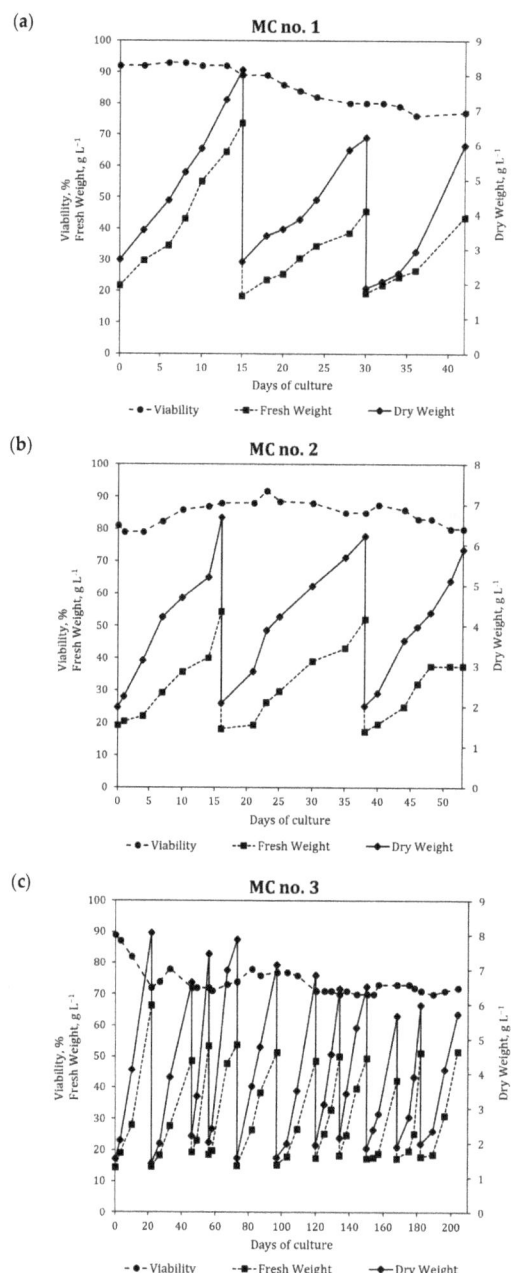

Figure 3. Growth curves of *Polyscias fruticosa* suspension cell culture during cultivation in 20 L bioreactor in a semi-continuous cultivation regime: (**a**) Multi-cycle (MC) no. 1; (**b**) MC no. 2; (**c**) MC no. 3.

2.3. Analysis of the Two Major Secondary Metabolites in the Suspension Cell Culture of Polyscias fruticosa during Cultivation in Flasks and Bioreactors

Triterpene glycosides ladyginoside A and PFS (28-O-β-D-glucopyranosyl ester of oleanolic acid 3-O-β-D-glucopyranosyl-(1→4)-β-D-glucuronopyranoside) in dried cell biomass of P. fruticosa were identified as described earlier [40–42] using UPLC-ESI-MS based on the interpretation of corresponding mass-spectra, comparison of chromatographic behavior, and mass-spectra with standard samples previously isolated from the leaves of Polyscias filicifolia [40–42] and the literature data [45,46]. Structural formulas of the two compounds are presented in Figure 4. The results of qualitative and quantitative analysis of triterpene glycosides (negative ion mode) in cell biomass are presented in Tables 3 and 4. Ladyginoside A and PFS were detected in cell biomass during cultivation in flasks and bioreactors (Table 4). Interestingly, the production of both compounds in bioreactors was significantly higher compared to flasks (p = 0.01 for PFS and p = 0.0002 for LadA). In flasks, PFS accumulation in cell biomass was higher at day 20 (end of exponential phase) compared to day 27 (end of stationary phase–beginning of degradation phase). At the same time, the content of LadA did not change significantly between these growth phases. The content of LadA remained relatively stable within each multi-cycle bioreactor run, ranging from 0.65 to 0.75 mg gDW^{-1} depending on a subcultivation cycle. The content of PFS varied in a broader range (0.70–1.17 mg gDW^{-1}) between subcultivations within one MC run. However, the average content of both triterpene glycosides was similar between the three MC rungs in bioreactors (Table 4), indicating relatively good reproducibility of biosynthesis of TG in the bioreactor cultivation system.

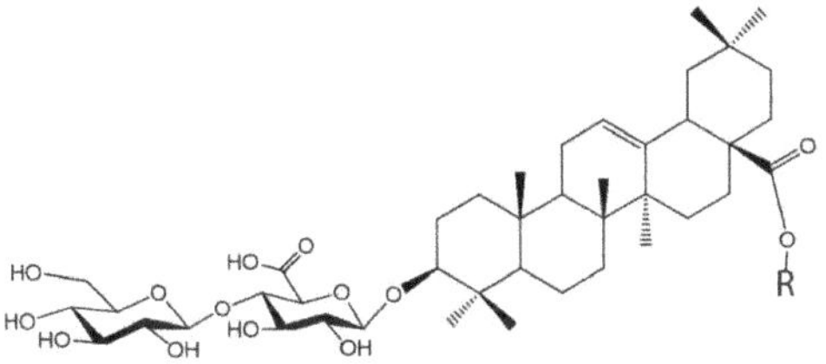

Figure 4. Chemical structure of triterpene glycosides found in cell biomass of Polyscias fruticosa: Ladyginoside A: R = H; PFS: R = β-D-glucopyranoside.

Table 3. Results of UPLC–ESI–MS analysis (negative ion mode) of extracts from cell biomass of Polyscias fruticosa suspension cell culture grown in flasks.

t_R, min *	[M-H]$^-$, m/z **	Identification Results
7.906	955.6	PFS ***
13.440	793.6	Ladyginoside A (LadA)

* Retention time on the chromatographic column (min); ** The m/z values for the ions detected in the mass spectra; *** 28-O-β-D-glucopyranosyl ester of oleanolic acid 3-O-β-D-glucopyranosyl-(1→4)-β-D-glucuronopyranoside.

Table 4. Content of PFS and ladyginoside A, as determined using UPLC–ESI–MS, in cell biomass samples of Polyscias fruticosa suspension cell culture grown in flasks and 20 L bioreactors.

Cultivation System	Variant	Triterpene Glycoside Content, mg gDW^{-1}	
		PFS	Ladyginoside A
Flasks	Day 20	0.50 ± 0.05 [bc]	0.22 ± 0.01 [b]
	Day 27	0.34 ± 0.12 [c]	0.20 ± 0.01 [b]
Bioreactors *	MC no. 1	0.78 ± 0.01 [a]	0.66 ± 0.01 [a]
	MC no. 2	1.03 ± 0.37 [a]	0.75 ± 0.08 [a]
	MC no. 3	0.92 ± 0.15 [ab]	0.79 ± 0.21 [a]
	Average of MC 1-3	0.91 ± 0.12	0.73 ± 0.07

* For each MC, values are means of triterpene glycoside content measured during individual subcultivation cycles at the time points of maximum biomass accumulation; PFS—28-O-β-D-glucopyranosyl ester of oleanolic acid 3-O-β-D-glucopyranosyl-(1→4)-β-D-glucuronopyranoside; MC—multi-cycle. Mean values in columns followed by the same letter are not significantly different at $p < 0.05$, according to Duncan's multiple range test.

2.4. Antioxidant and Antimicrobial Activities of the Extracts from Cell Biomass of P. fruticosa Grown in Bioreactors

For rapid assessment of the biological activity of the cell biomass extracts, their antimicrobial and antioxidant activities were determined and compared to those of leaves of the greenhouse plants.

The minimum inhibitory concentrations (MICs) of cell biomass extracts against the most common pathogens, *Staphylococcus aureus*, including methicillin-resistant strain MRSA, *Pseudomonas aeruginosa*, and *Escherichia coli*, are given in Table 5. Cell biomass was most effective against *E. coli*, followed by *S. aureus* and *P. aeruginosa*, and less effective against MRSA. However, for all pathogens tested, the antimicrobial activity of cell biomass extracts exceeded those exhibited by extracts from greenhouse plant leaves.

Table 5. The minimum inhibitory concentration (MICs) of extracts from *Polyscias fruticosa* cell biomass grown in bioreactors and greenhouse plant leaves.

Extracts	MICs, µg mL^{-1}			
	Escherichia coli ATCC 25922	*Staphylococcus aureus* ATCC 25923	MRSA ATCC 33591	*Pseudomonas aeruginosa* ATCC 27853
Cell biomass	250	500	2000	500
Plant leaves	4000	4000	4000	4000

MRSA—methicillin-resistant *Staphylococcus aureus*.

The antiradical activity of cell biomass and greenhouse plant leaf extracts was determined by two spectrophotometric tests, DPPH and TEAC, based on the ability of plant material samples to bind stable radicals. The *P. fruticosa* leaf extract showed higher antiradical activity in both tests, and the obtained absolute effective concentrations causing a 50% effect (EC$_{50}$) of extracts from leaves were lower than the EC$_{50}$ of extracts from cell culture biomass (Table 6).

Table 6. Antioxidant activities of the extracts from *P. fruticosa* cell biomass and greenhouse plant leaves expressed as EC$_{50}$ values (absolute effective concentrations causing a 50% effect), based on DPPH and TEAC assays.

Extracts	Assay *	
	DPPH EC$_{50}$, mg mL^{-1}	TEAC EC$_{50}$, mg mL^{-1}
Cell biomass	17.85 ± 0.44 [a]	11.10 ± 0.98 [a]
Plant leaves	10.69 ± 0.09 [b]	8.81 ± 0.13 [a]
Positive control	0.150 ± 0.002	0.036 ± 0.001

* DPPH—2,2-diphenyl-1-picrylhydrazyl free radical scavenging assay; TEAC—Trolox equivalent antioxidant capacity. Ascorbic acid and Trolox were positive controls for DPPH and TEAC assays, respectively. In columns, values followed by the same letter are not significantly different at $p < 0.05$ according to Student's unpaired *t*-test with Welch correction.

3. Discussion

Our results demonstrated that the suspension cell culture of *P. fruticosa* grew well in flasks (Table 1). The maximum dry weight accumulation in flasks achieved in our study (8.2 g L^{-1} at the inoculum density 1.1 g L^{-1}) was similar to what was previously recorded for the same cell strain [40,44] and to those reported for the suspension cell cultures of *P. fruticosa* by Kim et al. [31] (0.33–0.45 g per flask or 6.5–9.0 g L^{-1} depending on the media composition at inoculum density 3.3–3.4 g L^{-1}). The cell culture showed a lag phase when subcultured with inoculum density about 1.0 g L^{-1} [44]. The growth parameters of the *P. fruticosa* suspension cell culture were good enough to consider its bioreactor cultivation. For example, the economic coefficient of 0.27 suggests that almost one third of sucrose supplied to culture medium is utilized for building cell structures and compartments.

The size of cell aggregates in the suspension culture is essential for bioreactor cultivation. Large cell aggregates can lead to an inhomogeneous distribution of nutrients, oxygen, light, and other environmental factors between the cells, thus reducing the suspension growth. In addition, large cell aggregates create technological problems in bioreactor operation, such as excessive sedimentation and adhesion of cells to bioreactor walls and hose plugging [47,48]. In the present study, the suspension cell culture of *P. fruticosa* had no large aggregates (over 1 mm in diameter) throughout the growth cycle in flasks, which was a positive factor for further upscaling to bioreactors (Table 3). During bioreactor cultivation, the ratios of cell aggregate groups changed towards forming larger aggregates. However, such change had no impact on the cultivation process due to the small size of the cells.

For bioreactor cultivation, the bubble type of bioreactors and the semi-continuous cultivation regime were selected based on a review of the literature data on the classification of bioreactors and their technological parameters as well as on the results of our experience with bioreactor cultivation of the cell suspensions of other plant species [35,36,49–51]. The aeration regime for bioreactors was selected experimentally. The airflow rate varied depending on the cell suspension growth stage, taking into account the following requirements: (i) the concentration of dissolved oxygen (pO_2) should remain above 15%; (ii) there should be no stagnant ("dead") zones in the bioreactor, no cell sedimentation, and no intensive foaming [52,53]. Dissolved oxygen concentrations between 15 to 20% air saturation are usually recommended for plant cell cultivations [54].

Inoculum density is one of the main parameters for optimizing bioreactor cultivation for plant suspension cell cultures. Several studies revealed that variations in the inoculum size can lead to changes in culture growth kinetics and production of secondary metabolites and cell biomass, and this effect is specific for different cell cultures, even within the same plant species [55–58]. Increasing the inoculum density to a certain level leads to a reduction in or complete disappearance of the lag phase, an increment in the specific growth rate, earlier entrance to the stationary growth phase, and earlier synthesis of secondary metabolites. On the contrary, reducing the inoculum below a certain critical value, specific for each cell culture, may lead to a significant suppression and, in some cases, a complete arrestment of the culture growth. In the present study, higher inoculum density was used for bioreactor cultivation of *P. fruticosa* cell suspension compared to flask cultures (2.07 vs 1.10 gDW L^{-1}). This allowed omitting the lag phase and reducing the time required to reach the maximum biomass accumulation point, thus resulting in 25–30% shorter subculture cycles compared to flask culture. However, it also led to reduction in growth parameters (productivity, specific growth rate, and economic coefficient). A similar tendency was observed for flask cultures where inoculum size above 2.5 gDW L^{-1} led to a significant slowdown in cell culture growth, a decrease in viability and maximum biomass accumulation (Supplementary Figure S1), and medium darkening during 2–3 subcultivation cycles. Therefore, in the present study, inoculum size was controlled at ~1.8–2.4 gDW L^{-1}, and the fresh medium was fed into bioreactors at a cell suspension concentration of 6.5–8.5 gDW L^{-1}. Factors limiting the growth of cell cultures during cultivation with high inoculum density may include rapid depletion of nutrients in the culture medium, accumulation of toxic byproducts, tissue metabolites, or dead cell residues.

The maximum biomass accumulation and the average specific growth rate decreased in bubble-type bioreactors compared to flask culture (Table 2). It is likely that the growth limitations may be associated with higher stress levels in the bubble-type bioreactors in comparison with flasks [37,54,59]. However, the type of bioreactors used in the present study was selected based on our previous successful experience with suspension cell cultures of a number of medicinal species, such as *Dioscorea deltoidea, Panax japonicus, Stephania glabra, Taxus wallichiana,* and a relative species *Polyscias filicifolia* [35,60–62]. These cultures retained their main characteristics during cultivation in bioreactors of this type. Unlike cell cultures mentioned, a cell suspension of *P. fruticosa* requires further optimization of the bioreactor conditions and a more comprehensive investigation of cell suspension physiology. At the same time, the growth and biomass accumulation parameters were stable and repetitive between multicycles performed during three years, which makes the future of bioreactor cultivation of this cell strain quite promising.

It is essential that the cell culture grown in bioreactors accumulate biologically active compounds and maintain active biosynthesis of the desired secondary metabolites. Qualitative and quantitative analysis of secondary metabolites confirmed the presence of TG of the oleanolic acid group in cell biomass samples from both flasks and bioreactors. These compounds were PFS and LadA, identified earlier in this suspension cell culture [40,42,44] (Table 3). During the bioreactor cultivation, the suspension cell culture of *P. fruticosa* retained active synthesis of these metabolites at a sufficiently high level, exceeding those recorded during cultivation in flasks. With increasing duration of bioreactor cultivation (MC 3 in Table 4), a relatively stable synthesis was observed for LadA. In contrast, PFS content showed fluctuations within a 0.20–0.40 mg gDW^{-1} range between the subculture cycles. These results are in accordance with other studies [1,7,8,16] reporting oleanolic acid saponins as major components of *P. fruticosa*. PFS and LadA were found in the leaves and roots of *P. fruticosa* plants [7–9]. Moreover, the PFS saponin is quantitatively the primary compound isolated from leaves and roots of *P. fruticosa* plants. Therefore, PFS can be potentially used as a marker for quality control of *P. fruticosa*-based products, particularly cell biomass [8]. It is worth noting that the amount of PFS in the cell culture in our study was comparable to its content in leaves (1.30 mg gDW^{-1}) and roots (0.57 mg gDW^{-1}) of *P. fruticosa* plants [8].

It is acknowledged that the oleanolic acid glycosides are primarily responsible for the biological activity of *P. fruticosa* [10–14], including antibacterial effects. In particular, *P. fruticosa* leaf extract exhibited antibacterial activity on Gram-positive Cocci *Staphylococcus aureus* (including MRSA), *Bacillus subtilis*, and *E. coli* [13,14]. It is considered that oleanolic acid saponins can increase and/or disrupt the permeability of the bacterial cell membrane, which leads to damage and death of the bacterial cells [13]. In our work, the antibacterial activity of extracts of cell suspension biomass from bioreactors was investigated in vitro by broth microdilution assay using cultures of the four most common pathogens. The highest antibacterial activity for cell biomass extract was observed against the Gram-negative strain *Escherichia coli* (MIC 250 µg mL^{-1}). The lowest activity was revealed against the Gram-positive strain MRSA ATCC 33591 (MIC 2000 µg mL^{-1}). Importantly, the antimicrobial activity of the extracts from cell biomass exceeded those of extracts from plant leaves (Table 5).

Several publications reported the in vitro and in vivo antioxidant activities of crude extracts of *P. fruticosa* leaves and roots [24,27,63,64]. There are different methodologies for evaluating the antioxidant activity of both synthetic and natural compounds. In our study, we performed DPPH and ABTS assays as rapid and low-cost methods for antioxidant activity screening of cell biomass extracts. These methods are frequently used for the evaluation of the antioxidative potential of plant extracts, including most of the published studies on the antioxidant activities of *P. fruticosa* [24,27,28,63,65]. In our study, cell biomass extracts demonstrated moderate antioxidant activities that were slightly lower than the activity measured for extracts from leaves of the greenhouse plants (Table 6). These differences can be attributed to the activity of both TG found in cell biomass and polyphenolic compounds (not analyzed in the present study). For example, Tran et al. [27] demonstrated strong correlations between the antioxidant activity of the *P. fruticosa* extract and the content of polyphenols and triterpenoid saponins.

4. Materials and Methods

4.1. Cell Suspension Cultivation in Flasks and Bioreactors

Polyscias fruticosa (L.) Harms suspension cell culture, strain 6a, was provided by the All-Russian Collection of Plant Cell Cultures at the Institute of Plant Physiology of the Russian Academy of Sciences (Moscow, Russia). The strain was induced from a leaf of a greenhouse plant, as described earlier [43]. Stock cell culture was maintained in 250 mL Erlenmeyer flasks filled with 30 mL of modified Murashige and Skoog liquid medium with 30 g L^{-1}

sucrose and plant growth regulators 2,4-D (2 mg L^{-1}) and BA (1 mg L^{-1}) [43,66] (Figure 5, Table 7). The initial density (inoculum) size was set at 0.9–1.2 gDW L^{-1}. Cultures were grown on an orbital shaker (95–100 rpm) at 26–27 °C and 70–75% relative air humidity in darkness. Subcultures were performed every 14 days. In the experiments on growth curve estimation, the culture period was extended to 34 days to capture the degradation phase.

(a)  (b)

Figure 5. Suspension cell culture of *Polyscias fruticosa*: (**a**) in 250 mL flasks; (**b**) in 20 L bubble-type bioreactors.

Table 7. Major characteristics of cultivation systems used for *Polyscias fruticosa* cell suspension (based on [60]).

Cultivation System	250 mL Flasks on an Orbital Shaker	20 L Bioreactors (Bubble-Type)
Mode of operation	Batch	Semi-continuous
Working volume	35 mL medium	15 L
Aeration device	No	Sparger * $n_h = 1$ $d_h = 6.0$ mm

* Air supply 0.5 to 1.0 v vpm^{-1}; n_h—number of holes in the sparger; d_h—diameter of the hole in the sparger.

Bubble-type 20 L glass bioreactors (Institute of Plant Physiology of RAS, Moscow, Russia) with 15 L working volume and air supply through a sparger were used to cultivate cell suspension (Table 7; Figure 5). The choice of bioreactor type was based on previous studies demonstrating that barbotage bioreactors produced minimum mechanical damage to plant cell suspensions [35,60,67]. The scheme of the bioreactor is given in Supplementary Figure S2. Cultivation was performed using a semi-continuous regime at 26 ± 0.5 °C in darkness. To maintain the semi-continuous cultivation, a portion of cell suspension was regularly harvested from bioreactors using a special sterile hosepipe at the beginning of the stationary growth phase of a subcultivation cycle, which was measured as cell suspension concentration reaching 6.5–8.5 gDW L^{-1}. Simultaneously, the sterile fresh nutrient medium was added to bioreactors until cell suspension was diluted to 1.5–2.0 gDW L^{-1}, and the cultivation process was continued. This process was repeated when the cell suspension reached the beginning of the stationary phase of the next subcultivation cycle. Each bioreactor cultivation under a semi-continuous regime thus consisted of multiple subcultivation cycles and is therefore designated in the text as a "multi-cycle (MC) run." Three independent MC runs of bioreactors were performed in this study. The first two MC runs consisted of three subculture cycles, and the third MC run had 11 subculture cycles (Figure 3). The total duration of multicycles 1, 2, and 3 was 42, 53, and 204 days, respectively.

The dissolved oxygen (pO_2) concentration was maintained at 10–40% of saturation volume without intense foaming. Air was supplied to bioreactors at a rate varying from 0.1 to 1.0 v vpm^{-1} depending on the growth phase of the cell culture. The minimum air flow rate was set at the beginning of subculture cycles to reduce the adverse effects of intensive suspension mixing while avoiding cell sedimentation. During the exponential growth phase, the air supply rate was increased to the maximum possible that did not cause cell destruction, as confirmed by microscopic observations.

4.2. Assessment of Growth and Physiological Characteristics of the Cell Suspension Culture

During cultivation in flasks and bioreactors, FW and DW of cell biomass, aggregation level, and viability were recorded every 2–3 days during the cultivation cycle [60]. For each time point, three flasks or three samples from bioreactors were taken ($n = 3$).

To estimate the FW, 10–15 mL of the cell suspension was collected on paper filters in a Büchner funnel, and the culture medium was removed under vacuum. Cells were washed three times with distilled water under vacuum, and FW of the biomass sample was recorded. To estimate DW, cell biomass sample was dried at 40 °C for 48 h to a constant weight. The biomass samples for chemical analysis and biological activity assessment (see below) were prepared following the same procedure.

Cell viability was calculated under microscope after staining with 0.025% Evans blue (modified from [68]) as the percentage of cell aggregates composed of colorless (living) cells. Aggregation was defined as the ratio of different types of aggregates expressed in percentages. To increase contrast, cells were stained with 0.1% phenosafranin, and the number of aggregates of different types was counted under a light microscope. The diameter of large aggregates was measured under a binocular microscope. The measurements were performed in three biological and two analytical repetitions (a fixed volume of cell suspension was taken twice from each of the three flasks or bioreactor samples, and each time at least 100 aggregates were measured).

In addition, the following growth parameters were calculated based on DW according to [35,69]:

Specific growth rate, indicating the rate of dry weight increase per day:

$$\mu = \text{maximum value of } \mu_i = \ln(X_i/X_{i-1})/(t_i - t_{i-1}), [\text{day}^{-1}]$$

where X_i and X_{i-1} are, respectively, dry cell biomass concentrations (g L^{-1}) at time points t_i and t_{i-1};

Doubling time, indicating the time requiring for doubling dry cell weight at constant μ (exponential growth phase):

$$\tau = \ln 2/\mu, [\text{day}]$$

Productivity on dry biomass, indicating the amount of biomass that can be harvested from one liter of the suspension per day:

$$P = \text{maximum value of } P_i = (X_i - X_0)/(t_i - t_0) [\text{g (L day)}^{-1}]$$

where X_i and X_0 are, respectively, dry cell biomass concentrations (g L^{-1}) at time points t_i and at the time of inoculation t_0

Economic coefficient, indicating the efficiency of substrate (sucrose) utilization for cell growth:

$$Y = (X_{max} - X_0)/S_0,$$

where: X_0 and X_{max} are, respectively, initial and maximum dry cell biomass concentrations (g L^{-1}); S_0—initial sucrose concentration in the medium (g L^{-1}).

4.3. Extract Preparation and High-Performance Liquid Chromatography-Electrospray Ionization–Mass Spectrometry (UPLC-ESI-MS) Analysis of Triterpene Glycosides in Cell Biomass

Dried cell biomass was powdered, and extraction and purification by solid-phase extraction were performed according to previously published procedures [70]. The evaporated extracts were dissolved in 2 mL of 70% (*v*/*v*) aqueous methanol and filtered through a nylon filter "Acrodisc" with 0.2 μm pores (Pall Corporation, NY, USA).

The analysis of triterpene glycosides was performed using ACQUITY UPLC H-Class PLUS chromatograph (Waters, Milford, MA, USA) coupled with electrospray ionization and a hybrid time-of-flight mass spectrometry detector Xevo G2-XS TOF (Waters, Milford, MA, USA). Samples (0.05 μL) were separated on the ACQUITY UPLC BEH C18 column (50 × 2.1 mm, 1.7 μm; Waters, Drinagh, County Wexford, Ireland) at 40 °C with mobile phase flow rate 0.4 mL min^{-1}. The composition of the mobile phase was: 0.1% (*v*/*v*) solution of formic acid in water (phase A) and 0.1% (*v*/*v*) solution of formic acid in 99.9% (*v*/*v*) acetonitrile (phase B). Phases were prepared using deionized water (Simplicity UV, Millipore, Molsheim, France) and acetonitrile of "LC-MS" grade (Panreac, Barcelona, Spain). Analytes were separated using a gradient (B, % by volume): 0–1 min—5→15%, 1–5 min—15→30%, 5–11 min—30→38%, 11–15 min—38→65%, 15–15.5 min—65→95%.

Analysis was performed in a negative ion detection mode in the *m/z* range of 100–1900. The ionization source temperature was set to 150 °C, desolvation temperature 650 °C, capillary voltage 3.0 kV, sample injection cone voltage 30 V; desolvation gas (nitrogen) flow rate 1101 L h^{-1}. Data analysis was performed using MassLynx software 4.2 (Waters, Milford, MA, USA).

Triterpene glycosides were identified using cell culture grown in flasks as described earlier [41]. Structural identification of ladyginoside A and PFS (28-*O*-β-*D*-glucopyranosyl ester of oleanolic acid 3-*O*-β-*D*-glucopyranosyl-(1→4)-β-*D*-glucuronopyranoside) was carried out by comparing chromatographic and mass spectrometric characteristics of the found compounds with standard samples of these glycosides, previously isolated from the leaves of *Polyscias filicifolia* and the structure of which was unambiguously confirmed by 1D and 2D NMR and high-resolution mass spectrometry [41]. The concentrations of individual compounds in cell biomass grown in flasks and bioreactors were determined from the calibration curves constructed using the external standard of ginsenoside R$_0$ (Sigma-Aldrich, MA, USA). Standard deviations for retention time and chromatographic peak squares were below 1% and 15%, respectively. In a working concentration range (0.02–50.00 μg mL^{-1}), linear models approximated the calibration curve with correlation coefficients $R^2 > 0.995$.

4.4. Test Systems for Rapid Assessment of the Biological Activity of Cell Biomass Extracts: Antioxidant and Antimicrobial Tests

4.4.1. Preparation of Extracts from Plant Leaves and Cell Culture Biomass for Antioxidant and Antimicrobial Activity Tests

Antioxidant and antimicrobial activities of cell biomass harvested from a 20 L bioreactor were compared to those of greenhouse *P. fruticosa* plant leaves. Mature leaves were randomly collected from two pot plants (five per plant) growing in a greenhouse of the Main Botanic Garden (Moscow, Russia) and air-dried at 40 °C under the same conditions as cell biomass. Dried cell and leaf biomass samples were ground to a fine powder in a mortar. Samples weighing 2 g were extracted in a 20-fold (1:20) volume of 100% and 80% aqueous methanol in an ultrasonic bath (Sapfir, Russia) for 15 min each. The total extract was filtered, evaporated under vacuum at 40 °C, freeze-dried, and stored at 5 °C. For antimicrobial activity tests, samples were re-dissolved in a 10% dimethyl sulphoxide (DMSO) solution. The resulting extracts (20 mg mL^{-1}) were diluted by distilled water to a final concentration of 4 mg mL^{-1} and used in the tests. For the antioxidant activity assay, dry extracts prepared as described above were re-dissolved in 4% DMSO to obtain a stock solution of 100 mg mL^{-1}. These solutions were further diluted to a range of concentrations (from 7 to 12 mg mL^{-1} for the leaf extract and 5 to 20 mg mL^{-1} for the cell culture extract) and used to determine the EC$_{50}$ values in DPPH and TEAC assays.

4.4.2. Determination of Antioxidant (Antiradical) Activities Using DPPH and TEAC Assays

The antioxidant (antiradical) activity of the suspension cell culture and leaf extracts were determined by 2,2-diphenyl-1-picrylhydrazyl (DPPH) free radical scavenging and Trolox equivalent antioxidant capacity (TEAC) assays. The analyses were performed according to Masci et al. [71], with minor modifications as follows. All chemicals were analytical grade and purchased from Sigma. Absorbance was measured using a spectrophotometer Genesys 20 (Thermo Scientific, USA).

DPPH Assay

An amount of 10 µL of extract solutions diluted as described in 4.4.1. was added to 1 mL of 40 µmol L^{-1} methanolic solution of stable nitrogen-centered free radical DPPH$^{\cdot}$. The absorbance was monitored spectrophotometrically at 517 nm after one hour of incubation at room temperature in the dark. The capacity for scavenging the DPPH-radical was estimated from the difference between the absorbance measured with and without the addition of the extract.

TEAC Assay

A 7 mmol L^{-1} solution of the 2,2′-azinobis(3-ethylbenzothiazoline-6-sulfonic acid) diammonium salt (ABTS) was prepared by dissolving ABTS in water. Then, ABTS was oxidized to a radical cation (ABTS$^{+\cdot}$) by 2.45 mmol L^{-1} potassium persulfate during 16 h incubation at room temperature in the dark. The radical cation reagent (ABTS$^{+\cdot}$) was diluted with ethanol to obtain an absorbance of 0.70 ± 0.02 at 734 nm. To initiate the reaction, 10 µL of extract solutions diluted in 4% DMSO, as described above, were added to 1 mL of the ABTS$^{+\cdot}$ solution. The mixture was incubated for six minutes in the dark and at room temperature, and absorbance was measured at 734 nm. The antioxidant capacity of the samples was expressed in Trolox equivalents.

Estimation of the Effective Concentration (EC$_{50}$)

Antiradical activity curves were plotted referring to sample concentrations on the X axis and their relative radical scavenging capacity on the Y axis. Effective concentration (EC$_{50}$) was defined as a sample concentration that gives a 50% percentage effect. The EC$_{50}$ values were determined using a linear regression algorithm in GraphPad Prism version 8.0.1. The experiments were performed in triplicate, and the actual EC$_{50}$ was estimated by calculating the average value from three different EC$_{50}$ extrapolations with correlation coefficients $R^2 > 0.80$.

4.4.3. Determination of Antibacterial Activities

The antibacterial activity analysis was modified from Orlova et al. [72]. Bacterial strains used were *Escherichia coli* ATCC 25922, *Pseudomonas aeruginosa* ATCC 27853, and *Staphylococcus aureus* ATCC 25923, MRSA ATCC 33591. Strain MRSA ATCC 33591 was provided by Prof. R. Lehrer (University of California Los Angeles, CA, USA); *Escherichia coli* ATCC 25922, *Pseudomonas aeruginosa* ATCC 27853, *Staphylococcus aureus* ATCC 25923 were provided by the Department of Molecular Microbiology, IEM. Bacterial strains were cultured under aerobic conditions according to the approved standard protocols. The minimum inhibitory concentrations (MIC) of the cell and leaf extracts were determined by the broth microdilution method, as recommended by the Clinical Laboratory Standards Institute, USA [73]. The microorganisms were transferred from an agar plate culture to 2.1% (*w/v*) Mueller–Hinton broth (MHB, HiMedia, Mumbai, India) and incubated on an orbital shaker (150 rpm) at 37 °C for 2–6 h. After adjusting the turbidity to 0.5 McFarland

(1.5×10^8 CFU mL^{-1}), suspensions were diluted in sterile 2.1% (w/v) MHB until achieving a final bacterial concentration of 1.0×10^6 CFU mL^{-1}.

The cell suspension and leaf extracts were serially two-fold diluted (starting with the initial concentration 4 mg mL^{-1}) with sterile 2.1% (w/v) MHB. Then, 25 µL aliquots were placed in the wells of a 96-well sterile U-shaped plate (GreinerBio-one, Austria). Afterward, bacterial suspension (25 µL) was added to each well. The microtiter plates were incubated in a thermostat at 37 °C for 18 h. MICs were defined as the lowest extract concentrations that inhibited the visual growth of microorganisms. The experiments were performed in triplicate, and the final results were calculated as the medians based on the data from three independent experiments, each accompanied by the complete set of the controls of bacterial growth and viability, and medium sterility.

4.5. Statistical Analysis

Data on suspension cell growth are presented as the mean values and standard deviations recorded for the triplicates (three flasks or three fixed-size samples of cell suspension collected from bioreactors) for each data point. For viability assessment, a minimum of 250 cell aggregates were examined in each of the three replicates for every experimental condition. Growth parameters and triterpene glycosides' content during bioreactor cultivation were determined as the average of several subcultivation cycles within each multi-cycle bioreactor run. Standard deviations that constitute less than 10% of mean values are not displayed in the graphs. Antioxidant and antimicrobial tests were performed in triplicate.

The data analysis for this paper was generated using SAS software. Copyright© 2023 SAS Institute Inc. SAS and all other SAS Institute Inc. product or service names are registered trademarks or trademarks of SAS Institute Inc., Cary, NC, USA. Statistical significance of differences was estimated using the one-way ANOVA test followed by the Duncan Multiple Range test at $p < 0.05$ for growth parameters and Dunnett's test at $p < 0.05$ for triterpene glycoside content.

5. Conclusions

This is the first study on bioreactor cultivation of *P. fruticosa* suspension cell culture. The results demonstrated that suspension cell culture of this species can be grown in laboratory (20 L) bubble-type bioreactors under a semi-continuous regime for at least 200 days. Cell culture retained satisfactory growth and biosynthetic abilities and accumulated oleanolic acid triterpene glycosides (ladyginoside A and PFS). Cell biomass extracts exhibited antibacterial activity against four cultures of the most common pathogens. In antibacterial tests, the minimum inhibitory concentrations (MICs) of cell biomass extracts against the most common pathogens, *Staphylococcus aureus*, methicillin-resistant strain MRSA, *Pseudomonas aeruginosa*, and *Escherichia coli* varied within 250–2000 µg mL^{-1} which was higher than in leaves of greenhouse plants (MICs = 4000 µg mL^{-1}). Cell biomass extracts also exhibited antioxidant activity, as confirmed by DPPH and TEAC assays. The results of this study will be useful for further optimization of bioreactor cultivation for this species. Our data also suggest that cell biomass of *P. fruticosa* produced in bioreactors has excellent potential to become a sustainable source of the vegetative raw material of this species for different purposes, for example, for functional foods, cosmetics, and natural health products and therefore, help reduce overexploitation of wild plants in their native habitats.

Supplementary Materials: The following supporting information can be downloaded at: https://www.mdpi.com/article/10.3390/plants12203641/s1, Figure S1: Growth curves (dry weight) of the suspension cell culture of *Polyscias fruticosa* in 250 mL flasks plotted in semi-logarithmic coordinates. X_0—initial dry cell biomass concentration (inoculum size); X—dry cell biomass concentration at time of sampling. Figure S2: Scheme of a bubble-type bioreactor used for the cultivation of *Polyscias fruticosa* cell suspension. 1—exhaust air, 2—sterile compressed air input (sparger diameter 6 mm), 3—sterile nutrient medium input, 4—sterile compressed air input for clearing sample collection pipeline, 5—pipeline for collecting cell suspension samples. A, B, C—air sterilization filters; D—container with sterile liquid medium. Bioreactor dimensions: larger diameter—225 mm, height—480 mm, mouth outer/inner diameter—60/46 mm.

Author Contributions: Conceptualization—A.M.N. and M.V.T.; cell suspension cultivation in flasks and bioreactors—M.V.T., E.N.G., I.M.I. and E.S.S.; Phytochemical analysis conceptualization, triterpene glycosides identification—D.V.K.; quantitative analysis of triterpene glycosides—D.V.K. and T.M.T.; antimicrobial and antioxidant activity assays—M.K.L., T.M.T., A.O. and E.V.T.; data analysis—M.V.T., D.V.K., E.V.P. and M.K.L.; writing (manuscript preparation)—M.V.T., E.V.P. and A.O.; writing (revision and comments)—M.K.L., T.M.T., A.O., E.V.T., A.M.N. and D.V.K.; fund acquisition—A.M.N., M.V.T. and E.V.P. All authors have read and agreed to the published version of the manuscript.

Funding: Cultivation of *P. fruticosa* cell culture in flasks was performed using the equipment of the large-scale research facility "All-Russian Collection of cell cultures of higher plants" (ARCCC HP IPPRAS). Bioreactor cultivation of plant cell suspensions was performed using the equipment of the large-scale research facility "Experimental biotechnological facility" (EBF IPPRAS). Maintenance of the original cell strain and cell cultivation in flasks were financially supported within the state assignments of the Ministry of Science and Higher Education of the Russian Federation, theme No. 122042700045-3. Cell strain bioreactor cultivation and biochemical analysis were financially supported within the state assignments of the Ministry of Science and Higher Education of the Russian Federation, theme No. 122042600086-7. Biological activity of the extracts (antioxidant and antimicrobial tests) were performed under financial support of Russian Science Foundation project No. 19-14-00387.

Data Availability Statement: Raw data are available from authors upon request.

Conflicts of Interest: The authors declare no conflict of interest.

References

1. Ashmawy, N.S.; Gad, H.A.; Ashour, M.L.; El-Ahmady, S.H.; Singab, A.N.B. The Genus *Polyscias* (Araliaceae): A Phytochemical and Biological Review. *J. Herb. Med.* **2020**, *23*, 100377. [CrossRef]
2. Thach, B.D.; Le Nguyen Tu Linh, D.T.; Cang, T.T.B.; Giang, T.T.L.; Uyen, N.P.A.; Suong, N.K. Protocol Establishment for Multiplication and Regeneration of *Polyscias fruticosa* L. Harms. an Important Medicinal Plant in Vietnam. *Eur. J. Biotechnol. Genet. Eng.* **2016**, *3*, 31–37.
3. Pandya, D.; Mankad, A.; Pandya, H. Cost Effective Micropropagation of *Polyscias fruicosa* (L.) Harm. *Int. Assoc. Biol. Comput. Dig.* **2022**, *1*, 58–62. [CrossRef]
4. Nguyen, T.-T.-D.; Nguyen, Q.-D.; Nguyen, T.-V.-L. Kinetic Study on Chlorophyll and Antioxidant Activity from *Polyscias fruticosa* (L.) Harms Leaves via Microwave-Assisted Extraction. *Molecules* **2021**, *26*, 3761. [CrossRef] [PubMed]
5. Lutomski, J.; Luân, T.C.; Hoa, T.T. Polyacetylenes in the Araliaceae Family. Part IV. *Herba Pol.* **1992**, *38*, 137–140.
6. Tram, N.T.T.; Tuyet, H.D.; Minh, Q.N. One Unusual Sterol from *Polyscias fruticosa* (L.) Harms (Araliaceae). *Can Tho Univ. J. Sci.* **2017**, *7*, 33–36.
7. Huan, V.D.; Yamamura, S.; Ohtani, K.; Kasai, R.; Yamasaki, K.; Nham, N.T.; Chau, H.M. Oleanane Saponins from *Polyscias fruticosa*. *Phytochemistry* **1998**, *47*, 451–457. [CrossRef]
8. Tran, V.T.; Tran, T.H.H.; Nguyen, T.D. Validated High Performance Liquid Chromatography Method for Quantification of a Major Saponin in *Polyscias fruticosa*. *J. Multidiscip. Eng. Sci. Technol.* **2016**, *3*, 4880–4882.
9. Do, V.M.; Tran, C.L.; Nguyen, T.P. Polysciosides J and K, Two New Oleanane-Type Triterpenoid Saponins from the Leaves of *Polyscias fruticosa* (L.) Harms. Cultivating in an Giang Province, Viet Nam. *Nat. Prod. Res.* **2020**, *34*, 1250–1255. [CrossRef]
10. Koffuor, G.A.; Boye, A.; Ofori-Amoah, J.; Kyei, S.; Abokyi, S.; Nyarko, R.A.; Naalukyem, B.R. Anti-Inflammatory and Safety Assessment of *Polyscias fruticosa* (L.) Harms (Araliaceae) Leaf Extract in Ovalbumin-Induced Asthma. *J. Phytopharm.* **2014**, *3*, 337–342. [CrossRef]
11. Divakar, M.C.; Bensita, M.B. Screening of Various Leaf Extracts of *Polyscias fruticosa* Harms for Their Antidiabetic Activity. *Indian J. Nat. Prod.* **1998**, *14*, 24–28.
12. Varadharajan, R.; Rajalingam, D. Diuretic Activity of *Polyscias fruticosa* (L.) Harms. *Int. J. Innov. Drug Discov.* **2011**, *1*, 15–18.
13. Mawea, F.; Indrayati, A.; Rahmawati, I. Review: Antibacterial Activity of *Polyscias fruticosa* and Molecular Mechanism of Active Compounds. *Int. J. Health Med. Curr. Res.* **2021**, *6*, 2062–2087. [CrossRef]
14. Bensita, M.B.; Nilani, P.; Sandhya, S.M. Studies on the Adaptogenic and Antibacterial Properties of *Polyscias fructicosa* (L) Harms. *Anc. Sci. Life* **1999**, *18*, 231–246. [PubMed]
15. Luyen, N.T.; Dang, N.H.; Binh, P.T.X.; Hai, N.T.; Dat, N.T. Hypoglycemic Property of Triterpenoid Saponin PFS Isolated from *Polyscias fruticosa* Leaves. *An. Acad. Bras. Ciências* **2018**, *90*, 2881–2886. [CrossRef]
16. Tran, T.H.H.; Nguyen, H.D.; Nguyen, T.D. α-Amylase and α-Glucosidase Inhibitory Saponins from *Polyscias fruticosa* Leaves. *J. Chem.* **2016**, *2016*, 2082946. [CrossRef]

17. Le, Q.-U. A Science Opinion on *Polyscias fruticosa* and *Morus alba* L. Combination: Better Anti-Diabetic and Late Complication Inhibitory Properties? *Curr. Res. Diabetes Obes. J.* **2019**, *10*, 555798. [CrossRef]
18. Asumeng Koffuor, G.; Boye, A.; Kyei, S.; Ofori-Amoah, J.; Akomanin Asiamah, E.; Barku, A.; Acheampong, J.; Amegashie, E.; Kumi Awuku, A. Anti-Asthmatic Property and Possible Mode of Activity of an Ethanol Leaf Extract of *Polyscias fruticosa*. *Pharm. Biol.* **2016**, *54*, 1354–1363. [CrossRef]
19. Yen, T.T. Improvement of Learning Ability in Mice and Rats with the Root Extract of Dinh Lang (*Policias fruticosum* L.). *Acta Physiol. Hung.* **1990**, *75*, 69–76.
20. Yen, T.T. Stimulation of Sexual Performance in Male Rats with the Root Extract of Dinh Lang (*Policias fruticosum* L.). *Acta Physiol. Hung.* **1990**, *75*, 61–67.
21. Boye, A.; OseiOwusu, A.; Koffuor, G.; Barku, V.; Asiamah, E.; Asante, E. Assessment of *Polyscias fruticosa* (L.) Harm (Araliaceae) Leaf Extract on Male Fertility in Male Wistar Rats. *J. Intercult. Ethnopharmacol.* **2018**, *7*, 45–56. [CrossRef]
22. Yen, T.T.; Knoll, J. Extension of Lifespan in Mice Treated with Dinh Lang (*Policias fruticosum* L.) and (−)Deprenyl. *Acta Physiol. Hung.* **1992**, *79*, 119–124. [PubMed]
23. Selvaraj, B.; Le, T.T.; Kim, D.W.; Jung, B.H.; Yoo, K.-Y.; Ahn, H.R.; Thuong, P.T.; Tran, T.T.T.; Pae, A.N.; Jung, S.H.; et al. Neuroprotective Effects of Ethanol Extract of *Polyscias fruticosa* (EEPF) against Glutamate-Mediated Neuronal Toxicity in HT22 Cells. *Int. J. Mol. Sci.* **2023**, *24*, 3969. [CrossRef]
24. Ly, H.T.; Nguyen, T.T.H.; Le, V.M.; Lam, B.T.; Mai, T.T.T.; Dang, T.P.T. Therapeutic Potential of *Polyscias fruticosa* (L.) Harms Leaf Extract for Parkinson's Disease Treatment by Drosophila Melanogaster Model. *Oxidative Med. Cell. Longev.* **2022**, *2022*, 5262677. [CrossRef] [PubMed]
25. Nhu-Trang, T.-T.; Phan, T.-K.-T.; Lam, T.-M.-T.; Nguyen, P.-B.-D.; Nguyen, T.-V.-L. Drying Kinetics and Energy Consumption in Hot-Air and Microwave Drying of *Polyscias fruticosa* (L.) Harms Leaves. *AIP Conf. Proc.* **2022**, *2610*, 070004.
26. MinhThu, N.; Son, H.L. Effect of Storage Temperature and Preservatives on the Stability and Quality of *Polyscias fruticosa* (L.) Harms Herbal Health Drinks. *J. Pharm. Res. Int.* **2019**, *26*, 1–7. [CrossRef]
27. Tran, C.H.; Nguyen, H.A.; Nguyen, T.N.T.; Ha, T.T.N.; Le, T.H.A. Effects of Storage Conditions on Polyphenol and Triterpenoid Saponin Content and the Antioxidant Capacity of Ethanolic Extract from Leaves of *Polyscias fruticosa* (L.) Harms. *J. Sci. Technol. Food* **2020**, *3*, 47–53.
28. Nguyen, M.P. Impact of Roasting to Total Phenolic, Flavonoid and Antioxidant Activities in Root, Bark and Leaf of *Polyscias fruticosa*. *J. Pharm. Res. Int.* **2020**, *32*, 13–17. [CrossRef]
29. Sakr, S.S.; Melad, S.S.; El-Shamy, M.A.; Elhafez, A.E. In Vitro Propagation of *Polyscias fruticosa* Plant. *Int. J. Plant Soil Sci.* **2014**, *3*, 1254–1265. [CrossRef]
30. Vinh, D.T.; Phuong Hoa, M.T.; Nhu Thao, L.T.; Trang Nha, N.H.; Minh, T. Van Saponin Production by Cell Culture Techniques of *Polyscias fruticosa* L. Harms. *Acad. J. Biol.* **2015**, *37*, 135–141. [CrossRef]
31. Kim, P.T.A.; An, L.T.B.; Chung, N.T.; Truong, N.T.; Thu, L.T.A.; Huan, L.V.T.; Loc, N.H. Growth and Oleanolic Acid Accumulation of *Polyscias fruticosa* Cell Suspension Cultures. *Curr. Pharm. Biotechnol.* **2021**, *22*, 1266–1272. [CrossRef] [PubMed]
32. Phuong, T.T.B.; Trung, V.P.; An, N.H.; Tuan, N.D.; Nguyen, P.T.T. The Effects of 2,4-Dichlorophenoxyacetic Acid and α-Naphthaleneacetic Acid on Biomass Increment, Rhizogenesis and Somatic Embryogenesis of Suspension-Cultured Dinh Lang Cells [*Polyscias fruticosa* (L.) Harms]. *Indian J. Agric. Res.* **2021**, *56*, 70–75. [CrossRef]
33. Hau, N.T.; Nhu Thao, L.T.; Minh, T. Van Cultivation of Leaf-Tissue of *Polyscias fruticosa* (L.) Harms for Quantity of Saponin Accumulation. *Acad. J. Biol.* **2015**, *37*, 184–189. [CrossRef]
34. Lộc, P.V. A Study on Adventitious Root Formation of Ming Aralia (*Polyscias fruticosa* L. Harms) by in Vitro Culture. *Univ. Danang J. Sci. Technol.* **2014**, *3*, 106–108.
35. Titova, M.V.; Popova, E.V.; Konstantinova, S.V.; Kochkin, D.V.; Ivanov, I.M.; Klyushin, A.G.; Titova, E.G.; Nebera, E.A.; Vasilevskaya, E.R.; Tolmacheva, G.S.; et al. Suspension Cell Culture of *Dioscorea deltoidea*—A Renewable Source of Biomass and Furostanol Glycosides for Food and Pharmaceutical Industry. *Agronomy* **2021**, *11*, 394. [CrossRef]
36. Georgiev, M.I.; Weber, J. Bioreactors for Plant Cells: Hardware Configuration and Internal Environment Optimization as Tools for Wider Commercialization. *Biotechnol. Lett.* **2014**, *36*, 1359–1367. [CrossRef] [PubMed]
37. Valdiani, A.; Hansen, O.K.; Nielsen, U.B.; Johannsen, V.K.; Shariat, M.; Georgiev, M.I.; Omidvar, V.; Ebrahimi, M.; Tavakoli Dinanai, E.; Abiri, R. Bioreactor-Based Advances in Plant Tissue and Cell Culture: Challenges and Prospects. *Crit. Rev. Biotechnol.* **2019**, *39*, 20–34. [CrossRef]
38. Thanh, N.T.; Murthy, H.N.; Paek, K.Y. Optimization of Ginseng Cell Culture in Airlift Bioreactors and Developing the Large-Scale Production System. *Ind. Crops Prod.* **2014**, *60*, 343–348. [CrossRef]
39. Kubica, P.; Szopa, A.; Kokotkiewicz, A.; Miceli, N.; Taviano, M.F.; Maugeri, A.; Cirmi, S.; Synowiec, A.; Gniewosz, M.; Elansary, H.O.; et al. Production of Verbascoside, Isoverbascoside and Phenolic Acids in Callus, Suspension, and Bioreactor Cultures of *Verbena officinalis* and Biological Properties of Biomass Extracts. *Molecules* **2020**, *25*, 5609. [CrossRef]
40. Kochkin, D.V.; Sukhanova, E.S.; Nosov, A.M. The Accumulation of Triterpene Glycosides in the Growh Cycle of Cell Suspension Cultures of Polyscias Fruticosa. *Vestn. Volga State Univ. Technol. Ser For. Ecol. Nat. Manag.* **2014**, *4*, 67–73.

41. Kochkin, D.V.; Sukhanova, E.S.; Nosov, A.M. Triterpene Glycosides in Suspension-Cell Culture of *Polyscias fruticosa* (L.) Harms. In *Proceedings of the International Conference with Elements of Scientific School for Young Scientists "Perspectives of Phytobiotechnology to Improving the Quality of Life in the North"*, October 10–16, 2010; North-Eastern Federal University in Yakutsk: Yakutsk, Russia, 2020; pp. 201–204.
42. Kochkin, D.V.; Sukhanova, E.S.; Sergeev, R.V.; Nosov, A.M. Triterpene Glycosides of *Polyscias* spp. Cell Cultures. *Vestn. Volga State Univ. Technol. Ser For. Ecol. Nat. Manag.* **2014**, *1*, 69–76.
43. Sukhanova, E.S.; Chernyak, N.D.; Nosov, A.M. Obtaining and Description of *Polyscias filicifolia* and *Polyscias fruticosa* Calli and Suspension Cell Cultures. *Biotekhnologiia* **2010**, *4*, 44–50.
44. Sukhanova, E.S.; Kochkin, D.V.; Titova, M.V.; Nosov, A.M. Growth and Biosynthetic Characteristics of Different Polyscias Plant Cell Culture Strains. *Vestn. Volga State Univ. Technol. Ser For. Ecol. Nat. Manag.* **2012**, *2*, 57–66.
45. Yang, W.; Ye, M.; Qiao, X.; Liu, C.; Miao, W.; Bo, T.; Tao, H.; Guo, D. A Strategy for Efficient Discovery of New Natural Compounds by Integrating Orthogonal Column Chromatography and Liquid Chromatography/Mass Spectrometry Analysis: Its Application in Panax Ginseng, Panax Quinquefolium and Panax Notoginseng to Characterize 4. *Anal. Chim. Acta* **2012**, *739*, 56–66. [CrossRef] [PubMed]
46. Sun, T.-T.; Liang, X.-L.; Zhu, H.-Y.; Peng, X.-L.; Guo, X.-J.; Zhao, L.-S. Rapid Separation and Identification of 31 Major Saponins in Shizhu Ginseng by Ultra-High Performance Liquid Chromatography–Electron Spray Ionization–MS/MS. *J. Ginseng Res.* **2016**, *40*, 220–228. [CrossRef]
47. Huang, T.-K.; McDonald, K.A. Bioreactor Engineering for Recombinant Protein Production in Plant Cell Suspension Cultures. *Biochem. Eng. J.* **2009**, *45*, 168–184. [CrossRef]
48. Motolinía-Alcántara, E.A.; Castillo-Araiza, C.O.; Rodríguez-Monroy, M.; Román-Guerrero, A.; Cruz-Sosa, F. Engineering Considerations to Produce Bioactive Compounds from Plant Cell Suspension Culture in Bioreactors. *Plants* **2021**, *10*, 2762. [CrossRef]
49. Su, R.; Sujarani, M.; Shalini, P.; Prabhu, N. A Review on Bioreactor Technology Assisted Plant Suspension Culture. *Asian J. Biotechnol. Bioresour. Technol.* **2019**, *5*, 1–13. [CrossRef]
50. Esperança, M.N.; Mendes, C.E.; Rodriguez, G.Y.; Cerri, M.O.; Béttega, R.; Badino, A.C. Sparger Design as Key Parameter to Define Shear Conditions in Pneumatic Bioreactors. *Biochem. Eng. J.* **2020**, *157*, 107529. [CrossRef]
51. Seidel, S.; Maschke, R.W.; Werner, S.; Jossen, V.; Eibl, D. Oxygen Mass Transfer in Biopharmaceutical Processes: Numerical and Experimental Approaches. *Chem. Ing. Tech.* **2021**, *93*, 42–61. [CrossRef]
52. Garcia-Ochoa, F.; Gomez, E. Bioreactor Scale-up and Oxygen Transfer Rate in Microbial Processes: An Overview. *Biotechnol. Adv.* **2009**, *27*, 153–176. [CrossRef]
53. Garcia-Ochoa, F.; Gomez, E.; Santos, V.E. Fluid Dynamic Conditions and Oxygen Availability Effects on Microbial Cultures in STBR: An Overview. *Biochem. Eng. J.* **2020**, *164*, 107803. [CrossRef]
54. Kieran, P.; MacLoughlin, P.; Malone, D. Plant Cell Suspension Cultures: Some Engineering Considerations. *J. Biotechnol.* **1997**, *59*, 39–52. [CrossRef]
55. Lee, C.W.; Shuler, M.L. The Effect of Inoculum Density and Conditioned Medium on the Production of Ajmalicine and Catharanthine from Immobilized *Catharanthus roseus* Cells. *Biotechnol. Bioeng.* **2000**, *67*, 61–71. [CrossRef]
56. Sakurai, M.; Mori, T.; Seki, M.; Furusaki, S. Changes of Anthocyanin Composition by Conditioned Medium and Cell Inoculum Size Using Strawberry Suspension Culture. *Biotechnol. Lett.* **1996**, *18*, 1149–1154. [CrossRef]
57. Thanh, N.T.; Murthy, H.N.; Yu, K.W.; Jeong, C.S.; Hahn, E.J.; Paek, K.Y. Effect of Inoculum Size on Biomass Accumulation and Ginsenoside Production by Large-Scale Cell Suspension Cultures of Panax Ginseng. *J. Plant Biotechnol.* **2004**, *6*, 265–268.
58. Tan, S.H.; Maziah, M.; Ariff, A. Synergism Effect between Inoculum Size and Aggregate Size on Flavonoid Production in *Centella asiatica* (L.) Urban (Pegaga) Cell Suspension Cultures. *Int. J. Res. Eng. Technol.* **2013**, *2*, 244–253.
59. Georgiev, M.I.; Eibl, R.; Zhong, J.-J. Hosting the Plant Cells in Vitro: Recent Trends in Bioreactors. *Appl. Microbiol. Biotechnol.* **2013**, *97*, 3787–3800. [CrossRef]
60. Titova, M.V.; Popova, E.V.; Shumilo, N.A.; Kulichenko, I.E.; Chernyak, N.D.; Ivanov, I.M.; Klushin, A.G.; Nosov, A.M. Stability of Cryopreserved *Polyscias filicifolia* Suspension Cell Culture during Cultivation in Laboratory and Industrial Bioreactors. *Plant Cell Tissue Organ Cult.* **2021**, *145*, 591–600. [CrossRef]
61. Demidova, E.; Globa, E.; Klushin, A.; Kochkin, D.; Nosov, A. Effect of Methyl Jasmonate on the Growth and Biosynthesis of C13- and C14-Hydroxylated Taxoids in the Cell Culture of Yew (*Taxus wallichiana* Zucc.) of Different Ages. *Biomolecules* **2023**, *13*, 969. [CrossRef]
62. Titova, M.V.; Reshetnyak, O.V.; Osipova, E.A.; Osip'yants, A.I.; Shumilo, N.A.; Oreshnikov, A.V.; Nosov, A.M. Submerged Cultivation of *Stephania glabra* (Roxb.) Miers Cells in Different Systems: Specific Features of Growth and Accumulation of Alkaloid Stepharine. *Appl. Biochem. Microbiol.* **2012**, *48*, 645–649. [CrossRef]
63. Nguyen, N.Q.; Nguyen, M.T.; Nguyen, V.T.; Le, V.M.; Trieu, L.H.; Le, X.T.; Khang, T.V.; Giang, N.T.; Thach, N.Q.; Hung, T.T. The Effects of Different Extraction Conditions on the Polyphenol, Flavonoids Components and Antioxidant Activity of *Polyscias fruticosa* Roots. *IOP Conf. Ser. Mater. Sci. Eng.* **2020**, *736*, 022067. [CrossRef]
64. Nguyen, T.T.H. Studies on Antioxidant Activity of *Polyscias fruticosa* Harms. (Araliaceae). *J. Med. Mater. Hanoi.* **2003**, 142–146.
65. Nguyen, M.P. Changes of Phytochemical, Antioxidant Characteristics of *Polyscias fruticosa* Rhizomes during Convective and Freeze Drying. *Biosci. Res.* **2020**, *17*, 323–326.

66. Murashige, T.; Skoog, F. A Revised Medium for Rapid Growth and Bio Assays with Tobacco Tissue Cultures. *Physiol. Plant.* **1962**, *15*, 473–497. [CrossRef]
67. Titova, M.V.; Shumilo, N.A.; Kulichenko, I.E.; Ivanov, I.M.; Sukhanova, E.S.; Nosov, A.M. Features of Respiration and Formation of Steroidal Glycosides in *Dioscorea deltoidea* Cell Suspension Culture Grown in Flasks and Bioreactors. *Russ. J. Plant Physiol.* **2015**, *62*, 557–563. [CrossRef]
68. Dixon, R.A.; Gonzales, R.A. (Eds.) *Plant Cell Culture: A Practical Approach*, 2nd ed.; Practical Approach Series, 145; Oxford University Press: New York, NY, USA, 1995.
69. Pirt, S.J. *Principles of Microbe and Cell Cultivation*; Blackwell Scientific: Oxford, UK, 1975; ISBN 0632081503.
70. Kochkin, D.V.; Galishev, B.A.; Glagoleva, E.S.; Titova, M.V.; Nosov, A.M. Rare Triterpene Glycoside of Ginseng (Ginsenoside Malonyl-Rg1) Detected in Plant Cell Suspension Culture of *Panax japonicus* Var. Repens. *Russ. J. Plant Physiol.* **2017**, *64*, 649–656. [CrossRef]
71. Masci, A.; Mattioli, R.; Costantino, P.; Baima, S.; Morelli, G.; Punzi, P.; Giordano, C.; Pinto, A.; Donini, L.M.; D'Erme, M.; et al. Neuroprotective Effect of Brassica Oleracea Sprouts Crude Juice in a Cellular Model of Alzheimer's Disease. *Oxid. Med. Cell. Longev.* **2015**, *2015*, 781938. [CrossRef]
72. Orlova, A.; Kysil, E.; Tsvetkova, E.; Meshalkina, D.; Whaley, A.; Whaley, A.O.; Laub, A.; Francioso, A.; Babich, O.; Wessjohann, L.A.; et al. Phytochemical Characterization of Water Avens (*Geum rivale* L.) Extracts: Structure Assignment and Biological Activity of the Major Phenolic Constituents. *Plants* **2022**, *11*, 2859. [CrossRef]
73. *CLSI Standard M07*; Methods for Dilution Antimicrobial Susceptibility Tests for Bacteria That Grow Aerobically. 11th ed. Clinical and Laboratory Standards Institute: Wayne, PA, USA, 2018.

Disclaimer/Publisher's Note: The statements, opinions and data contained in all publications are solely those of the individual author(s) and contributor(s) and not of MDPI and/or the editor(s). MDPI and/or the editor(s) disclaim responsibility for any injury to people or property resulting from any ideas, methods, instructions or products referred to in the content.

Article

Unveiling the Dual Nature of Heavy Metals: Stressors and Promoters of Phenolic Compound Biosynthesis in *Basilicum polystachyon* (L.) Moench In Vitro

Sumanta Das [1,*], Kaniz Wahida Sultana [1], Moupriya Mondal [1], Indrani Chandra [1,*] and Ashwell R. Ndhlala [2]

1. Department of Biotechnology, The University of Burdwan, Burdwan 713104, West Bengal, India; kwsultana@gmail.com (K.W.S.); moupriyamondal652@gmail.com (M.M.)
2. Department of Plant Production, Soil Science and Agricultural Engineering, Green Biotechnologies Research Centre of Excellence, University of Limpopo, Private Bag X1106, Sovenga 0727, South Africa
* Correspondence: sumantad1991@gmail.com (S.D.); ichandrabiotech@gmail.com (I.C.)

Abstract: The global industrial revolution has led to a substantial rise in heavy metal levels in the environment, posing a serious threat to nature. Plants synthesize phenolic compounds under stressful conditions, which serve as protective agents against oxidative stress. *Basilicum polystachyon* (L.) Moench is an herbaceous plant of the Lamiaceae family. Some species within this family are recognized for their capacity to remediate sites contaminated with heavy metals. In this study, the effects of mercury (II) chloride and lead (II) nitrate on the in vitro propagation of *B. polystachyon* were investigated. Shoot tips from in vitro plantlets were cultured in Murashige and Skoog's (MS) media with heavy metals ranging from 1 to 200 µM to induce abiotic stress and enhance the accumulation of phenolic compounds. After three weeks, MS medium with 1 µM of lead (II) supported the highest shoot multiplication, and the maximum number of roots per explant was found in 100 µM of lead (II), whereas a higher concentration of heavy metals inhibited shoot multiplication and root development. The plantlets were hardened in a greenhouse with a 96% field survival rate. Flame atomic absorption spectroscopy (FAAS) was used to detect heavy metal contents in plant biomass. At both 200 µM and 50 µM concentrations, the greatest accumulation of mercury (II) was observed in the roots (16.94 ± 0.44 µg/g) and shoots (17.71 ± 0.66 µg/g), respectively. Similarly, lead (II) showed the highest accumulation in roots (17.10 ± 0.54 µg/g) and shoots (7.78 ± 0.26 µg/g) at 200 µM and 50 µM exposures, respectively. Reverse-phase high-performance liquid chromatography (RP-HPLC) identified and quantified various phenolic compounds in *B. polystachyon* leaves, including gallic acid, caffeic acid, vanillic acid, *p*-coumaric acid, ellagic acid, rosmarinic acid, and *trans*-cinnamic acid. These compounds were found in different forms, such as free, esterified, and glycosylated. Mercury (II)-exposed plants exhibited elevated levels of vanillic acid (1959.1 ± 3.66 µg/g DW), ellagic acid (213.55 ± 2.11 µg/g DW), and rosmarinic acid (187.72 ± 1.22 µg/g DW). Conversely, lead (II)-exposed plants accumulated higher levels of caffeic acid (42.53±0.61 µg/g DW) and *p*-coumaric acid (8.04 ± 0.31 µg/g DW). *Trans*-cinnamic acid was the predominant phenolic compound in control plants, with a concentration of 207.74 ± 1.45 µg/g DW. These results suggest that sublethal doses of heavy metals can act as abiotic elicitors, enhancing the production of phenolic compounds in *B. polystachyon*. The present work has the potential to open up new commercial opportunities in the pharmaceutical industry.

Keywords: abiotic stress; lamiaceae family; lead; medicinal plant; mercury; phenolic acids; propagation

Citation: Das, S.; Sultana, K.W.; Mondal, M.; Chandra, I.; Ndhlala, A.R. Unveiling the Dual Nature of Heavy Metals: Stressors and Promoters of Phenolic Compound Biosynthesis in *Basilicum polystachyon* (L.) Moench In Vitro. *Plants* **2024**, *13*, 98. https://doi.org/10.3390/plants13010098

Academic Editors: Kee-Yoeup Paek and Hosakatte Niranjana Murthy

Received: 30 November 2023
Revised: 25 December 2023
Accepted: 26 December 2023
Published: 28 December 2023

Copyright: © 2023 by the authors. Licensee MDPI, Basel, Switzerland. This article is an open access article distributed under the terms and conditions of the Creative Commons Attribution (CC BY) license (https://creativecommons.org/licenses/by/4.0/).

1. Introduction

Heavy metal contamination is a serious environmental threat that substantially limits crop productivity [1]. Essential metals like cobalt (Co^{2+}), copper (Cu^{2+}), iron (Fe^{3+}), manganese (Mn^{2+}), nickel (Ni^{2+}) and zinc (Zn^{2+}) are crucial for plant growth and development [2]. In contrast, non-essential heavy metals such as cadmium (Cd^{2+}), mercury (Hg^{2+}), lead (Pb^{2+}), chromium (Cr^{2+}), arsenic (As^{2+}), etc. are extremely noxious for plants and animals [3]. High concentrations of these metals can be harmful to plants, as they interfere with plant metabolism and development [4]. The excessive accumulation of heavy metals in plant tissues can interfere with photosynthesis, respiration and nutrient uptake [5]. Both lead and mercury are the most toxic elements that can disrupt plant growth, as recognized by the United States Environmental Protection Agency (EPA) and the US Agency for Toxic Substances and Disease Registry (ATSDR) [6,7]. Mercury exists in various forms, including elemental, inorganic, and organic compounds, all of which are toxic. Lead, on the other hand, forms particularly harmful organometallic compounds when combined with carbon [8]. They can cause oxidative stress in cells, leading to increased production of reactive oxygen species (ROS), lipid peroxidation, and signaling compounds that can interrupt the defense system of plants [5,9,10].

Muszynska, et al. [11] and Demarco, et al. [12] found that some plants can survive in stressful conditions, even when high concentrations of heavy metals can cause alterations in photosynthesis by reducing the chlorophyll content in leaves. Elevated levels of heavy metals in the growth medium can impact mineral uptake, leading to imbalances between essential and trace elements, as demonstrated by Gatti [13] and Okem, et al. [14]. Interestingly, some studies have reported that low concentrations of heavy metals can promote plant growth [15–19]. However, prolonged exposure to heavy metals can generate free radicals like superoxide anion (O^{2-}), hydroxyl radicals (.OH), and non-free radicals such as hydrogen peroxide (H_2O_2), organic peroxide (ROOH), and singlet oxygen ($^1O_2$) [20]. Plants respond to heavy metal stress by upregulating genes that encode proteins involved in the production of secondary metabolites [21]. The production of secondary metabolites in plants can be either stimulated or inhibited by exposure to heavy metal contamination [22]. Heavy metal exposure reduces the production of secondary metabolites in plants as per previous studies [23–25]. There is some evidence that plants can produce phenolic compounds in the presence of metals by increasing their metabolic activity [26–31].

The Industrial Revolution transformed global perspectives on addressing environmental waste management, as traditional remediation technologies were often costly and could have adverse effects on the environment. In contrast, plant-based remediation is a more sustainable and efficient approach that requires low energy and minimal expenses and can effectively remove metal pollutants from the environment or transform them into non-toxic forms. Non-edible aromatic plants are often a more appropriate choice for remediation efforts due to their inherent aromatic properties, minimizing the risk of contaminating the food chain with harmful substances [32]. Importantly, these plants are neither consumed by humans nor animals, effectively halting the transmission of heavy metals from soil to the food chain, and subsequently, to the human body [33]. Plant tissue culture is a powerful tool for selecting metal-tolerant plants, which can be used for phytoremediation [34–39]. A diverse range of plants, including shrubs, ornamental perennials, and annuals, can take up and degrade pollutants. Among the species of the Lamiaceae family, *Ocimum basilicum* L. holds the potential for phytoremediation [40,41]. Many other plant species can also clean up heavy metals from soil. These include *Bidens pilosa* L., *Tagetes minuta* L. [42], *Salix alba* L. [43], *Helianthus annuus* L. [44,45] for lead and *Brassica juncea* L. [41], *Caladium bicolor* L., *Cyperus kyllingia* L., *Digitaria radicosa* (Presl) Miq, *Lindernia crustacea* L., *Paspalum conjugatum* L. and *Zingiber purpureum* (Roxb.) [46] for mercury.

Furthermore, other plant species recognized for their ability to accumulate heavy metals comprise *Basilicum polystachyon* (L.) Moench, the plant selected for this study, which is a fast-growing Lamiaceae species. This aromatic herb is found in Asia, Africa, and India [47]. The present study investigated the potential for in vitro propagation of *B. polystachyon* under heavy metal stress, as well as the accumulation of heavy metals in plant biomass and the effect of heavy metal stress on the enhancement of phenolic compounds.

2. Materials and Methods

2.1. Chemicals and Solvents

Murashige and Skoog's (MS) basal medium, agar-agar, and diluent solution for DNA were obtained from Hi-media, India. Bavestien® (Carbendazimpowder) was procured from ASF India Limited, New Delhi, India. Tween-20 (Polysorbate 20), mercury (II) chloride, lead (II) nitrate, methanol, n-hexane, acetonitrile, acetone, ethyl acetate, diethyl ether, nitric acid (HNO_3), perchloric acid ($HClO_4$), sucrose, sodium hydroxide (NaOH), ethylenediaminetetraacetic acid (EDTA) and phytagel were obtained from Merck, Merck-Sigma Aldrich, St. Louis, MO, USA. All standard phenolic compounds viz., gallic acid, caffeic acid, *p*-coumaric acid, ellagic acid, rosmarinic acid, *trans*-cinnamic acid and vanillic acid were procured from Sigma Aldrich, Merck-Sigma Aldrich, St. Louis, MO, USA. All the solvents used in these experiments were HPLC grade.

2.2. Source of Plant Material and Sterilization Grade

The *B. polystachyon* sample was obtained from 2-month-old plants grown ex vitro in the field at the Department of Biotechnology, The University of Burdwan, Burdwan 713104, West Bengal, India (23°15′25.2″ N, 87°51′01.7″ E). The plant specimen was identified and verified by K. Karthigeyan, Scientist—'F', at the Botanical Survey of India in Kolkata, and a voucher specimen (BU/SD-01) was deposited at the Department of Biotechnology at The University of Burdwan. The use of this plant in the present study complies with institutional, national, and international guidelines and legislation. All experiments were performed in accordance with relevant guidelines and regulations. Shoot tips were excised from the plant and used as the explant source. Explants were washed in Milli-Q water (Millipore system, Merck, Rahway, NJ, USA) for 5 min, then disinfected in 70% ethanol v/v for 10 s followed by washing with 0.01% v/v Tween-20 for 4 min [48,49]. Surface sterilization was performed using 0.1% w/v mercuric chloride for 45 s in a Biosafety Cabinet A2 (Biobase Inc., Jinan, China) under aseptic conditions. The explant was thoroughly rinsed three times with Milli-Q water after sterilization [34,50].

2.3. Media Preparation and Culture Condition

A shoot tip measuring 6–7 mm in length was placed in a culture vessel (25 × 150 mm) containing 20 mL of Murashige and Skoog's (MS) basal media, as developed by Murashige and Skoog [51]. The media was enriched with various concentrations (0, 1, 25, 50, 100, and 200 µM) of heavy metals, specifically mercury (II) chloride and lead (II) nitrate. Additionally, the media contained 30 g/L sucrose and 0.15% w/v phytagel as supporting substances. A control culture was also established using MS medium without the inclusion of heavy metals. To ensure the best environment for plant regeneration, the plant growth chamber was maintained at a constant temperature of 25 ± 2 °C, 55% humidity, and subjected to a 16 h photoperiod with 2000 lux of light intensity [34].

2.4. Shoot Multiplication, Rooting and Acclimatization

The multiplication of regenerated in vitro shoots was achieved by transferring mother explants and subculturing in vitro raised plantlets on fresh culture media at a regular interval of three weeks. Each experiment was repeated three times and the data were recorded after three weeks using 20 replicates. Experimental data were recorded considering parameters such as the number of shoots per explant, shoot length, number of roots per shoot, and root length. In vitro plantlets (three weeks old) were removed from culture tubes, thoroughly washed with Milli-Q water to remove agar, and then transferred to plastic pots (100 × 80 mm) containing a sterilized mixture of sand and soil (1:1 w/w). A 0.1% w/v solution of Bavestien® was applied to the surface of the plastic pots to inhibit fungal growth. The pots were covered with transparent poly bags (300 × 220 mm) and placed in a plant growth chamber (Thermo Fisher Scientific, Waltham, MA, USA) at a temperature of 25 ± 2 °C with a 16 h photoperiod. After one week, the plastic covers were removed, and the plantlets were maintained under the same conditions for two weeks. The partially acclimatized plantlets were then moved to a net house environment. After two weeks, acclimatized plants were transferred to their natural habitat [34].

2.5. In Vitro Selection and Analysis of Heavy Metal Contents

To evaluate heavy metal accumulation in *B. polystachyon* under different concentrations, in vitro regenerated plantlets (three weeks old) were collected, washed, and subsequently dried at room temperature. The plantlets were then divided into two parts i.e., shoot and root. A 0.1 g of dried tissue sample was transferred to a Teflon-lined vessel and digested with HNO_3-$HClO_4$ (3:1 v/v) in a microwave (LG, Delhi, India) digester. After dilution with distilled water up to 25 mL, the samples were filtered and analyzed by flame atomic absorption spectroscopy (PerkinElmer Inc., Waltham, MA, USA). The linearity and range of different concentrations of reference heavy metals were evaluated precisely. The determination of metals used a specific hollow cathode lamp (HCL) and air-acetylene flame, with a slit width set to 0.7 nm and detection wavelengths of 283.3 nm and 253.7 nm for lead (II) and mercury (II), respectively [52]. The heavy metal contents in plant parts were expressed as µg/g DW.

Determination of Tolerance Index (*TI*) and Translocation Factor (*TF*)

The *TI* can be calculated according to the following equation [53,54]:

$$TI\ (\%) = \frac{Dry\ weight\ of\ treated\ plant}{Dry\ weight\ of\ control\ plant} \times 100$$

The *TF* was determined according to the formula followed by Yoon, et al. [55]:

$$TF = \frac{Heavy\ metal\ content\ in\ shoot}{Heavy\ metal\ content\ in\ root}$$

2.6. Extraction, Identification, Quantification and Assessment of Phenolic Compounds

Heavy-metal-induced and control plant cultures (50 µM) were collected. Leaves were cut, air-dried for 72 h, and ground into fine powder. Then, 0.5 g of each sample powder was dissolved in 10 mL of an extraction buffer (methanol: water: acetone in a 5:3:2 ratio) and incubated for 48 h in a shaker incubator (Spac-N-Service, Kolkata, India) at room temperature. The samples were subjected to ultrasonic extraction (PIEZO-U-SONIC Ultrasonic Processor, Kolkata, India) at two distinct time intervals, i.e., 5 min and 10 min. The frequency of the ultrasound was 40 kHz. The extracted solutions were centrifuged (CPR-30 Plus, Remi Lab World, Mumbai, India) at $5600 \times g$ for 5 min, and the supernatant was collected and filtered through a PTFE membrane filter (Hi-media, Thane, India). The resulting supernatant was used to separate the different forms of phenolic compounds, i.e., free, esterified and glycosylated using the method described by Das, et al. [56] and Arruda, et al. [57].

2.6.1. Extraction of Free Form Phenolic Compounds

The previously obtained supernatant was subjected to rotary vacuum evaporation (RE 100 Pro, Biobase Inc., Jinan, China) at 40 °C to remove organic solvents. The resulting aqueous phase was acidified to pH 2 and transferred to a separating funnel. The clear supernatant was extracted three times with an equal volume of n-hexane (1:1, v/v) to eliminate interfering lipid molecules. The organic and aqueous phases were separated, and the aqueous phase was collected. An equal volume of diethyl ether and ethyl acetate (1:1, v/v) was added to the aqueous phase, and the mixture was transferred to a separating funnel. The organic phase was collected, dehydrated, and filtered through anhydrous sodium sulfate using Whatman No. 1 filter paper. The solvent was then removed under vacuum rotary evaporation at 35 °C. The resulting dry residue of free form phenolics was dissolved in 1 mL of methanol for further use.

2.6.2. Extraction of Esterified Form Phenolic Compounds

The remaining aqueous phase, obtained from the extraction of the esterified phenolics, was hydrolyzed with a mixture of 4 M NaOH, 10 mM EDTA, and 1% ascorbic acid (using a solvent to aqueous phase ratio of 2:1, v/v). The solution was incubated for 3 h at room temperature in a water bath shaker (120 rpm) to release the esterified phenolics. The pH of the solution was adjusted to 2 and transferred to a separating funnel, where it was mixed with an equal volume of diethyl ether and ethyl acetate (1:1, v/v). The organic phase was collected, dehydrated, and filtered through anhydrous sodium sulfate using Whatman No. 1 filter paper. The resulting solution was evaporated under a vacuum using a rotary evaporator at 35 °C. The dry residue of esterified form phenolics was dissolved in 1 mL of methanol for further use.

2.6.3. Extraction of Glycosylated Form Phenolic Compounds

The aqueous solution left over from the extraction of esterified phenolic compounds was subjected to hydrolysis with twice the volume of 6 M HCl and incubated for 30 min in a shaker incubator (120 rpm) to release glycosylated phenolic compounds. The pH of the resulting solution was adjusted to 2 and then mixed with an equal volume of diethyl ether and ethyl acetate (1:1, v/v) in a separating funnel. The organic phase was collected, dehydrated, and filtered using Whatman No. 1 filter paper with anhydrous sodium sulfate. The resulting solution was evaporated under a vacuum rotary evaporator at 35 °C. The dry residue of glycosylated form phenolics was dissolved in 1 mL of methanol for future use.

2.7. Instrumentation

An RP-HPLC system (Chromaster, Hitachi Corporation, Tokyo, Japan) equipped with a UV detector and a quaternary gradient pump was utilized for the analysis of phenolic compounds. The C18 reversed-phase column (5C18-MS-II, 4.6 ID 250 mm, cosmosil-Nacalai Tesque INC., Kyoto, Japan) was maintained at a temperature of 25 °C. RP-HPLC is a widely used technique for analyzing phenolic compounds, and most of these compounds can be detected in the UV range [58]. The elution gradient for the sample was achieved by using two solvents: solvent A (2% glacial acetic acid in water) and solvent B (acetonitrile: water, 70:30) (Supplementary Table S1). A 20 µL sample was injected, and the flow rate was set to 1 mL/min. The detection wavelength was 280 nm [59].

Identification and Quantification of Phenolic Compounds Using RP-HPLC

Phenolic compounds were identified by comparing the retention times of extracted compounds with those of standard phenolic acids, including gallic acid, caffeic acid, p-coumaric acid, rosmarinic acid, *trans*-cinnamic acid, vanillic acid, and ellagic acid. Standard solutions of each acid were prepared by dissolving 1 mg in 1 mL of methanol and were used to quantify the phenolic compounds present in the leaves of *B. polystachyon*. The quantitative determination of phenolic compounds was carried out following the proposed method [50,60].

$$\text{Sample concentration}(\mu g/gDW) = \frac{\text{Sample area}}{\text{Standard area}} \times \frac{\text{Standard weight}}{\text{Standard dilution}} \times \frac{\text{Sample dilution}}{\text{Sample weight}}$$

2.8. Data Collection and Statistical Analysis

The data were analyzed using one-way analysis of variance (ANOVA) and the results were presented as mean ± standard error. The significance of differences among means was determined by Duncan's multiple range test (DMRT) [61,62] at a significance level of $p \leq 0.05$ using SPSS 26.0 version software (SPSS Inc., Armonk, NY, USA). RP-HPLC data and statistical graphs were generated using Origin 2022 and GraphPad Prism 9.5, respectively.

3. Results and Discussion

3.1. Effect of Heavy Metal on In Vitro Propagation and Tolerance Index

Shoot tips from field-grown plants of *B. polystachyon* were cultured in MS media that contained various concentrations of mercury (II) and lead (II) for plant regeneration (Tables 1 and 2). Multiple shoots with roots were induced after 3 weeks of culture incubation in MS media that was supplemented with various concentrations of mercury (II) (Figure 1B–F). Plants showed tolerance to all concentrations of heavy metals (1, 25, 50, 100, and 200 µM), as evidenced by the lack of chlorosis and necrosis. Heavy metal concentration significantly impacted plant growth in *B. polystachyon*, with lower concentrations promoting better shoot and root development (Table 1 and Figure 1). Among the concentrations tested, 1 µM mercury (II) induced the highest number and length of shoots in *B. polystachyon* (Figure 1A,B). Increasing the concentration reduced overall growth, resulting in a gradual decline in both shoot and root lengths. Surprisingly, at 50 µM, root number was maximized while maintaining moderate shoot growth, as depicted in Figure 1D. However, the shoot length was greatly reduced in 100 µM concentration of mercury (II), and root formation was generally poor (Figure 1E). In addition, stout and hairy adventitious roots were observed in MS media supplemented with 200 µM of mercury (II) (Figure 1F). This highlights the complex interplay between heavy metal concentration and plant growth, suggesting the existence of a critical balance for optimal development. The present results are consistent with [63] reports that higher concentrations of mercury (II) inhibit the growth of *Pisum sativum* L. Passow and Rothstein [64] and Shieh and Barber [65] also reported that mercury (II) affected cell membranes, leading to a breakdown in the transport mechanism of plants. Higher concentrations of heavy metal stress resulted in increased leaf wilting, significantly reduced shoot multiplication, and shorter shoots and roots compared to control plants. This could potentially be due to a reduction in cell division and differentiation [66]. Several studies on basil plants suggest that the presence of heavy metals in the growth medium has negative effects on various physiological processes [67,68].

As depicted in Figure 1H–L, the results suggest that higher concentrations of lead (II) had a more significant impact on in vitro regeneration. Lead (II) at 1 µM in MS media maximized both shoot multiplication and root induction (Figure 1H). Although 50 µM proved optimal for both, exceeding this level generally inhibited root development, with a notable exception at 100 µM, where root number unexpectedly increased (Figure 1K). Intriguingly, at 200 µM, the mean shoot length plummeted to 1.00 ± 0.40 cm (Figure 1L). Significantly, shoot multiplication was completely suppressed at 200 µM of both mercury (II) and lead (II), whereas robust adventitious root formation persisted. Intriguingly, the presence of these metals also induced *B. polystachyon* to flower, as shown in Figure 1D. This finding aligns with the previously reported observation that heavy metals can promote flowering, suggesting a more complex interplay between these elements and plant development [69]. Similar to *Cyamopsis tetragonoloba* L. and *Sesamum indicum* L. [70], which exhibited tolerance to lead (II), *B. polystachyon* also demonstrated the ability to tolerate low levels of heavy metals, according to this study. Heavy metal stress, particularly at high levels, significantly reduced plant growth and development, likely due to interference with nutrient uptake pathways. The present study demonstrates that in vitro-regenerated *B. polystachyon* plantlets exhibit a noteworthy degree of tolerance to heavy metals.

Acclimatization of in vitro-grown plantlets was essential for their successful transplantation to in vivo climatic conditions. By the end of the experiment, the result established that *B. polystachyon* plantlets exposed to heavy metals were successfully propagated. Subsequently, the plantlets were acclimatized to the greenhouse environment, and the hardened plants were transplanted to the field with a 96% survival rate (Figure 1G–M). Based on the findings, it can be demonstrated that the in vitro propagation technique offers a promising approach for selecting heavy-metal-tolerant plants.

Table 1. Effect of Hg (II) on in vitro shoot multiplication, root induction, and plant tolerance index.

Hg (II) (µM)	No. of Shoots Per Explant Mean ± SE	Shoot Length (cm) Mean ± SE	No. of Roots Per Shoot Mean ± SE	Root Length (cm) Mean ± SE	TI (%)
Control	8.25 ± 0.21 [a]	4.25 ± 0.27 [a]	8.00 ± 0.75 [d]	5.25 ± 0.18 [a]	0.00
1	7.50 ± 0.18 [b]	3.25 ± 0.27 [b]	8.25 ± 0.54 [c]	4.75 ± 0.17 [b]	96.87
25	6.25 ± 0.27 [c]	2.75 ± 0.45 [c]	8.5 ± 0.64 [b]	4.00 ± 0.24 [c]	90.31
50	5.25 ± 0.25 [d]	2.25 ± 0.44 [d]	9.0 ± 0.85 [a]	3.25 ± 0.35 [d]	78.75
100	2.00 ± 0.16 [e]	2.0 ± 0.85 [e]	4.50 ± 0.34 [e]	2.75 ± 0.40 [e]	54.68
200	1.00 ± 0.25 [f]	1.75 ± 0.27 [f]	2.50 ± 0.25 [f]	1.75 ± 0.81 [f]	40.00

Values represent mean ± SE of 20 replicates per experiment, with each experiment repeated three times. Mean followed by the same letter is not significantly different ($p \leq 0.05$) using Duncan's multiple range test.

Table 2. Effect of Pb (II) on in vitro shoot multiplication, root induction, and plant tolerance index.

Pb (II) (µM)	No. of Shoots Per Explant Mean ± SE	Shoot Length (cm) Mean ± SE	No. of Roots Per Shoot Mean ± SE	Root Length (cm) Mean ± SE	TI (%)
Control	8.25 ± 0.21 [a]	4.25 ± 0.27 [a]	8.00 ± 0.75 [e]	5.25 ± 0.18 [a]	0.00
1	7.75 ± 0.22 [b]	3.50 ± 0.28 [b]	8.25 ± 0.74 [d]	4.0 ± 0.27 [b]	97.18
25	6.50 ± 0.27 [c]	2.75 ± 0.24 [c]	9.00 ± 0.34 [c]	3.25 ± 0.22 [c]	90.00
50	5.75 ± 0.24 [d]	2.25 ± 0.55 [d]	9.5 ± 0.24 [b]	2.50 ± 0.25 [d]	79.37
100	2.75 ± 0.36 [e]	1.25 ± 0.65 [e]	10.25 ± 0.66 [a]	2.0 ± 0.20 [e]	70.86
200	1.75 ± 0.25 [f]	1.0 ± 0.40 [f]	4.00 ± 0.28 [f]	1.0 ± 0.11 [f]	42.18

Values represent mean ± SE of 20 replicates per experiment, with each experiment repeated three times. Mean followed by the same letter is not significantly different ($p \leq 0.05$) using Duncan's multiple range test.

Figure 1. Effect of heavy metal stress on in vitro propagation of *B. polystachyon* after three weeks of culture incubation. (**A**) Control, (**B**) MS medium containing 1 µM Hg (II), (**C**) 25 µM Hg (II), (**D**) 50 µM Hg (II), (**E**) 100 µM Hg (II), (**F**) 200 µM Hg (II), (**G**) acclimatization of plantlets, (**H**) 1 µM Pb (II), (**I**) 25 µM Pb (II), (**J**) 50 µM Pb (II), (**K**) 100 µM Pb (II), (**L**) 200 µM Pb (II), (**M**) hardened plantlets.

As shown in Tables 1 and 2, the results demonstrated that as the plant was exposed to different concentrations of heavy metals, the TI (tolerance index) values markedly decreased at higher concentrations. Specifically, the plant had a 70.86% TI at 100 µM of lead (II) and a 42.18% TI at 200 µM of lead (II). Similarly, the plant had a 78.75% TI at 50 µM of mercury (II) and a 40% TI at 200 µM of mercury (II). Interestingly, low concentrations of mercury (II) and lead (II) had a partially negative effect on shoot and root growth, but *B. polystachyon* was capable of accumulating high levels of heavy metals and had a survival rate of 96%. Cano-Ruiz, et al. [71] found similar results for cadmium, nickel, lead, and copper. This study supports the work of Youssef [72] on *Ocimum basilicum* L., which confirmed the effects of heavy metals on the plant. Other studies have shown that plants from the Lamiaceae family, such as *Mentha crispa* L., *Mentha piperita* L, *Ocimum basilicum* L. and *Ocimum sanctum* L. can exhibit resistance against the harmful effects of heavy metal toxicity [73–75]. The results of the study suggest that plantlets regenerated in vitro could be used to accumulate heavy metals in contaminated sites.

The effect of varying concentrations of heavy metals on *B. polystachyon* was assessed by measuring the fresh and dry weights of plantlets regenerated using the in vitro method. As the concentrations of mercury (II) and lead (II) increased in the MS media, the fresh and dry weights of the plantlets gradually decreased, as shown in Figure 2. The highest fresh weight (3.12 ± 0.27 g) and dry weight (0.86 ± 0.18 g) of *B. polystachyon* were noted in the MS media supplemented with 1 µM of mercury (II) compared to lead (II), but the difference was not statistically significant. The fresh and dry weights of plantlets decreased as the concentration of heavy metals increased in MS media. The plant biomass declined significantly at 100 µM and 200 µM, after reaching 50 µM. Figure 2 illustrates that, despite the control plants exhibiting marginally greater mean fresh weight (3.2 ± 0.34 g) and dry weight (0.9 ± 0.13 g), there was no significant difference in biomass between the control and the 1 µM heavy metal treatments. These findings suggest that low concentrations of heavy metals did not significantly affect the plant biomass.

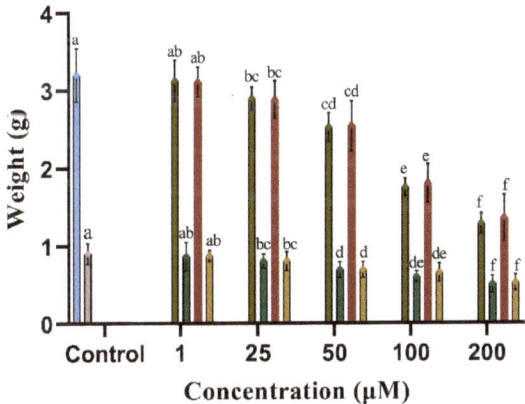

Figure 2. Effect of heavy metals on the fresh and dry weights of in vitro-raised plantlets. Values represent mean ± SE of 10 replicates per experiment, with each experiment repeated three times. Mean followed by the same letter is not significantly different ($p \leq 0.05$) using Duncan's multiple range test.

Moreover, a reduction in the biomass of in vitro regenerated plantlets was noted in the regenerating media with higher concentrations of heavy metals. The growth of the plant was significantly affected by the addition of 200 µM of mercury (II) and lead (II) in the medium, as evidenced by the significant reduction in shoot and root length. The toxic nature of heavy metals interferes with cellular levels, disrupting the plant metabolic pathways and ultimately reducing growth and development. Previous research demonstrated that high concentrations of heavy metals in the growth medium can disrupt the ability of plants to take up water and nutrients, leading to decreased biomass in *Ocimum basilicum* L. and *Mentha piperita* L [33,76,77].

3.2. Potential for Heavy Metal Accumulation

In Table 3, it was demonstrated that the plant exhibited the ability to accumulate a particular level of heavy metals in both its root and shoot. Notably, the highest accumulation of mercury (II) was observed at concentrations of 200 µM and 50 µM, with the root and shoot values of 16.94 ± 0.44 µg/g and 17.68 ± 0.66 µg/g, respectively. There was no significant difference observed in the accumulation of heavy metal in the shoot between concentrations of 1 µM and 100 µM. Figure 3 shows the TF values for heavy metals, where values greater than one indicate efficient metal translocation from root to shoot.

Table 3. Translocation of heavy metals in various plant tissues.

Concentration (µM)	Hg (II) (µg/g)		Pb (II) (µg/g)	
	Root	Shoot	Root	Shoot
Control	0 ± 0	0 ± 0	0 ± 0	0 ± 0
1	7.34 ± 0.47 [e]	8.76 ± 0.28 [cd]	7.2 ± 0.19 [e]	5.27 ± 0.43 [de]
25	9.17 ± 0.38 [d]	12.45 ± 0.45 [b]	9.56 ± 0.33 [d]	6.89 ± 0.55 [b]
50	11.56 ± 0.56 [c]	17.68 ± 0.66 [a]	13.66 ± 0.64 [c]	7.78 ± 0.26 [a]
100	14.43 ± 0.58 [b]	8.28 ± 0.47 [cd]	16.48 ± 0.72 [b]	6.24 ± 0.32 [c]
200	16.94 ± 0.44 [a]	7.78 ± 0.26 [d]	17.10 ± 0.54 [a]	5.56 ± 0.14 [de]

Values represent mean ± SE of 3 replicates per experiment, with each experiment repeated three times. Mean followed by the same letter is not significantly different ($p \leq 0.05$) using Duncan's multiple range test.

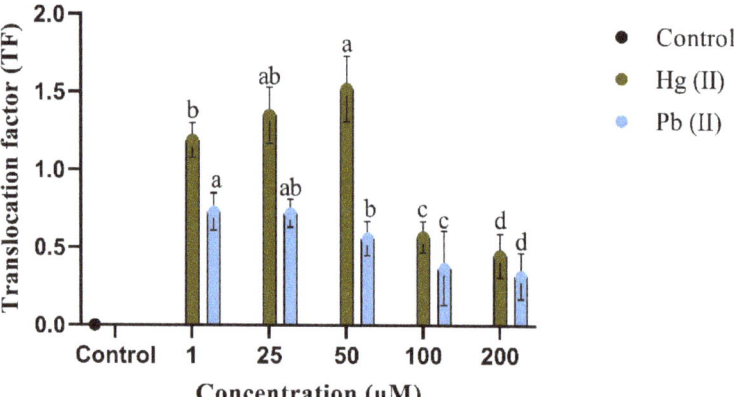

Figure 3. TF of Hg (II) and Pb (II) in *B. polystachyon*. Values represent mean ± SE of 3 replicates per experiment, with each experiment repeated three times. Mean followed by the same letter is not significantly different ($p \leq 0.05$) using Duncan's multiple range test.

The present study found that the roots of *B. polystachyon* accumulated the highest level of mercury (II) at a concentration of 200 µM, with a translocation factor of 0.45. However, at a concentration of 50 µM, the TF value was 1.52, suggesting that the mercury (II) was able to move from the root into the shoot. Lone, et al. [78] reported that the accumulation of high levels of metals in aerial parts of the plant led to decreased plant height, which is consistent with a previous study on *Lindernia crustacea* L., where the maximum mercury accumulation occurred in the shoots [46]. Additionally, it was stated that the TF can be influenced by various factors, such as plant species, root uptake efficiency, water absorption, element type, and soil nutrient availability.

The accumulation capacity of *B. polystachyon* showed that the plant could survive heavy metal stress up to 200 µM. The root accumulated significantly higher lead (II) than the shoot, with concentrations of 17.10 ± 0.54 µg/g and 7.78 ± 0.26 µg/g, respectively, at 200 µM and 50 µM. The results also revealed that increasing the concentration of lead (II) led to a decrease in TF values, suggesting that the plant was less able to transport lead (II) from the roots into the shoots at higher concentrations. The lowest TF value (0.32) was found in lead (II) enriched plants, indicating that the majority of the lead (II) remained in the roots and was not translocated to the shoots. Previous studies have shown that lead (II) accumulates mostly in the roots, gradually moving to the shoots [79,80]. The study found that the plant had the potential to accumulate and translocate lead (II) and mercury (II) in root and shoot. Purohit, et al. [81] and Yan, et al. [82] demonstrated that the transporter of P1B-type ATPases is involved in the transport of heavy metals from root to shoot. The results of this study suggest that *B. polystachyon* is a potential plant for phytoremediation, even though it accumulates high levels of lead (II) in its roots.

Several studies have shown that lead (II) primarily accumulates in the roots of plants due to its binding to ion-exchangeable sites on the cell wall, which prevents it from moving into the cells [83,84]. The present results are consistent with previous studies that have shown that lead (II) is poorly translocated from root to shoot, likely due to the presence of a physical barrier in the root zone. Plants have developed effective mechanisms to deal with high levels of heavy metal exposure, which are dependent on their biochemical processes. The present findings align with earlier studies conducted by Dinu, et al. [85] and Youssef [72], indicating that species of the Lamiaceae family, such as *Ocimum* sp., are capable of accumulating heavy metals. Additionally, these plants exhibit a higher concentration of secondary metabolites, which offer protection against oxidative damage and prevent cell oxidation. The results suggest that the plant holds promising potential for utilization in heavy metal remediation purposes.

3.3. Effect of Heavy Metal on the Enhancement of Production of Phenolic Compounds

Table 4 shows the contents of phenolic compounds in the leaves of plants exposed to heavy metals and the control. Supplementary Figure S1 displays the chromatograms depicting the standard phenolic acids. The results showed that there were significant variations in the contents of phenolic compounds, depending on their form. These forms included free, esterified, and glycosylated phenolics. Remarkably, the results suggest that plants grown in vitro in the presence of heavy metals may increase their production of phenolic compounds. Quantitative analysis of free, esterified, and glycosylated phenolic compounds revealed a substantial upregulation in lead (II)-exposed plants. Table 4 illustrates a significant increase in the content of free phenolics, which constitute the primary form of phenolic compounds in plants exposed to lead (II). Rosmarinic acid (30.82 ± 0.31 µg/g DW), caffeic acid (25.57 ± 0.54 µg/g DW), gallic acid (16.23 ± 0.43 µg/g DW), and ellagic acid (10.71 ± 0.22 µg/g DW) were identified as the most abundant free phenolics, whereas caffeic acid (42.53 ± 0.61 µg/g DW), gallic acid (15.95 ± 0.38 µg/g DW), and rosmarinic acid (31.13 ± 0.47 µg/g DW) dominated the esterified and glycosylated fractions, respectively (Supplementary Figure S2A–C). In addition to rosmarinic acid, *p*-coumaric acid was also found in free and glycosylated forms. Among phenolic fraction, caffeic acid (42.53 ± 0.61 µg/g DW) emerged as the most predominant phenolic compound, followed by rosmarinic acid (31.13 ± 0.47 µg/g DW), gallic acid (16.23 ± 0.43 µg/g DW), ellagic acid (12.86 ± 0.28 µg/g DW), *p*-coumaric acid (8.04 ± 0.31 µg/g DW), *trans*-cinnamic acid (7.57 ± 0.20 µg/g DW), and vanillic acid (0.54 ± 0.7 µg/g DW).

Table 4. Comparison of phenolic compound content in *B. polystachyon* leaves under heavy metal exposure and in control plants.

Phenolic Compound	Hg (II)			Pb (II)			Control		
	Free Form Phenolics (µg/g DW)	Esterified Form Phenolics (µg/g DW)	Glycosylated Form Phenolics (µg/g DW)	Free Form Phenolics (µg/g DW)	Esterified Form Phenolics (µg/g DW)	Glycosylated Form Phenolics (µg/g DW)	Free Form Phenolics (µg/g DW)	Esterified Form Phenolics (µg/g DW)	Glycosylated Form Phenolics (µg/g DW)
Gallic acid	5.18 ± 0.56 [f]	0 ± 0	33.16 ± 0.68 [a]	16.23 ± 0.43 [b]	15.95 ± 0.38 [bc]	0 ± 0	7.46 ± 0.24 [d]	7.06 ± 0.45 [de]	15.52 ± 0.20 [bc]
Caffeic acid	0 ± 0	0 ± 0	18.51 ± 0.44 [c]	25.57 ± 0.54 [b]	42.53 ± 0.61 [a]	5.98 ± 0.37 [d]	0 ± 0	0 ± 0	0 ± 0
Vanillic acid	77.74 ± 1.08 [c]	1959.1 ± 3.66 [a]	0 ± 0	0 ± 0	0.54 ± 0.07 [d]	0 ± 0	143.57 ± 1.7 [b]	0 ± 0	0 ± 0
p-Coumaric acid	0 ± 0	0 ± 0	2.13 ± 0.15 [b]	8.04 ± 0.31 [a]	0 ± 0	0.94 ± 0.05 [c]	0 ± 0	0 ± 0	0 ± 0
Ellagic acid	213.55 ± 2.11 [a]	14.94 ± 0.35 [b]	7.17 ± 0.22 [e]	10.71 ± 0.22 [d]	12.86 ± 0.28 [c]	4.45 ± 0.17 [f]	0 ± 0	0 ± 0	0 ± 0
Rosmarinic acid	0 ± 0	187.72 ± 1.22 [a]	45.09 ± 0.78 [b]	30.82 ± 0.45 [cd]	0 ± 0	31.13 ± 0.47 [cd]	0 ± 0	0 ± 0	0 ± 0
Trans-cinnamic acid	33.046 ± 0.69 [c]	9.30 ± 0.66 [e]	33.32 ± 0.32 [c]	7.57 ± 0.20 [f]	1.61 ± 0.12 [g]	2.15 ± 0.11 [f]	11.52 ± 0.29 [d]	82.31 ± 1.13 [b]	207.74 ± 1.45 [a]

Values represent mean ± SE of 3 replicates per experiment, with each experiment repeated three times. Mean followed by the same letter is not significantly different ($p \leq 0.05$) using Duncan's multiple range test.

In plants exposed to mercury (II), vanillic acid was the most abundant phenolic compound, followed by ellagic acid, rosmarinic acid, *trans*-cinnamic acid, gallic acid, caffeic acid, and *p*-coumaric acid (Supplementary Figure S3). The esterified phenolics were the most abundant, followed by the glycosylated form and then the free form (Table 4). Within the esterified phenolics, vanillic acid exhibited the highest concentration (1959.1 ± 3.66 µg/g DW), followed by rosmarinic acid (187.72 ± 1.22 µg/g DW). Lower concentrations were observed for *trans*-cinnamic acid (9.30 ± 0.66 µg/g DW) and ellagic acid (14.94 ± 0.35 µg/g DW) (Supplementary Figure S3B). In the free phenolic compounds, ellagic acid emerged as the most abundant with a concentration of 213.55 ± 0.15 µg/g DW, followed by vanillic acid (77.74 ± 1.08 µg/g DW). *Trans*-cinnamic acid and gallic acid were also present but at lower concentrations of 33.046 ± 0.69 µg/g DW and 5.183 ± 0.56 µg/g DW, respectively (Supplementary Figure S3A). The glycosylated form exhibited the highest phenolic compound diversity, identifying a total of six phenolic acids, including rosmarinic acid (45.09 ± 0.78 µg/g DW), *trans*-cinnamic acid (33.32 ± 0.32 µg/g DW), gallic acid (33.16 ± 0.68 µg/g DW), caffeic acid (18.51 ± 0.44 µg/g DW), ellagic acid (7.17 ± 0.22 µg/g DW), and *p*-coumaric acid (2.13 ± 0.15 µg/g DW) (Supplementary Figure S3C). In the free and glycosylated forms, rosmarinic acid emerged as the predominant phenolic compound, with the exception of the esterified form, where ellagic acid was present. Notably, *p*-coumaric acid was exclusively detected in its glycosylated form. These findings are consistent with previous research, which has shown that plants grown in media containing heavy metals accumulate higher levels of phenolic acids [86,87].

Mercury (II) was the most effective metal in stimulating the production of vanillic acid, ellagic acid and rosmarinic acid, whereas lead (II)-induced plants exhibited different types of phenolic compounds, with similar concentrations of some phenolic acids. The results suggest that the decrease in the amount of *trans*-cinnamic acid that plants accumulate may be attributed to their exposure to lead (II) and mercury (II), which is a precursor to other phenolic compounds. Therefore, its reduction may have led to a decrease in the production of these compounds, including caffeic acid and *p*-coumaric acid. The findings of this study suggest that heavy metal exposure can lead to an increase in the production of phenolic compounds in plants which could be an adaptive response that helps plants to protect themselves from the harmful effects of heavy metals.

Mercury (II) exposure significantly enhanced the synthesis of gallic acid, vanillic acid, ellagic acid, rosmarinic acid and *trans*-cinnamic acid in *B. polystachyon* plantlets. Interestingly, lead (II) exposure resulted in lower levels of these phenolic compounds compared to mercury (II) treatment, with the exception of caffeic acid and *p*-coumaric acid, even though both metals induce the biosynthesis of phenolic acid. Lead (II)-induced plants exhibited a diverse range of phenolic compounds, although some phenolic acids were found in similar concentrations. The present findings are in line with earlier research, suggesting that elicitors, known for triggering an immune response in plants, can also enhance the production of phenolic compounds [88,89]. Kisa, et al. [90] demonstrated that heavy metal stress can alter the expression of genes involved in the production of phenylpropanoids, leading to the accumulation of large amounts of phenolic acids in stressed plants. To the best of our knowledge, this is the first report to establish that heavy metal stress significantly increases the accumulation of phenolic compounds in *B. polystachyon*.

In the control plants, the glycosylated form emerged as the predominant type of phenolic compound. Notably, *trans*-cinnamic acid (207.74 ± 1.45 µg/g DW) and gallic acid (15.52 ± 0.20 µg/g DW) were identified as the major compounds within this form (Supplementary Figure S4C). Comparatively, the free form of phenolics was found in lower concentrations than the glycosylated form, as outlined in Table 4. In the free form, vanillic acid (143.57 ± 1.7 µg/g DW), *trans*-cinnamic acid (11.52 ± 0.29 µg/g DW), and gallic acid (7.46 ± 0.24 µg/g DW) were detected (Supplementary Figure S4A), whereas the esterified form contained *trans*-cinnamic acid (82.31 ± 1.13 µg/g DW) and gallic acid (7.06 ± 0.45 µg/g DW) (Supplementary Figure S4B). *Trans*-cinnamic acid remained the major phenolic compound in both the esterified and glycosylated forms, whereas vanillic acid dominated the free form. Notably, the results indicated an increase in vanillic acid content in plants exposed to heavy metals compared to the control. This phenomenon could be attributed to the upregulation of enzymes involved in the conversion of benzoic acid to vanillic acid under stressful environmental conditions [91]. The results of this study are in line with the previous report on *Zea mays* L. plants where the content of vanillic acid increased with exposure to mercury (II) and lead (II) [90]. It is worth noting that cinnamic acid and its derivatives have applications in the pharmaceutical industry [92]. Previous research has demonstrated that rosmarinic acid processes anti-inflammatory, antiviral, antibacterial, antioxidant, and antimutagenic properties [93–95]. Caffeic acid has been identified as an anti-inflammatory agent [96], and a study conducted on diabetic mice in 2009 found that it could potentially combat diabetes and increase blood insulin levels [97,98]. Vanillic acid is used as a flavoring agent and also exhibits beneficial biological activities, particularly in chemo-protection, anti-inflammation, and antimicrobial activities [99]. The identified phenolic compounds from this study have the possibility for beneficial applications in the pharmaceutical and cosmetic industries.

4. Conclusions

To the best of our knowledge, this is the first report that presents a novel and promising approach for regenerating *B. polystachyon* in vitro in the presence of mercury (II) and lead (II). It establishes the plant's capacity to efficiently accumulate, translocate, and adapt to these heavy metal contaminants. The findings of this study suggest that *B. polystachyon* may have the potential for heavy metal remediation, emphasizing the need for further investigation. Furthermore, the study highlights the plant's response to heavy metal stress, which triggers an increase in the accumulation of various phenolic compounds, such as gallic acid, caffeic acid, vanillic acid, *p*-coumaric acid, ellagic acid, and rosmarinic acid. Based on the findings of the present study, it can be concluded that *B. polystachyon* possesses substantial potential for the large-scale production of a diverse range of phenolic compounds, with promising applications in various industries.

Supplementary Materials: The following supporting information can be downloaded at: https://www.mdpi.com/article/10.3390/plants13010098/s1.

Author Contributions: S.D.: Conceptualization, Methodology, Investigation, Data Analysis, Writing—original draft; K.W.S.: Data analysis, Writing—Editing, M.M.: Data analysis; A.R.N.: Visualization; I.C.: Supervision. All authors have read and agreed to the published version of the manuscript.

Funding: The fellowship grant (Senior Research Fellowship, Award letter No. FC(Sc)/RS/SF/BIOTECH/2020-21/32) provided by the Government of West Bengal and The University of Burdwan is gratefully acknowledged by the authors.

Data Availability Statement: All data generated or analyzed during this study are available in the published article and its Supplementary Materials files.

Acknowledgments: The authors would also like to express their gratitude to the Department of Biotechnology at The University of Burdwan for providing the necessary research facilities. The authors gratefully acknowledge the Department of Plant Production, Soil Science and Agricultural Engineering, Green Biotechnologies Research Centre of Excellence, University of Limpopo, South Africa. Additionally, the authors extend their sincere appreciation to Kaushik Sarkar, Technical Assistant Gr-I at the Department of Biotechnology, The University of Burdwan.

Conflicts of Interest: The authors declare no conflict of interest.

References

1. Sivarajasekar, N.; Baskar, R. Adsorption of basic red 9 on activated waste Gossypium hirsutum seeds: Process modeling, analysis and optimization using statistical design. *J. Ind. Eng. Chem.* **2014**, *20*, 2699–2709. [CrossRef]
2. Ernst, W.H. Evolution of metal tolerance in higher plants. *For. Snow Landsc. Res.* **2006**, *80*, 251–274.
3. Janicka-Russak, M.; Kabała, K.; Burzyński, M.; Kłobus, G. Response of plasma membrane H+-ATPase to heavy metal stress in Cucumis sativu s roots. *J. Exp. Bot.* **2008**, *59*, 3721–3728. [CrossRef] [PubMed]
4. Ernst, W.H.; KRAUSS, G.J.; Verkleij, J.A.; Wesenberg, D. Interaction of heavy metals with the sulphur metabolism in angiosperms from an ecological point of view. *Plant Cell Environ.* **2008**, *31*, 123–143. [CrossRef] [PubMed]
5. Xu, J.; Yin, H.X.; Li, X. Protective effects of proline against cadmium toxicity in micropropagated hyperaccumulator, *Solanum nigrum* L. *Plant Cell Rep.* **2009**, *28*, 325–333. [CrossRef]
6. Martinez-Fernandez, D.; Walker, D.J.; Romero-Espinar, P.; Flores, P.; del Rio, J.A. Physiological responses of *Bituminaria bituminosa* to heavy metals. *J. Plant Physiol.* **2011**, *168*, 2206–2211. [CrossRef]
7. Singh, A.; Prasad, S.M. Remediation of heavy metal contaminated ecosystem: An overview on technology advancement. *Int. J. Environ. Sci. Technol.* **2015**, *12*, 353–366. [CrossRef]
8. Khafouri, A.; Talbi, E.; Abdelouas, A. Assessment of Heavy Metal Contamination of the Environment in the Mining Site of Ouixane (North East Morocco). *Water Air Soil Poll.* **2021**, *232*, 398. [CrossRef]
9. DalCorso, G.; Manara, A.; Furini, A. An overview of heavy metal challenge in plants: From roots to shoots. *Metallomics* **2013**, *5*, 1117–1132. [CrossRef]
10. Benyo, D.; Horvath, E.; Nemeth, E.; Leviczky, T.; Takacs, K.; Lehotai, N.; Feigl, G.; Kolbert, Z.; Ordog, A.; Galle, R.; et al. Physiological and molecular responses to heavy metal stresses suggest different detoxification mechanism of *Populus deltoides* and *P.-x canadensis*. *J. Plant Physiol.* **2016**, *201*, 62–70. [CrossRef]
11. Muszynska, E.; Hanus-Fajerska, E.; Kozminska, A. Differential Tolerance to Lead and Cadmium of Micropropagated *Gypsophila fastigiata* Ecotype. *Water Air Soil Poll.* **2018**, *229*, 42. [CrossRef] [PubMed]
12. Demarco, C.F.; Afonso, T.F.; Pieniz, S.; Quadro, M.S.; Camargo, F.A.D.; Andreazza, R. Phytoremediation of heavy metals and nutrients by the *Sagittaria montevidensis* into an anthropogenic contaminated site at Southern of Brazil. *Int. J. Phytoremediat.* **2019**, *21*, 1145–1152. [CrossRef] [PubMed]
13. Gatti, E. Micropropagation of *Ailanthus altissima* and in vitro heavy metal tolerance. *Biol. Plant.* **2008**, *52*, 146–148. [CrossRef]
14. Okem, A.; Moyo, M.; Stirk, W.A.; Finnie, J.F.; Van Staden, J. Investigating the effect of cadmium and aluminium on growth and stress-induced responses in the micropropagated medicinal plant *Hypoxis hemerocallidea*. *Plant Biol.* **2016**, *18*, 805–815. [CrossRef] [PubMed]
15. Calabrese, E.J.; Baldwin, L.A. Hormesis as a biological hypothesis. *Environ. Health Perspect.* **1998**, *106*, 357–362.
16. Calabrese, E.J.; Baldwin, L.A. Chemical hormesis: Its historical foundations as a biological hypothesis. *Toxicol. Pathol.* **1999**, *27*, 195–216. [CrossRef] [PubMed]
17. Calabrese, E.J.; Baldwin, L.A. Radiation hormesis: Its historical foundations as a biological hypothesis. *Hum. Exp. Toxicol.* **2000**, *19*, 41–75. [CrossRef]
18. Liu, Z.L.; Chen, W.; He, X.Y.; Jia, L.; Yu, S.; Zhao, M.Z. Hormetic Responses of *Lonicera Japonica Thunb.* to Cadmium Stress. *Dose-Response* **2015**, *13*, 14-033.He. [CrossRef]
19. Velini, E.D.; Alves, E.; Godoy, M.C.; Meschede, D.K.; Souza, R.T.; Duke, S.O. Glyphosate applied at low doses can stimulate plant growth. *Pest. Manag. Sci.* **2008**, *64*, 489–496. [CrossRef]
20. Pena, L.B.; Barcia, R.A.; Azpilicueta, C.E.; Mendez, A.A.E.; Gallego, S.M. Oxidative post translational modifications of proteins related to cell cycle are involved in cadmium toxicity in wheat seedlings. *Plant Sci.* **2012**, *196*, 1–7. [CrossRef]
21. Eriksen, R.L.; Padgitt-Cobb, L.K.; Townsend, M.S.; Henning, J.A. Gene expression for secondary metabolite biosynthesis in hop (*Humulus lupulus* L.) leaf lupulin glands exposed to heat and low-water stress. *Sci. Rep.* **2021**, *11*, 5138. [CrossRef] [PubMed]
22. Lajayer, B.A.; Ghorbanpour, M.; Nikabadi, S. Heavy metals in contaminated environment: Destiny of secondary metabolite biosynthesis, oxidative status and phytoextraction in medicinal plants. *Ecotoxicol. Environ. Saf.* **2017**, *145*, 377–390. [CrossRef] [PubMed]
23. Thangavel, P.; Sulthana, A.S.; Subburam, V. Interactive effects of selenium and mercury on the restoration potential of leaves of the medicinal plant, *Portulaca oleracea* Linn. *Sci. Total Environ.* **1999**, *243*, 1–8. [CrossRef]

24. Murch, S.J.; Haq, K.; Rupasinghe, H.V.; Saxena, P.K. Nickel contamination affects growth and secondary metabolite composition of St. John's wort (*Hypericum perforatum* L.). *Environ. Exp. Bot.* **2003**, *49*, 251–257. [CrossRef]
25. Pandey, N.; Pathak, G.C.; Pandey, D.K.; Pandey, R. Heavy metals, Co, Ni, Cu, Zn and Cd, produce oxidative damage and evoke differential antioxidant responses in spinach. *Braz. J. Plant Physiol.* **2009**, *21*, 103–111. [CrossRef]
26. Kasparova, M.; Siatka, T. Abiotic elicitation of the explant culture of *Rheum palmatum* L. by heavy metals. *Ceska A Slov. Farm. Cas. Ceske Farm. Spol. A Slov. Farm. Spol.* **2004**, *53*, 252–255.
27. Zhang, C.H.; Yan, Q.; Cheuk, W.K.; Wu, J.Y. Enhancement of tanshinone production in *Salvia miltiorrhiza* hairy root culture by Ag+ elicitation and nutrient feeding. *Planta Med.* **2004**, *70*, 147–151.
28. Kim, D.-I.; Pedersen, H.; Chin, C.-K. Stimulation of berberine production in *Thalictrum rugosum* suspension cultures in response to addition of cupric sulfate. *Biotechnol. Lett.* **1991**, *13*, 213–216. [CrossRef]
29. Michalak, A. Phenolic compounds and their antioxidant activity in plants growing under heavy metal stress. *Pol. J. Environ. Stud.* **2006**, *15*, 523–530.
30. Rai, V.; Khatoon, S.; Bisht, S.S.; Mehrotra, S. Effect of cadmium on growth, ultramorphology of leaf and secondary metabolites of *Phyllanthus amarus* Schum. and Thonn. *Chemosphere* **2005**, *61*, 1644–1650. [CrossRef]
31. Gad, N.; Aziz, E.E.; Kandil, H. Effect of cobalt on growth, herb yield and essential quantity and quality in dill (*Anethum graveolens*). *Middle East. J. Agric. Res.* **2014**, *3*, 536–542.
32. Gupta, A.K.; Verma, S.K.; Khan, K.; Verma, R.K. Phytoremediation Using Aromatic Plants: A Sustainable Approach for Remediation of Heavy Metals Polluted Sites. *Environ. Sci. Technol.* **2013**, *47*, 10115–10116. [CrossRef] [PubMed]
33. Zheljazkov, V.D.; Craker, L.E.; Xing, B. Effects of Cd, Pb, and Cu on growth and essential oil contents in dill, peppermint, and basil. *Environ. Exp. Bot.* **2006**, *58*, 9–16. [CrossRef]
34. Das, S.; Sultana, K.W.; Chandra, I. In vitro micropropagation of *Basilicum polystachyon* (L.) Moench and identification of endogenous auxin through HPLC. *Plant Cell Tiss. Org.* **2020**, *141*, 633–641. [CrossRef]
35. Waoo, A.A.; Khare, S.; Ganguly, S. Toxic effect of different lead concentrations on in-vitro culture of *Datura inoxia*. *J. Sci. Innov. Res.* **2014**, *3*, 532–535. [CrossRef]
36. Rout, G.R.; Samantaray, S.; Das, P. In vitro selection and biochemical characterisation of zinc and manganese adapted callus lines in *Brassica* spp. *Plant Sci.* **1999**, *146*, 89–100. [CrossRef]
37. Watmough, S.A.; Dickinson, N.M. Multiple metal resistance and co-resistance in *Acer pseudoplatanus* L. (sycamore) callus cultures. *Ann. Bot.* **1995**, *76*, 465–472. [CrossRef]
38. Fourati, E.; Vogel-Mikus, K.; Bettaieb, T.; Kavcic, A.; Kelemen, M.; Vavpetic, P.; Pelicon, P.; Abdelly, C.; Ghnaya, T. Physiological response and mineral elements accumulation pattern in *Sesuvium portulacastrum* L. subjected in vitro to nickel. *Chemosphere* **2019**, *219*, 463–471. [CrossRef]
39. Doran, P.M. Application of Plant Tissue Cultures in Phytoremediation Research: Incentives and Limitations. *Biotechnol. Bioeng.* **2009**, *103*, 60–76. [CrossRef]
40. Zahedifar, M.; Moosavi, A.A.; Zarei, Z.; Shafigh, M.; Karimian, F. Heavy metals content and distribution in basil (*Ocimum basilicum* L.) as influenced by cadmium and different potassium sources. *Int. J. Phytoremediat.* **2019**, *21*, 435–447. [CrossRef]
41. Szczyglowska, M.; Piekarsaka, A.; Konieczka, P.; Namiesnik, J. Use of Brassica Plants in the Phytoremediation and Biofumigation Processes. *Int. J. Mol. Sci.* **2011**, *12*, 7760–7771. [CrossRef] [PubMed]
42. Salazar, M.J.; Pignata, M.L. Lead accumulation in plants grown in polluted soils. Screening of native species for phytoremediation. *J. Geochem. Explor.* **2014**, *137*, 29–36. [CrossRef]
43. Milan, B.; Slobodanka, N.; Nataša, N.; Borivoj, K.; Milan, Ž.; Marko, K.; Andrej, P.; Saša, O. Response of *Salix alba* L. to heavy metals and diesel fuel contamination. *Afr. J. Biotechnol.* **2012**, *11*, 14313–14319. [CrossRef]
44. Riza, M. Phytoremediation of Pb and Cd contaminated soils by using sunflower (*Helianthus annuus*) plant. *Ann. Agr. Sci* **2018**, *63*, 123–127.
45. Tejeda-Agredano, M.; Gallego, S.; Vila, J.; Grifoll, M.; Ortega-Calvo, J.; Cantos, M. Influence of the sunflower rhizosphere on the biodegradation of PAHs in soil. *Soil Biol. Biochem.* **2013**, *57*, 830–840. [CrossRef]
46. Muddarisna, N.; Krisnayanti, B. Selection of mercury accumulator plants for gold mine tailing contaminated soils. *J. Degrad. Min. Lands Manag.* **2015**, *2*, 341.
47. Das, S.; Sultana, K.W.; Chandra, I. In vitro propagation, phytochemistry and pharmacology properties of *Basilicum polystachyon* (L.) Moench (Lamiaceae): A short review. *S. Afr. J. Bot.* **2023**, *155*, 178–186. [CrossRef]
48. Nehra, N.S.; Kartha, K.K. Meristem and shoot tip culture: Requirements and applications. In *Plant Cell and Tissue Culture*; Springer: Berlin/Heidelberg, Germany, 1994; pp. 37–70.
49. Sultana, K.W.; Chandra, I.; Roy, A. Callus induction and indirect regeneration of *Thunbergia coccinea* Wall. *Plant Physiol. Rep.* **2020**, *25*, 58–64. [CrossRef]
50. Das, S.; Sultana, K.W.; Chandra, I. Adventitious rhizogenesis in *Basilicum polystachyon* (L.) Moench callus and HPLC analysis of phenolic acids. *Acta Physiol. Plant* **2021**, *43*, 146. [CrossRef]
51. Murashige, T.; Skoog, F. A revised medium for rapid growth and bio assays with tobacco tissue cultures. *Physiol. Plant.* **1962**, *15*, 473–497. [CrossRef]
52. Hseu, Z.-Y. Evaluating heavy metal contents in nine composts using four digestion methods. *Bioresour. Technol.* **2004**, *95*, 53–59. [CrossRef] [PubMed]

53. Mattina, M.I.; Lannucci-Berger, W.; Musante, C.; White, J.C. Concurrent plant uptake of heavy metals and persistent organic pollutants from soil. *Environ. Pollut.* **2003**, *124*, 375–378. [CrossRef] [PubMed]
54. Wilkins, D. The measurement of tolerance to edaphic factors by means of root growth. *New Phytol.* **1978**, *80*, 623–633. [CrossRef]
55. Yoon, J.; Cao, X.; Zhou, Q.; Ma, L.Q. Accumulation of Pb, Cu, and Zn in native plants growing on a contaminated Florida site. *Sci. Total Environ.* **2006**, *368*, 456–464. [CrossRef] [PubMed]
56. Das, S.; Sultana, K.W.; Chandra, I. Characterization of polyphenols by RP-HPLC in *Basilicum polystachyon* (L.) Moench with their antioxidant and antimicrobial properties. *S. Afr. J. Bot.* **2022**, *151*, 926–940. [CrossRef]
57. Arruda, H.S.; Pereira, G.A.; de Morais, D.R.; Eberlin, M.N.; Pastore, G.M. Determination of free, esterified, glycosylated and insoluble-bound phenolics composition in the edible part of araticum fruit (*Annona crassiflora* Mart.) and its by-products by HPLC-ESI-MS/MS. *Food Chem.* **2018**, *245*, 738–749. [CrossRef] [PubMed]
58. Spagnuolo, P.A.; Ahmed, N.; Buraczynski, M.; Roma, A.; Tait, K.; Tcheng, M. Analytical Methods–Functional Foods and Dietary Supplements. *Compr. Biotechnol.* **2019**, *4*, 519–531.
59. Baltas, N.; Pakyildiz, S.; Can, Z.; Dincer, B.; Kolayli, S. Biochemical properties of partially purified polyphenol oxidase and phenolic compounds of *Prunus spinosa* L. subsp. dasyphylla as measured by HPLC-UV. *Int. J. Food Prop.* **2017**, *20*, 1377–1391. [CrossRef]
60. Subiramani, S.; Sundararajan, S.; Govindarajan, S.; Sadasivam, V.; Ganesan, P.K.; Packiaraj, G.; Manickam, V.; Thiruppathi, S.K.; Ramalingam, S.; Narayanasamy, J. Optimized in vitro micro-tuber production for colchicine biosynthesis in *Gloriosa superba* L. and its anti-microbial activity against Candida albicans. *Plant Cell Tiss. Org.* **2019**, *139*, 177–190. [CrossRef]
61. Harter, H.L. Critical values for Duncan's new multiple range test. *Biometrics* **1960**, *16*, 671–685. [CrossRef]
62. Duncan, D.B. Multiple range and multiple F tests. *Biometrics* **1955**, *11*, 1–42. [CrossRef]
63. Beauford, W.; Barber, J.; Barringer, A. Uptake and distribution of mercury within higher plants. *Physiol. Plant.* **1977**, *39*, 261–265. [CrossRef]
64. Passow, H.; Rothstein, A. The binding of mercury by the yeast cell in relation to changes in permeability. *J. Gen. Physiol.* **1960**, *43*, 621–633. [CrossRef] [PubMed]
65. Shieh, Y.; Barber, J. Uptake of mercury by Chlorella and its effect on potassium regulation. *Planta* **1973**, *109*, 49–60. [CrossRef] [PubMed]
66. Amirmoradi, S.; Moghaddam, P.R.; Koocheki, A.; Danesh, S.; Fotovat, A. Effect of cadmium and lead on quantitative and essential oil traits of peppermint (*Mentha piperita* L.). *Not. Sci. Biol.* **2012**, *4*, 101–109. [CrossRef]
67. Sharma, P.; Dubey, R.S. Lead toxicity in plants. *Braz. J. Plant Physiol.* **2005**, *17*, 35–52. [CrossRef]
68. Fattahi, B.; Arzani, K.; Souri, M.K.; Barzegar, M. Effects of cadmium and lead on seed germination, morphological traits, and essential oil composition of sweet basil (*Ocimum basilicum* L.). *Ind. Crop Prod.* **2019**, *138*, 111584. [CrossRef]
69. Atanassova, B.; Zapryanova, N. Influence of Heavy Metal Stress on Growth and Flowering of *Salvia Splendens* Ker. -Gawl. *Biotechnol. Biotechnol. Equip.* **2009**, *23*, 173–176. [CrossRef]
70. Amin, H.; Arain, B.A.; Jahangir, T.M.; Abbasi, M.S.; Amin, F. Accumulation and distribution of lead (Pb) in plant tissues of guar (*Cyamopsis tetragonoloba* L.) and sesame (*Sesamum indicum* L.): Profitable phytoremediation with biofuel crops. *Geol. Ecol. Landsc.* **2018**, *2*, 51–60.
71. Cano-Ruiz, J.; Galea, M.R.; Amoros, M.C.; Alonso, J.; Mauri, P.V.; Lobo, M.C. Assessing *Arundo donax* L. in vitro-tolerance for phytoremediation purposes. *Chemosphere* **2020**, *252*, 126576. [CrossRef]
72. Youssef, N.A. Changes in the morphological traits and the essential oil content of sweet basil (*Ocimum basilicum* L.) as induced by cadmium and lead treatments. *Int. J. Phytoremediation* **2021**, *23*, 291–299. [CrossRef] [PubMed]
73. Zemiani, A.; Boldarini, M.T.B.; Anami, M.H.; de Oliveira, E.F.; da Silva, A.F. Tolerance of *Mentha crispa* L. (garden mint) cultivated in cadmium-contaminated oxisol. *Environ. Sci. Pollut. Res.* **2021**, *28*, 42107–42120. [CrossRef] [PubMed]
74. Dinu, C.; Gheorghe, S.; Tenea, A.G.; Stoica, C.; Vasile, G.G.; Popescu, R.L.; Serban, E.A.; Pascu, L.F. Toxic Metals (As, Cd, Ni, Pb) Impact in the Most Common Medicinal Plant (*Mentha piperita*). *Int. J. Env. Res. Pub He* **2021**, *18*, 3904. [CrossRef] [PubMed]
75. Sreelakshmi, C. Heavy Metal Removal from Wastewater Using Ocimum Sanctum. *Int. J. Latest Technol. Eng. Manag. Appl. Sci.* **2017**, *6*, 85–90.
76. Patra, M.; Bhowmik, N.; Bandopadhyay, B.; Sharma, A. Comparison of mercury, lead and arsenic with respect to genotoxic effects on plant systems and the development of genetic tolerance. *Environ. Exp. Bot.* **2004**, *52*, 199–223. [CrossRef]
77. Hatamian, M.; Nejad, A.R.; Kafi, M.; Souri, M.K.; Shahbazi, K. Growth characteristics of ornamental Judas tree (*Cercis siliquastrum* L.) seedling under different concentrations of lead and cadmium in irrigation water. *Acta Sci. Pol. Hortorum Cultus* **2019**, *18*, 87–96. [CrossRef]
78. Lone, M.; Saleem, S.; Mahmood, T.; Saifullah, K.; Hussain, G. Heavy metal contents of vegetables irrigated by sewage/tubewell water. *Int. J. Agric. Biol.* **2003**, *5*, 533–535.
79. Brunet, J.; Repellin, A.; Varrault, G.; Terryn, N.; Zuily-Fodil, Y. Lead accumulation in the roots of grass pea (*Lathyrus sativus* L.): A novel plant for phytoremediation systems? *Comptes Rendus Biol.* **2008**, *331*, 859–864. [CrossRef]
80. Yabanli, M.; Yozukmaz, A.; Sel, F. Heavy metal accumulation in the leaves, stem and root of the invasive submerged macrophyte *Myriophyllum spicatum* L. (Haloragaceae): An example of Kadin Creek (Mugla, Turkey). *Braz. Arch. Biol. Technol.* **2014**, *57*, 434–440. [CrossRef]

81. Purohit, R.; Ross, M.O.; Batelu, S.; Kusowski, A.; Stemmler, T.L.; Hoffman, B.M.; Rosenzweig, A.C. Cu(+)-specific CopB transporter: Revising P1B-type ATPase classification. *Proc. Natl. Acad. Sci. USA* **2018**, *115*, 2108–2113. [CrossRef]
82. Yan, A.; Wang, Y.M.; Tan, S.N.; Yusof, M.L.M.; Ghosh, S.; Chen, Z. Phytoremediation: A Promising Approach for Revegetation of Heavy Metal-Polluted Land. *Front. Plant Sci.* **2020**, *11*, 359. [CrossRef] [PubMed]
83. Wierzbicka, M. Comparison of lead tolerance in *Allium cepa* with other plant species. *Environ. Pollut.* **1999**, *104*, 41–52. [CrossRef]
84. Geebelen, W.; Vangronsveld, J.; Adriano, D.C.; Van Poucke, L.C.; Clijsters, H. Effects of Pb-EDTA and EDTA on oxidative stress reactions and mineral uptake in *Phaseolus vulgaris*. *Physiol. Plant.* **2002**, *115*, 377–384. [CrossRef] [PubMed]
85. Dinu, C.; Vasile, G.-G.; Buleandra, M.; Popa, D.E.; Gheorghe, S.; Ungureanu, E.-M. Translocation and accumulation of heavy metals in *Ocimum basilicum* L. plants grown in a mining-contaminated soil. *J. Soils Sediments* **2020**, *20*, 2141–2154. [CrossRef]
86. Cai, Z.Z.; Kastell, A.; Mewis, I.; Knorr, D.; Smetanska, I. Polysaccharide elicitors enhance anthocyanin and phenolic acid accumulation in cell suspension cultures of *Vitis vinifera*. *Plant Cell Tiss. Org.* **2012**, *108*, 401–409. [CrossRef]
87. Guru, A.; Dwivedi, P.; Kaur, P.; Pandey, D.K. Exploring the role of elicitors in enhancing medicinal values of plants under in vitro condition. *S. Afr. J. Bot.* **2021**, *149*, 1029–1043. [CrossRef]
88. Zhao, J.L.; Zhou, L.G.; Wu, J.Y. Effects of biotic and abiotic elicitors on cell growth and tanshinone accumulation in *Salvia miltiorrhiza* cell cultures. *Appl. Microbiol. Biot.* **2010**, *87*, 137–144. [CrossRef]
89. Ghorbanpour, M. Major essential oil constituents, total phenolics and flavonoids content and antioxidant activity of *Salvia officinalis* plant in response to nano-titanium dioxide. *Indian J. Plant Physiol.* **2015**, *20*, 249–256. [CrossRef]
90. Kisa, D.; Elmastas, M.; Ozturk, L.; Kayir, O. Responses of the phenolic compounds of *Zea mays* under heavy metal stress. *Appl. Biol. Chem.* **2016**, *59*, 813–820. [CrossRef]
91. Metsamuuronen, S.; Siren, H. Bioactive phenolic compounds, metabolism and properties: A review on valuable chemical compounds in Scots pine and Norway spruce. *Phytochem. Rev.* **2019**, *18*, 623–664. [CrossRef]
92. Gunia-Krzyzak, A.; Sloczynska, K.; Popiol, J.; Koczurkiewicz, P.; Marona, H.; Pekala, E. Cinnamic acid derivatives in cosmetics: Current use and future prospects. *Int. J. Cosmet. Sci.* **2018**, *40*, 356–366. [CrossRef] [PubMed]
93. Fan, Y.-T.; Yin, G.-J.; Xiao, W.-Q.; Qiu, L.; Yu, G.; Hu, Y.-L.; Xing, M.; Wu, D.-Q.; Cang, X.-F.; Wan, R. Rosmarinic acid attenuates sodium taurocholate-induced acute pancreatitis in rats by inhibiting nuclear factor-κB activation. *Am. J. Chin. Med.* **2015**, *43*, 1117–1135. [CrossRef] [PubMed]
94. Gautam, R.K.; Gupta, G.; Sharma, S.; Hatware, K.; Patil, K.; Sharma, K.; Goyal, S.; Chellappan, D.K.; Dua, K. Rosmarinic acid attenuates inflammation in experimentally induced arthritis in Wistar rats, using Freund's complete adjuvant. *Int. J. Rheum. Dis.* **2019**, *22*, 1247–1254. [CrossRef] [PubMed]
95. Elufioye, T.O.; Habtemariam, S. Hepatoprotective effects of rosmarinic acid: Insight into its mechanisms of action. *Biomed. Pharmacother.* **2019**, *112*, 108600. [CrossRef] [PubMed]
96. Natarajan, K.; Singh, S.; Burke, T.R.; Grunberger, D.; Aggarwal, B.B. Caffeic acid phenethyl ester is a potent and specific inhibitor of activation of nuclear transcription factor NF-kappa B. *Proc. Natl. Acad. Sci. USA* **1996**, *93*, 9090–9095. [CrossRef]
97. Chao, P.C.; Hsu, C.C.; Yin, M.C. Anti-inflammatory and anti-coagulatory activities of caffeic acid and ellagic acid in cardiac tissue of diabetic mice. *Nutr. Metab.* **2009**, *6*, 33. [CrossRef]
98. Jung, U.J.; Lee, M.K.; Park, Y.B.; Jeon, S.M.; Choi, M.S. Antihyperglycemic and antioxidant properties of caffeic acid in db/db mice. *J. Pharmacol. Exp. Ther.* **2006**, *318*, 476–483. [CrossRef]
99. Calixto-Campos, C.; Carvalho, T.T.; Hohmann, M.S.; Pinho-Ribeiro, F.A.; Fattori, V.; Manchope, M.F.; Zarpelon, A.C.; Baracat, M.M.; Georgetti, S.R.; Casagrande, R.; et al. Vanillic Acid Inhibits Inflammatory Pain by Inhibiting Neutrophil Recruitment, Oxidative Stress, Cytokine Production, and NFkappaB Activation in Mice. *J. Nat. Prod.* **2015**, *78*, 1799–1808. [CrossRef]

Disclaimer/Publisher's Note: The statements, opinions and data contained in all publications are solely those of the individual author(s) and contributor(s) and not of MDPI and/or the editor(s). MDPI and/or the editor(s) disclaim responsibility for any injury to people or property resulting from any ideas, methods, instructions or products referred to in the content.

MDPI
St. Alban-Anlage 66
4052 Basel
Switzerland
www.mdpi.com

Plants Editorial Office
E-mail: plants@mdpi.com
www.mdpi.com/journal/plants

Disclaimer/Publisher's Note: The statements, opinions and data contained in all publications are solely those of the individual author(s) and contributor(s) and not of MDPI and/or the editor(s). MDPI and/or the editor(s) disclaim responsibility for any injury to people or property resulting from any ideas, methods, instructions or products referred to in the content.

www.ingramcontent.com/pod-product-compliance
Lightning Source LLC
LaVergne TN
LVHW070358100526
838202LV00014B/1337